P9-CCC-128

COLLEGE ALGEBRA

COLLEGE ALGEBRA

Margaret L. Lial
Charles D. Miller
American River College
Sacramento, California

Scott, Foresman and Company
Glenview, Illinois   Brighton, England

Cover Photograph by JEROME KRESCH: geometric design created by photographing light in motion.

Library of Congress Catalog Card Number: 72-91385.
ISBN: 0-673-07788-8.
AMS 1970 Subject Classification 98A10.

Printed in the United States of America.

Regional offices of Scott, Foresman and Company are located in Dallas, Texas; Glenview, Illinois; Oakland, New Jersey; Palo Alto, California; Tucker, Georgia; and Brighton, England.

PREFACE

This book is designed for students who need a good, solid foundation in algebra before going on to further work in mathematics, such as calculus, statistics, or mathematics for the business or social science student. We assume that students come to this text with a background of at least an introductory course in algebra. The book progresses in a gradually increasing level of sophistication so that a student taking a course from this text is thoroughly and completely prepared to enter a more advanced course. Since many of the students studying from this book want to go on to calculus, and since there are currently several different approaches to the teaching of calculus, we have been careful to ensure that the text provides good preparation for all of these different calculus courses.

Too often college algebra texts maintain a high level of rigor throughout the entire text. We, however, prefer to begin at a reasonable level, and increase in rigor so that at the end of the course the student has reached the level expected in the next course.

There is now a trend away from the excesses of rigor, sophistication, and notational complexities that were beginning to creep into mathematics texts. The trend now is to intuition and understanding—appeal to the intuition of the student and guide him carefully so that understanding follows. Mathematics is a lively, exciting, and fascinating subject, and it seems a shame to hide this behind a cloud of logical connectives, set symbols, and complicated axiom structures.

There is a substantial amount of analytic geometry in this text—virtually enough to prepare a student for calculus. Functions are emphasized throughout. The $a + bi$ definition of a complex number is used, rather than the (a, b) definition that is so confusing to many students. Much material is included on inequalities, which are so important in later work. There are many, many problem sets and exercises, enough for almost any kind of course.

The student-oriented style of writing that is used has now been completely tested in many, many classes at a very large number of schools of all sizes and types. We are highly gratified at the student reaction to our efforts. We have tried to write for students, in a way that they can understand. Topics have been explained as simply and clearly as possible and many examples of each new topic are included. Judging by the reactions to our earlier books, we feel that we have succeeded in writing just about the most readable texts on the market.

Several people have helped us to produce this text. Vern Heeren of American River College provided a thorough, careful reading of the manuscript. Pamela Conaghan of Scott, Foresman was very helpful. Zallia Todd and Sharlene Wills helped in the technical preparation of the manuscript.

Sacramento, California

Margaret L. Lial

Charles D. Miller

CONTENTS

Chapter 1. THE REAL NUMBER SYSTEM 1

1.1 Sets of Numbers 1
1.2 Axioms of the Real Numbers 4
1.3 Consequences of the Field Axioms 9
1.4 Axioms of Order and Completeness 15
1.5 Absolute Value 19

Chapter 2. ALGEBRAIC EXPRESSIONS AND EXPONENTS 24

2.1 Polynomials 24
2.2 Quotients of Polynomials 28
2.3 Special Products and Factoring 32
2.4 Rational Expressions 35
2.5 Integer Exponents 39
2.6 Rational Exponents 42
2.7 Simplifying Radicals 46

Chapter 3. FIRST DEGREE RELATIONS AND FUNCTIONS 50

3.1 Relations and Functions 50
3.2 Linear Functions 54
3.3 Straight Lines 57
3.4 Linear Equations and Inequalities 61
3.5 Linear Inequalities 65
3.6 Nonlinear First Degree Relations 68
3.7 Arithmetic Progressions 71

Chapter 4. **SECOND DEGREE RELATIONS AND FUNCTIONS** **75**

4.1 The Quadratic Function 75
4.2 Zeros of Quadratic Functions 79
4.3 Equations Quadratic in Form 82
4.4 Quadratic Inequalities 86
4.5 Circles 89
4.6 Ellipses and Hyperbolas 92
4.7 Variation 99

Chapter 5. **EXPONENTIAL AND LOGARITHMIC FUNCTIONS** **103**

5.1 Inverse Relations 103
5.2 Exponential Functions 107
5.3 Logarithmic Functions 110
5.4 Common Logarithms 113
5.5 Exponential and Logarithmic Equations 118
5.6 Natural Logarithms 120
5.7 Geometric Progressions 126
5.8 Sums of Infinite Geometric Sequences 129

Chapter 6. **MATRIX THEORY** **134**

6.1 Basic Properties of Matrices 134
6.2 Multiplication of Matrices 137
6.3 The Algebra of Matrices 142
6.4 Determinants 148
6.5 Properties of Determinants 152

Chapter 7. **SYSTEMS OF EQUATIONS AND INEQUALITIES** **159**

7.1 Linear Systems 159
7.2 Solution of Linear Systems by Determinants—Cramer's Rule 164
7.3 Solution of Linear Systems by Matrices 169
7.4 Solutions of Linear Systems of Equations by Inverses 173
7.5 Nonlinear Systems of Equations 175
7.6 Systems of Inequalities 180

Chapter 8. POLYNOMIAL FUNCTIONS 183

8.1 Operations on Complex Numbers 183
8.2 Complex Zeros of Quadratic Polynomials 188
8.3 The Factor Theorem 190
8.4 Complex Zeros of General Polynomial Functions 193
8.5 Rational Zeros of Polynomial Functions 197
8.6 Approximate Zeros of Polynomial Functions 199
8.7 Graphing Polynomial Functions 202
8.8 Rational Functions 206

Chapter 9. COMBINATORICS 210

9.1 Permutations 210
9.2 Combinations 214
9.3 Basic Properties of Probability 218
9.4 Probabilities of Alternate and Successive Events 223
9.5 The Binomial Theorem 229
9.6 The Binomial Theorem and Probability 233
9.7 Mathematical Induction 236

APPENDIX 242

Table 1 Squares and Square Roots 243
Table 2 Common Logarithms 244

ANSWERS TO SELECTED EXERCISES 247

INDEX 273

1

THE REAL NUMBER SYSTEM

Algebra is primarily the study of numbers and the general properties of numbers. In this chapter, we introduce the idea of set and then concentrate on one of the most important sets of numbers, the set of real numbers. In Section 1.1 we discuss the necessary definitions needed to begin the study of the set of real numbers.

1.1 Sets of Numbers

A **set** is a collection of **elements** or **members.** We say a set is **well-defined** if its elements are clearly determined, that is, if we can decide whether or not a given element belongs to the set. We may describe a set by listing the elements of the set between braces (called **set braces**). For example, we can write the set containing the numbers 1, 2, 3, and 4 as

$$\{1, 2, 3, 4\}.$$

Since the order in which the elements are listed is unimportant, we could write instead $\{1, 2, 4, 3\}$ or any other arrangement of the four numbers. It is customary to use capital letters to refer to sets. Thus, we might refer to the set described above as S, where

$$S = \{1, 2, 3, 4\}.$$

To indicate that the number 1 is an element of the set S, we write $1 \in S$ (read "1 is an element of the set S" or "1 belongs to the set S"), while we write $5 \notin S$ to indicate that 5 is not an element of S.

It is sometimes more convenient to describe a set with a verbal phrase. Thus, we might describe the set S above as the set consisting of the first four counting numbers. In this example the notation $\{1, 2, 3, 4\}$, where we listed the elements between set braces, is briefer and clearer than the verbal description. However, the set F consisting of all the numbers between 0 and 1 could not be described by a listing of the elements. (Try it.) Set F is an example of an *infinite set,* one which has an unending list of distinct elements. On the other hand, a *finite set* is one which has a limited number of elements. Some infinite sets, unlike F, can be described by a listing process. For example, the set of numbers used for counting, called the **natural numbers,** can be written as

$$N = \{1, 2, 3, 4, \cdots\}.$$

A set whose elements are all elements of a second set is said to be a **subset** of that second set. Each element of the set $S = \{1, 2, 3, 4\}$ described above is an element of N, and so S is a subset of N, which we write as $S \subset N$. In this case, N is not a subset of S, written $N \not\subset S$, since there are elements in N which are not in S (the elements 5 and 6 for example).

Two sets, A and B, are said to be **equal** whenever $A \subset B$ and $B \subset A$. Thus, by $A = B$ we mean that the two sets contain exactly the same elements. If two sets contain the same number of elements (but not necessarily the same elements), they are called **equivalent** sets. From this we see that the two sets

$$S = \{1, 2, 3, 4\} \qquad \text{and} \qquad W = \{a, b, c, d\}$$

are equivalent sets, but not equal sets.

Two sets can be combined to form a third set. The set of all elements which belong to *either* of the sets A *or* B is called the **union** of A and B, written $A \cup B$. The set of all elements which belong to *both* A *and* B is called the **intersection** of A and B and is written $A \cap B$. For example, given

$$A = \{0, 1, 2, 3\} \qquad \text{and} \qquad B = \{0, 2, 4, 6\},$$

we have $A \cup B = \{0, 1, 2, 3, 4, 6\}$ while $A \cap B = \{0, 2\}$.

Example 1 Find $M \cup N$ and $M \cap N$ if $M = \{2, 4, 6\}$ and $N = \{1, 3\}$.

By the definitions, we have $M \cup N = \{1, 2, 3, 4, 6\}$. The set $M \cap N$ consists of all elements common to both M and N, but there are no such elements. To express this fact, we introduce a set called the **empty set** (or **null set**), written $\varnothing$. Thus, $M \cap N = \varnothing$.

Two sets, such as M and N in Example 1, whose intersection is the empty set, are called **disjoint sets.** Hence, the set of all even numbers and the set of all odd numbers are disjoint sets.

The set U of all elements to be used in a particular discussion is called the **universe of discourse** or **universal set.** For a given universe U and set A, the set of all elements in U which are not in set A is called the **complement** of A, written A'. For example, if U is the set of natural numbers N, and E is the set of even natural numbers, then E' is the set of odd natural numbers. Note that

$$E \cup E' = U \qquad \text{and} \qquad E \cap E' = \varnothing,$$

by the definition of complementary sets.

A convenient way to represent sets requires the use of a **variable,** a letter used to represent one unspecified element of a set of numbers. The set of numbers is called the **domain** or **replacement set** of the variable. For example, we can write $\{x \mid x \in N\}$ (read "set of all x such that x is an element of N") to represent the set of natural numbers. In this case N is the domain of the variable x.

At this point, let us list some of the sets of numbers which we shall be using throughout this book.

- ▶ *natural numbers* $\{1, 2, 3, 4, \cdots\}$
- ▶ *whole numbers* $\{0, 1, 2, 3, 4, \cdots\}$
- ▶ *integers* $\{\cdots, -2, -1, 0, 1, 2, \cdots\}$
- ▶ *rational numbers* $\left\{\frac{p}{q} \,\middle|\, p \text{ and } q \text{ are integers, } q \neq 0\right\}$ or $\{x \mid x \text{ is a repeating or terminating decimal}\}$
- ▶ *irrational numbers* $\{x \mid x \text{ is a nonrepeating, nonterminating decimal}\}$
- ▶ *real numbers* $\{x \mid x \text{ is a rational or irrational number}\}$
- ▶ *imaginary numbers* $\{x \mid x = a + bi \text{ where } a \text{ and } b \text{ are real numbers, } b \neq 0, i^2 = -1\}$
- ▶ *complex numbers* $\{x \mid x = a + bi, a \text{ and } b \text{ are real numbers}\}$

In this chapter we shall make a thorough study of the set of real numbers. Then we shall study the imaginary and complex numbers in a later chapter.

1.1 Exercises

List the elements of the following sets between set braces.

1. $\{x \mid x \text{ is less than 5 and } x \text{ is a natural number}\}$
2. $\{x \mid x \text{ is a whole number between } -2\frac{1}{2} \text{ and } 2\frac{1}{2}\}$
3. $\{x \mid x \text{ is an integer between } -2\frac{1}{2} \text{ and } 2\frac{1}{2}\}$
4. $\{x \mid x \text{ is a natural number between } -2\frac{1}{2} \text{ and } 2\frac{1}{2}\}$
5. $\{x \mid x \text{ is a positive odd integer greater than } 5\}$

Given $U = \{x | x \text{ is an integer}\}$, $M = \{0, 2, 4, 6, 8\}$, $N = \{x | x \text{ is an even integer}\}$, $P = \{x | x \text{ is an odd integer}\}$, $Q = \{-1, 0, 1, 2, 3, 4\}$, *and* $R = \{-3, -2, -1\}$, *find each of the following sets.*

6. $M \cap Q$
7. $M \cup Q$
8. $M \cap P$
9. $N \cup P$
10. $N \cap P$
11. $Q \cap R$
12. N'
13. $M' \cap P$
14. $Q \cap R'$
15. $U' \cup R$
16. $(M \cap Q) \cap N$
17. $(Q \cap R) \cup M$
18. $(R \cap N) \cup Q$
19. $R \cap (P \cap Q)$
20. $(M' \cap Q) \cup R$

Let N represent the set of natural numbers, W the set of whole numbers, I the integers, F the rational numbers, H the irrational numbers, and R the real numbers. Complete each of the following statements with the appropriate symbol: $\subset$, $=$, $\cap$, *or* $\cup$.

21. N _____ I
22. N _____ F
23. N _____ $W = W$
24. N _____ $R = R$
25. I _____ $R = I$
26. F _____ $H = \varnothing$
27. F _____ $H = R$
28. $\varnothing$ _____ N
29. F' _____ H
30. I _____ $(N \cup I)$

State the conditions on sets X, Y, and Z for which the following are true.

31. $X \cap Y = \varnothing$
32. $X \cup Y = \varnothing$
33. $X \cap Y = X$
34. $X \cup Y = X$
35. $X \cap \varnothing = X$
36. $X \cup \varnothing = X$
37. $X \cap Y' = \varnothing$
38. $X \cap Y = Y \cup X$
39. $X \cap (Y \cup Z) = (X \cap Y) \cup (X \cap Z)$
40. $X \cap (Y \cap Z) = (X \cap Y) \cap Z$

Use the definitions given in this section to show that the following are true for all sets X, Y, and Z.

41. $X \cap X' = \varnothing$
42. $X \cup X' = U$
43. $X \cap Y = Y \cap X$
44. $X \cup (Y \cup Z) = (X \cup Y) \cup Z$
45. If $X \subset Y$ and $Y \subset Z$, then $X \subset Z$.

1.2 Axioms of the Real Numbers

A **mathematical system** requires (1) a set of elements, (2) relations between the elements, (3) operations on the elements, and (4) properties, the rules which govern the manipulation of the elements. In the remainder of this chapter, we shall investigate the mathematical system which has the real numbers for elements.

The most important relation between two real numbers is that of equality. To indicate the relation **"*a* is equal to *b*"** we write $a = b$. The symbol $\neq$

represents the relation "is not equal to." (In a later section, we shall discuss another relation, "is less than.") Two expressions which represent the same real number are said to be **equal.** The axioms of equality which are stated below describe the assumed properties of the relation "equals."

Axioms of Equality

For all real numbers a, b, and c:

$a = a$.	*reflexive axiom*
If $a = b$, then $b = a$.	*symmetric axiom*
If $a = b$ and $b = c$, then $a = c$.	*transitive axiom*
If $a = b$, then a may replace b in any expression.	*substitution axiom*

A **binary** operation on a set is an operation on two elements of the set which produces a third element. (The three elements need not be distinct, nor need the third element be within the set.) There are two fundamental binary operations on real numbers, addition, indicated by $+$, and multiplication, indicated by $\cdot$, as in $2 \cdot 3$. When indicating a product with variables, the elements are written with no symbol between them, as ab or $2x$. Multiplication is also indicated by parentheses, as in the expression $2(x - y)$ which indicates that the quantity in the parentheses is to be multiplied by 2. Thus, grouping by parentheses determines the order in which operations must be performed.

The set of real numbers, together with the relation of equality and the operations of addition and multiplication form the **real number system.** This system is an example of a **field.** The rules which govern the operations of addition and multiplication of a field are called the **field axioms.**

Field Axioms for Addition and Multiplication

For all real numbers a, b, and c:

$a + b = b + a$.
$ab = ba$.
commutative axioms

$(a + b) + c = a + (b + c)$.
$(ab)c = a(bc)$.
associative axioms

There exists a unique real number 0, the **additive identity,** such that

$$a + 0 = 0 + a = a.$$

There exists a unique real number 1, the **multiplicative identity,** such that

$$a \cdot 1 = 1 \cdot a = a.$$

identity axioms

There exists a unique real number $-a$, the **additive inverse** of a, such that

$$a + (-a) = (-a) + a = 0.$$

If $a \neq 0$, there exists a unique real number $1/a$, the **multiplicative inverse** of a, such that *inverse axioms*

$$a \cdot \frac{1}{a} = \frac{1}{a} \cdot a = 1.$$

$a + b$ is a real number.
ab is a real number. *closure axioms*

$a(b + c) = ab + ac$. *distributive axiom*

The closure axioms guarantee that the addition or multiplication of two real numbers will be a real number. The commutative axioms assure us that we can add two numbers or multiply two numbers without regard to order, while the associative axioms show that the way in which we group, when using only one operation with more than two numbers, has no effect on the final result.

The identity axioms establish the existence of the identity elements, 0 for addition and 1 for multiplication, which preserve the identity of any element with which they are combined. The additive inverse axiom notes the existence of the *additive inverse* element $-a$ for each element a, which when added to a, yields the *additive identity*, 0. And the multiplicative inverse axiom notes the existence of the *multiplicative inverse* element $1/a$ for each nonzero element a, which when multiplied by a, yields the *multiplicative identity*, 1. The additive inverse element $-a$ is also called the **negative** of a, while the multiplicative inverse element $1/a$ is also called the **reciprocal** of a.

The distributive axiom describes a relationship between multiplication and addition, which, in effect, changes an indicated product to an indicated sum, or an indicated sum to an indicated product. By the commutative axiom, the distributive axiom can also be written as

$$(a + b)c = ac + bc.$$

Repeated application of the distributive axiom gives the **extended distributive axiom,**

$$a(b + c + d + \ldots + n) = ab + ac + ad + \ldots + an,$$

which can also be rewritten using the commutative axiom.

We shall define **subtraction** for any real numbers a and b as

▶ $$a - b = a + (-b).$$

Thus, the subtraction $a - b$ is defined as the addition of a and the additive inverse of b. **Division** of any real number a by any nonzero real number b

(denoted $a \div b$, $\frac{a}{b}$, or a/b) is defined analogously as follows.

$$a \div b = \frac{a}{b} = a \cdot \frac{1}{b}.$$

That is, to find the quotient a/b, we find the product of a and the multiplicative inverse of b.

1.2 Exercises

Answer true or false. If false, tell why.

1. The set of integers is closed with respect to addition.
2. The set of irrational numbers is closed with respect to multiplication.
3. The set $\{0, 1\}$ is closed with respect to subtraction.
4. The set $\{0, 1\}$ is closed with respect to multiplication.
5. The set $\{1, -1\}$ is closed with respect to division.
6. The set of irrational numbers contains an identity element for multiplication.
7. The set of natural numbers contains an identity element for addition.
8. The set of even numbers satisfies the inverse axiom for addition.
9. The set of rational numbers between 0 and 1 inclusive satisfies the inverse axiom for multiplication.
10. Any subset of the real numbers satisfies the commutative axioms.

Identify the axiom(s) which is illustrated in each of the following.

11. $3\left(\frac{1}{3}\right) = \left(\frac{1}{3}\right)3$
12. $\frac{3}{4} + \left(-\frac{3}{4}\right) = 0$
13. $2\left(\frac{1}{4}\right) + 2\left(\frac{3}{4}\right) = 2(1)$
14. $x + y = y + x$
15. $(y + x) + 2x = y + (x + 2x)$
16. If $(x + y) + 2x = y + (x + 2x)$, and $y + (x + 2x) = y + 3x$, then $(x + y) + 2x = y + 3x$.
17. $(x + 6)\left(\frac{1}{x + 6}\right) = 1, (x \neq -6)$
18. If x is a real number, $x + 2$ is a real number.
19. If $x + 2 = 3$ and $(x + 2) - y = z$, then $3 - y = z$.
20. $(7 - y) + 0 = 7 - y$

In this section we stated that the field of real numbers forms a mathematical system. Another example of a mathematical system is one in which the elements are sets. The following exercises formulate this system.

21. State two relations between sets.
22. State two binary operations on sets.
23. State five axioms for sets. (Hint: refer to Exercise Set 1.1, Exercises 41–45.)

Use the axioms and definitions of real numbers to complete the following proofs.

24. Prove: $a[b + (c + d)] = ab + ac + ad$.

$a[b + (c + d)] = ab + a(c + d)$ ______

$ab + a(c + d) = ab + ac + ad$ ______

$a[b + (c + d)] = ab + ac + ad$ ______

25. Prove: $0 - a = -a$.

$0 - a = 0 + (-a)$ ______

$0 + (-a) = -a$ ______

$0 - a = -a$ ______

26. Prove: $a - a = 0$.

$a - a = a + (-a)$ ______

$a + (-a) = 0$ ______

$a - a = 0$ ______

27. Prove: $\frac{1}{a}(ba) = b$.

$\frac{1}{a}(ba) = \frac{1}{a}(ab)$ ______

$\frac{1}{a}(ab) = \left(\frac{1}{a} \cdot a\right)b$ ______

$\left(\frac{1}{a} \cdot a\right)b = 1 \cdot b$ ______

$1 \cdot b = b$ ______

$\frac{1}{a}(ba) = b$ ______

28. Prove: $(a + b)c = ac + bc$.

$(a + b)c = c(a + b)$ ______

$c(a + b) = ca + cb$ ______

$(a + b)c = ca + cb$ ______

$ca + cb = ac + bc$ ______

$(a + b)c = ac + bc$ ______

29. Prove: $(m + n) + (p + q) = (m + p) + (n + q)$.

$(m + n) + (p + q) = [(m + n) + p] + q$ ____________

$[(m + n) + p] + q = [m + (n + p)] + q$ ____________

$[m + (n + p)] + q = [m + (p + n)] + q$ ____________

$[m + (p + n)] + q = [(m + p) + n] + q$ ____________

$[(m + p) + n] + q = (m + p) + (n + q)$ ____________

$(m + n) + (p + q) = (m + p) + (n + q)$ ____________

30. Prove: $(a + b) + (-a) = b$.

$(a + b) + (-a) = a + [b + (-a)]$ ____________

$a + [b + (-a)] = a + [-a + b]$ ____________

$a + [-a + b] = [a + (-a)] + b$ ____________

$[a + (-a)] + b = 0 + b$ ____________

$0 + b = b$ ____________

$(a + b) + (-a) = b$ ____________

1.3 Consequences of the Field Axioms

From the basic definitions and axioms stated in Section 1.2, we can demonstrate many further properties of the real numbers. These consequences of the axioms and definitions are called **theorems** and are distinguished by name from the axioms, since they follow deductively from the basic properties which we accept without proof. In this section, we shall demonstrate the proofs of several important and useful properties which are consequences of the field axioms and the definitions.

The first theorem we shall prove is used so much that we give it a name.

THEOREM 1.1 ADDITION PROPERTY OF EQUALITY
For all real numbers a, b, and c,
if $a = b$, then $a + c = b + c$.

Proof

$a + c$ is a real number	closure axiom for addition
$a + c = a + c$	reflexive axiom
$a = b$	given
$a + c = b + c$	substitution axiom

The conclusion of this theorem can also be stated as $c + a = c + b$, by the commutative axiom. There is a comparable theorem for multiplication.

THEOREM 1.2 MULTIPLICATION PROPERTY OF EQUALITY
For all real numbers, a, b, and c,
if $a = b$, then $ac = bc$.

The conclusion here could also be stated as $ca = cb$. The proof of the multiplication property of equality is similar to that for Theorem 1.1 and is left for the exercises.

THEOREM 1.3 CANCELLATION PROPERTY OF ADDITION
For all real numbers a, b, and c,
if $a + c = b + c$, then $a = b$.

Proof

$a + c = b + c$	given
$(a + c) + (-c) = (b + c) + (-c)$	addition property of equality
$a + [c + (-c)] = b + [c + (-c)]$	associative axiom
$c + (-c) = 0$	additive inverse axiom
$a + 0 = b + 0$	substitution axiom
$a = b$	additive identity axiom and substitution axiom

Note that Theorem 1.3 is the converse of Theorem 1.1—that is, the hypothesis and conclusion of Theorem 1.1 are respectively the conclusion and hypothesis of Theorem 1.3. It is convenient to combine these two theorems into one statement as follows.

THEOREM 1.3 (A) For all real numbers a, b, and c,
$a = b$ if and only if $a + c = b + c$.

The words "if and only if" here tell us that the theorem is true with either part as hypothesis and the other as conclusion. Thus, in general, the phrase "p if and only if q" means "if p, then q *and* if q, then p." Proof of such a theorem requires proof of both "if p, then q" and "if q, then p."

THEOREM 1.4 CANCELLATION PROPERTY OF MULTIPLICATION
For all real numbers a, b, and $c \neq 0$,
$a = b$ if and only if $ac = bc$.

To prove this theorem, we must prove two statements: (1) if $a = b$, then $ac = bc$, and (2) if $ac = bc$, then $a = b$. The proof is left for the exercises.

Another important and useful theorem is the multiplication property of zero.

THEOREM 1.5 MULTIPLICATION PROPERTY OF ZERO
For all real numbers a,
$a \cdot 0 = 0 \cdot a = 0$.

Proof

$a = a \cdot 1$	identity axiom of multiplication
$a = a(1 + 0)$	identity axiom of addition and substitution axiom
$a = a \cdot 1 + a \cdot 0$	distributive axiom
$a = a + a \cdot 0$	identity axiom of multiplication and substitution axiom
$-a + a = -a + (a + a \cdot 0)$	addition property of equality
$-a + a = (-a + a) + a \cdot 0$	associative axiom of addition
$0 = 0 + a \cdot 0$	additive inverse axiom and substitution
$0 = a \cdot 0$	additive identity axiom and substitution
$a \cdot 0 = 0$	symmetric axiom
$a \cdot 0 = 0 \cdot a$	commutative axiom of multiplication
$0 = 0 \cdot a$	transitive axiom of equality
$0 \cdot a = 0$	symmetric axiom

The rules for operating with quotients are summarized in Theorem 1.6.

THEOREM 1.6 For all real numbers a, $b \neq 0$, c, and $d \neq 0$:

(a) $\frac{a}{b} = \frac{c}{d}$ if and only if $ad = bc$

(b) $\frac{ad}{bd} = \frac{a}{b}$*

(c) $\frac{1}{b} \cdot \frac{1}{d} = \frac{1}{bd}$

(d) $\frac{a}{b} \cdot \frac{c}{d} = \frac{ac}{bd}$

(e) $\frac{a}{b} + \frac{c}{b} = \frac{a + c}{b}$.

Theorem 1.6(a) describes the useful **equality test for rational numbers,** while (b) states the **fundamental theorem of rational numbers** which is the basis for reducing fractions to lowest terms or building fractions to higher terms. Theorem 1.6(c) states that the product of the multiplicative inverses of b and d equals the multiplicative inverse of the product bd. Parts (d) and (e) show how to perform the operations of multiplication and addition for rational numbers. The proofs of these statements are included in the exercises.

Since rational numbers are real numbers, the definition of subtraction for rational numbers follows from the definition of subtraction of real numbers.

$$\frac{a}{b} - \frac{c}{d} = \frac{a}{b} + \left(-\frac{c}{d}\right)$$

* Here $bd \neq 0$ follows from the two restrictions $b \neq 0$ and $d \neq 0$.

Similarly, the definition of division for rational numbers follows from the definition of division of real numbers and the definition of multiplication of rational numbers.

$$\frac{\frac{a}{b}}{\frac{c}{d}} = \frac{a}{b} \cdot \frac{1}{\frac{c}{d}} = \frac{a}{b} \cdot \frac{d}{c} = \frac{ad}{bc} \qquad \left(\frac{c}{d} \neq 0\right)$$

The next theorem includes several facts about operations with real numbers.

THEOREM 1.7 For all real numbers a and b:

(a) $-(-a) = a$

(b) $(-a) + (-b) = -(a + b)$

(c) $(-a)b = -(ab)$

(d) $(-a)(-b) = ab$

(e) $\frac{-a}{b} = \frac{a}{-b} = -\frac{a}{b}$, $(b \neq 0)$

(f) $\frac{-a}{-b} = \frac{a}{b}$, $(b \neq 0)$.

Theorem 1.7(a) states that the additive inverse of the additive inverse of a is a, while (b) states that the sum of the additive inverses of a and b is the additive inverse of the sum of a and b. Theorem 1.7(c) and (d) deal with products of additive inverses. Theorem 1.7(e) and (f) are consequences of the definition of division and Theorem 1.7(c) and (d).

The proof of Theorem 1.7(a) follows from the fact that both a and $-(-a)$ are additive inverses of $-a$ by definition. That is, $a + (-a) = 0$ and $-a + [-(-a)] = 0$. However, by the additive inverse axiom, for each real number, there is a *unique* additive inverse, so that we must have $-(-a) = a$. The proofs for the remainder of Theorem 1.7 are left as exercises.

1.3 Exercises

Complete the following proofs by giving reasons for each step, using axioms, definitions, or previously proved theorems. Some steps require more than one reason. Assume a, b, c, and d represent real numbers.

1. Prove: $(-a) + (-b) = -(a + b)$.

$[(-a) + (-b)] + (a + b) = (-a + a) + (-b + b)$ __________

$= 0 + 0$ __________

$= 0$ __________

$(-a) + (-b)$ is the additive inverse of $(a + b)$ __________

$-(a + b)$ is the additive inverse of $(a + b)$ __________

$(-a) + (-b) = -(a + b)$ __________

2. Prove: $-(a - b) = b - a$.

$(b - a) + (a - b) = [b + (-a)] + [a + (-b)]$ ______

$= [a + (-a)] + [b + (-b)]$ ______

$= 0 + 0$ ______

$= 0$ ______

$b - a$ is the additive inverse of $a - b$ ______

$-(a - b)$ is the additive inverse of $a - b$ ______

$-(a - b) = b - a$ ______

3. Prove: If $x + a = b$, then $x = b - a$.

$x + a = b$ given

$(x + a) + (-a) = b + (-a)$ ______

$x + [a + (-a)] = b + (-a)$ ______

$x + 0 = b + (-a)$ ______

$x = b + (-a)$ ______

$x = b - a$ ______

4. Prove: $(-a)b = -(ab)$.

$[(-a)b] + ab = [-a + a]b$ ______

$= 0 \cdot b$ ______

$= 0$ ______

$(-a)b$ is the additive inverse of ab ______

$-(ab)$ is the additive inverse of ab ______

$(-a)b = -(ab)$ ______

5. Prove: If $b \neq 0$, $\frac{-a}{b} = -\frac{a}{b}$.

$\frac{-a}{b} = -a\left(\frac{1}{b}\right)$ ______

$= -\left(a \cdot \frac{1}{b}\right)$ ______

$= -\frac{a}{b}$ ______

6. Prove: For $b \neq 0$ and $d \neq 0$, $\frac{a}{b} = \frac{c}{d}$ if and only if $ad = bc$.

PART I

$\frac{a}{b} = \frac{c}{d}$ given

$\frac{a}{b}(bd) = \frac{c}{d}(bd)$ ______

$\left(a \cdot \frac{1}{b}\right)(bd) = \left(c \cdot \frac{1}{d}\right)(bd)$ ______

$ad\left(\frac{1}{b} \cdot b\right) = cb\left(\frac{1}{d} \cdot d\right)$ ______

$ad(1) = cb(1)$ ______

$ad = bc$ ______

PART II

$ad = cb$	given
$ad\left(\frac{1}{b} \cdot \frac{1}{d}\right) = cb\left(\frac{1}{b} \cdot \frac{1}{d}\right)$	________
$\left(a \cdot \frac{1}{b}\right)\left(d \cdot \frac{1}{d}\right) = \left(c \cdot \frac{1}{d}\right)\left(b \cdot \frac{1}{b}\right)$	________
$\left(a \cdot \frac{1}{b}\right)1 = \left(c \cdot \frac{1}{d}\right)1$	________
$a \cdot \frac{1}{b} = c \cdot \frac{1}{d}$	________
$\frac{a}{b} = \frac{c}{d}$	________

From parts I and II, we have $\frac{a}{b} = \frac{c}{d}$ if and only if $ad = bc$.

7. Prove: If $b \neq 0$ and $d \neq 0$, $\frac{ad}{bd} = \frac{a}{b}$.

$(ad)b = (ad)b$	________
$(ad)b = (bd)a$	________
$\frac{ad}{bd} = \frac{a}{b}$	________

8. Prove: If $a \neq 0$ and $b \neq 0$, then $\frac{1}{a} \cdot \frac{1}{b} = \frac{1}{ab}$.

$\left(\frac{1}{a} \cdot \frac{1}{b}\right)(ab) = \left(\frac{1}{a} \cdot a\right)\left(\frac{1}{b} \cdot b\right)$	________
$= 1 \cdot 1$	________
$= 1$	________
$\frac{1}{a} \cdot \frac{1}{b}$ is the multiplicative inverse of ab	________
$\frac{1}{ab}$ is the multiplicative inverse of ab	________
$\frac{1}{a} \cdot \frac{1}{b} = \frac{1}{ab}$	________

Prove the following theorems. Assume a, b, and c are real numbers.

9. $(-1)a = -a$
10. If $x = \frac{a}{b}$, $b \neq 0$, then $a = bx$.
11. $a = b$ if and only if $ac = bc$, $(c \neq 0)$
12. $(-a)(-b) = ab$
13. $\frac{-a}{-b} = \frac{a}{b}$, $(b \neq 0)$
14. $\frac{a}{b} \cdot \frac{c}{d} = \frac{ac}{bd}$, $(b \neq 0,\ d \neq 0)$
15. $\frac{a}{b} + \frac{c}{b} = \frac{a + c}{b}$, $(b \neq 0)$
16. $a(b - c) = ab - ac$

1.4 Axioms of Order and Completeness

We saw in Section 1.2 that "is equal to" and "is not equal to" are relations between real numbers. Another important relation between real numbers is the order relation "is less than." To indicate the relation "a is less than b" we write $a < b$.

To discuss the ordering of the real numbers it is convenient to consider the correspondence which can be established between the set of real numbers and the set of points on a line. By choosing some point on a line as the point which we shall identify as 0, and then choosing a unit of measure, we can identify the number 1 with the point one unit to the right of 0, -1 with the point one unit to the left, and so on. We shall call this line the **number line,** and abbreviate the phrase "the point corresponding to the number x" as "the point x."

On the number line, points to the right of 0 represent positive numbers, and points to the left of 0 represent negative numbers. Note that the positive numbers increase toward the right, so that for positive numbers a and b, $a < b$ means a is to the left of b on the number line. To maintain this order, we shall define the relation "is less than" as follows.

▶ For all real numbers a and b,
$a < b$ if and only if, for some positive real number c, $a + c = b$.

This relationship is illustrated in Figure 1.1.

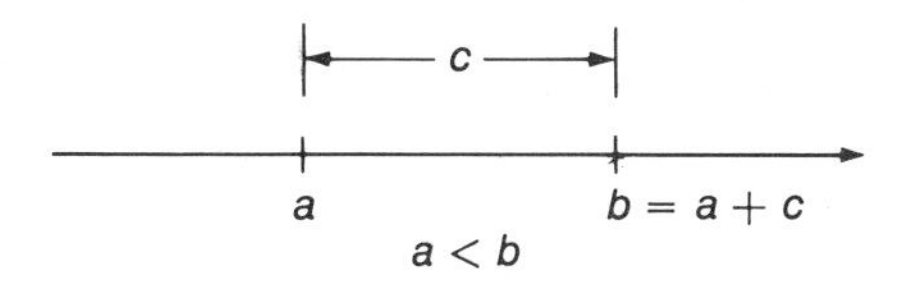

Figure 1.1

Several other order relations can be derived from the definitions of $<$ and $=$. We can restate the relation "a is less than b" as "b is greater than a," written $b > a$. The following variations of these relations are frequently used.

$\leq$	is less than or equal to
$\geq$	is greater than or equal to
$\nless$	is not less than
$\ngtr$	is not greater than

Sentences using these order relations, along with $<$, $>$, and $\neq$, are sometimes called inequalities.

From the definition of $a < b$, and by referring to the number line, we can see that the following order axioms hold.

Axioms of Order

For all real numbers a, b, and c:

Exactly one of these three conditions holds: $a = b$, $a < b$, or $b < a$.	*trichotomy axiom*
If $a > 0$ and $b > 0$, then $a + b > 0$ and $ab > 0$.	*closure axiom*

We can now restate the definition of positive and negative numbers in terms of inequalities.

▶ The number a is **positive** if and only if $a > 0$, and a is **negative** if and only if $a < 0$.

Several further properties of the relation $<$ are given in Theorem 1.8.

THEOREM 1.8 For all real numbers a, b, and c:

(a) if $a < b$, then $a + c < b + c$
(b) if $a < b$ and $c > 0$, then $ac < bc$
(c) if $a < b$ and $c < 0$, then $ac > bc$
(d) if $a < b$ and $b < c$, then $a < c$
(e) if $a > 0$, then $-a < 0$.

Proof of (a)

$a < b$	given
For some positive number m, $a + m = b$.	definition of $<$
$(a + m) + c = b + c$	addition property of equality
$(a + c) + m = b + c$	commutative and associative axioms
$a + c < b + c$	definition of $<$

The proofs of parts (b), (c), (d), and (e) are similar.

A nonempty set S of real numbers is said to have an **upper bound** b if b is a real number and for every element x in S, $x \leq b$. If b is less than every other upper bound of S, it is called the **least upper bound.** Similarly, S has a lower bound b', if for every element x of S, $x \geq b'$. Also, b' is the **greatest lower bound** of S if it is larger than every other lower bound of S. For example, the set $M = \{0, 1, 2\}$ has as upper bounds the numbers 2, 3.5, $\sqrt{15}$, and so on, with 0 and all negative numbers as lower bounds. The number 2 is the least upper bound while 0 is the greatest lower bound. The set $\{x \mid x \leq 2\}$ also has 2 as a least upper bound, but has no lower bound. The least upper bound and greatest lower bound need not belong to the set. For example, the set $\{x \mid x > 5\}$ has no upper bound, but has a greatest lower bound of 5, which is not an element of the set.

The last property needed to describe the set of real numbers is called the axiom of completeness. This axiom guarantees that the number line representing the set of real numbers will have no "gaps."

Axiom of Completeness

If a nonempty set S of real numbers has an upper bound, then S has a least upper bound. Also, if S has a lower bound, then S has a greatest lower bound.

With this axiom, we have completed the description of the real numbers as a mathematical system. The set of real numbers is a **complete ordered field,** since it satisfies the completeness axiom, the order axioms, and the field axioms. In fact, the set of real numbers is the only set which satisfies all of these axioms and thus is the only complete ordered field. The set of rational numbers is an ordered field but is not complete. To see why, consider the set $\{x \mid x \text{ is a rational number, } x^2 < 2\}$. This set has an upper bound in the set of rational numbers, for example, 1.5, but it has no least upper bound in the set of rational numbers. The field of complex numbers is not ordered, as is shown in a later chapter.

1.4 Exercises

Mark each of the following true or false.

1. $4 < -2$
2. $-7 < -5$
3. $-\frac{1}{2} > -\frac{1}{3}$
4. $\frac{1}{4} + \frac{1}{4} \leq \frac{1}{2}$
5. $\frac{2}{3} \geq -\frac{2}{3}$
6. $\frac{23}{4} \nless \frac{15}{4}$

Identify the axiom or theorem of this section which is illustrated by each of the following statements. Assume x and y represent real numbers.

7. If $x < 2$ and $2 < y$, then $x < y$.
8. If $x < 4$ then $x + 7 < 11$.
9. If $x < -2$ then $4x < -8$.
10. If $x < -2$ then $-4x > 8$.
11. If $x + 3 < y$, then $x < y - 3$.
12. Either $x < 0$, $x = 0$, or $x > 0$.

For each of the following sets of real numbers give both the greatest lower bound and the least upper bound.

13. $\{x \mid x \leq 6\}$
14. $\{x \mid -2 \leq x < 5\}$
15. $\{x \mid x < 5 \text{ and } x \text{ is a natural number}\}$
16. $\{x \mid x < 5 \text{ and } x \text{ is a rational number}\}$

17. $\{\cdots, 10, 1, .1, .01, .001, \cdots\}$

18. $\left\{1, \frac{1}{2}, \frac{1}{3}, \frac{1}{4}, \cdots\right\}$

Complete the following proofs for all real numbers a, b, c, and d.

19. Prove: If $a < b$ and $c > 0$, then $ac < bc$.

$a < b$	given
For some $m > 0$, $a + m = b$.	________
$(a + m)c = bc$	________
$ac + mc = bc$	________
$c > 0$	________
$mc > 0$	________
$ac < bc$	________

20. Prove: If $a < b$ and $c < 0$, then $ac > bc$.

$a < b$	given
For some $m > 0$, $a + m = b$.	________
$(a + m)c = bc$	________
$ac + mc = bc$	________
$c < 0$	________
$mc < 0$	________
$-(mc) > 0$	________
$(ac + mc) + (-mc) = bc + (-mc)$	________
$ac = bc + (-mc)$	________
$bc < ac$	________

21. Prove: $a > b$ if and only if $-a < -b$.

PART I

$a > b$	________
$a(-1) < b(-1)$	________
$a(-1) = -a$ and $b(-1) = -b$	________
$-a < -b$	________

PART II

$-a < -b$	given
________	________
________	________
________	________

22. Prove: If $a > 1$, then $a^2 > 1$.

$a > 1$	given
$a(a) > 1(a)$	________
$1(a) = a$	________
$1(a) > 1$	________
$a^2 = a(a) > 1$	________

23. Prove: If $a < b$ and $c < d$, then $a + c < b + d$.

$a < b$ and $c < d$	given
For some $m > 0$, $a + m = b$.	________
For some $n > 0$, $c + n = d$.	________
$(a + m) + (c + n) = b + d$	________
$(a + c) + (m + n) = b + d$	________
$m + n > 0$	________
$a + c < b + d$	________

24. Prove: For positive numbers a, b, c, and d, if $a < b$ and $c < d$, then $ac < bd$.

For some positive number m, $a + m = b$.	________
For some positive number n, $c + n = d$.	________
$(a + m)(c + n) = bd$	________
$a(c + n) + m(c + n) = bd$	________
$ac + an + mc + mn = bd$	________
$an + mc + mn > 0$	________
$ac < bd$	________

25. Prove: If $a > 1$, then $a^3 > 1$.

1.5 Absolute Value

Associated with each real number there is a nonnegative quantity which represents the number's distance from 0 on the number line. This is called the **absolute value** of the number, written $|x|$, and is defined as follows.

▶
$$|x| = \begin{cases} x \text{ if } x \geq 0 \\ -x \text{ if } x < 0 \end{cases}$$

For example, $|5| = 5$, $|-5| = 5$, and $-|-3| = -3$.

Since absolute value represents distance on the number line, to represent the distance between the numbers x and b, we can write $|x - b|$, where $|x - b|$ is defined as follows.

▶
$$|x - b| = \begin{cases} x - b \text{ if } x - b \geq 0 \\ -(x - b) \text{ if } x - b < 0 \end{cases}$$

The conditions $x - b \geq 0$ and $x - b < 0$ are equivalent to the simpler statements $x \geq b$ and $x < b$. When $b = 0$, this more general definition becomes our earlier definition of absolute value.

Example 2 Solve $|x| = 3$.

Since either $|3| = 3$ or $|-3| = 3$, the solution set is $\{3, -3\}$.

Example 3 Solve $|x - 4| = 2$.

Again we have two cases.

(1) If $x \geq 4$, $|x - 4| = x - 4$, so that

$$x - 4 = 2$$
$$x = 6.$$

(2) If $x < 4$, $|x - 4| = -(x - 4)$, and

$$-(x - 4) = 2$$
$$x = 2.$$

Thus, the solution set is $\{2, 6\}$.

In Section 1.4, we defined inequality relations. We can use these relations to describe intervals on the number line. For example, the set of numbers $\{x \mid 1 < x < 5\}$ corresponds to the set of points on the number line between 1 and 5 as shown in Figure 1.2. The fact that 1 and 5 are not included in the set is shown by the open circles at those points. When we write an inequality such as $1 < x < 5$ we must keep in mind that this represents the compound sentence $1 < x$ *and* $x < 5$, where a **compound sentence** refers to two simple sentences joined with a connective, such as *and* or *or*.

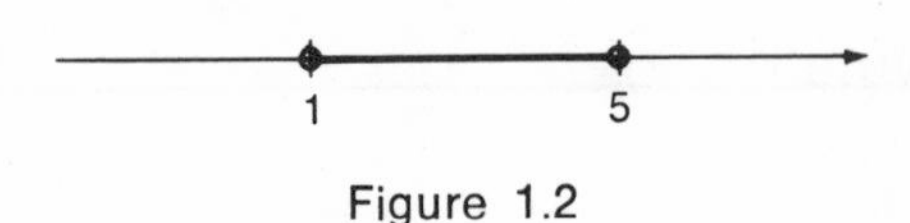

Figure 1.2

We can use the same method to describe other intervals on the number line, as shown in Figure 1.3, where a solid circle indicates that the point belongs to the set.

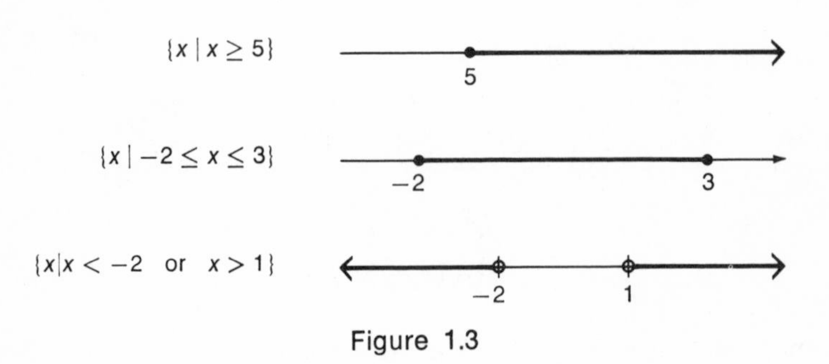

Figure 1.3

Now, let us consider absolute value inequalities. To solve the inequality $|x| < 5$, we use the definition of $|x|$. There are two cases:

(1) $x < 0$ and $|x| < 5$

If $x < 0$, $|x| = -x$, and so $|x| < 5$ becomes $-x < 5$, or $x > -5$. Thus, $x < 0$ and $|x| < 5$ becomes $x < 0$ and $x > -5$, or $-5 < x < 0$.

(2) $x \geq 0$ and $|x| < 5$
If $x \geq 0$, $|x| = x$, and so $|x| < 5$ becomes $x < 5$. Thus, $x \geq 0$ and $|x| < 5$ becomes $x \geq 0$ and $x < 5$, or $0 \leq x < 5$.

From the first case, we know every number x satisfying $-5 < x < 0$ is a solution, and from the second case all numbers x such that $0 \leq x < 5$ are solutions. The final solution set is the *union* of the solution sets from the two cases. It can be written $\{x| -5 < x < 0 \text{ or } 0 \leq x < 5\}$, or simply

$$\{x\,|\,-5 < x < 5\}.$$

The solution of the inequality $|x| > 5$ again requires two cases:

(1) $x < 0$ and $|x| > 5$
Here $|x| = -x$ so $|x| > 5$ becomes $-x > 5$, or $x < -5$. Now $x < 0$ and $|x| > 5$ becomes $x < 0$ and $x < -5$, or simply $x < -5$.

(2) $x \geq 0$ and $|x| > 5$
Here $|x| = x$ so $|x| > 5$ becomes $x > 5$. Now $x \geq 0$ and $|x| > 5$ becomes $x \geq 0$ and $x > 5$, or simply $x > 5$.

Again the final solution set is the union of the solution sets from case (1) and case (2). It can be written $\{x\,|\,x < -5 \text{ or } x > 5\}$. Generalizing, we have the following.

THEOREM 1.9 For all real numbers a and b, $b \geq 0$:
(a) $|a| \leq b$ if and only if $-b \leq a \leq b$
(b) $|a| \geq b$ if and only if $a \leq -b$ or $a \geq b$.

Example 4 Solve $|x| \geq 3$.
From Theorem 1.9(b), if $|x| \geq 3$, then $x \leq -3$ or $x \geq 3$, as shown on the number line of Figure 1.4. The solution set here is written $\{x|x \leq -3 \text{ or } x \geq 3\}$.

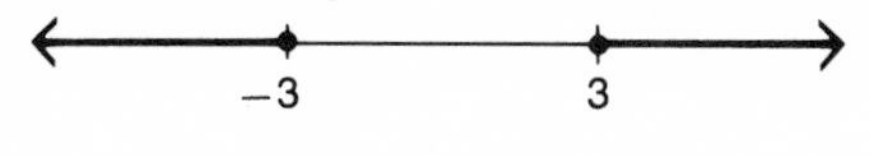

Figure 1.4

Example 5 Solve $|x - 2| < 5$.
From Theorem 1.9(a), letting $a = x - 2$ and $b = 5$, we have

$$|x - 2| < 5$$
$$-5 < x - 2 < 5$$
$$-3 < x < 7.$$

The solution set is written $\{x\,|\,-3 < x < 7\}$.

Properties of operations with absolute value are given in the following theorem.

THEOREM 1.10 For all real numbers a and b:

(a) $|ab| = |a| \cdot |b|$

(b) $|a + b| \leq |a| + |b|$

(c) $|a - b| \geq |a| - |b|$

(d) $\left|\frac{a}{b}\right| = \frac{|a|}{|b|}$, $(b \neq 0)$.

We omit the proof of these theorems. The proofs depend on the definitions of addition and multiplication for real numbers, and the definition of absolute value. Some examples of such proofs are included in the exercises.

1.5 Exercises

Complete the following statements by choosing the correct relation: $=$, $\leq$, *or* $\geq$.

1. $|5|$ ______ $|-5|$
2. $|x|$ ______ $|-x|$
3. $-|x|$ ______ $|x|$
4. $-|-x|$ ______ $|x|$
5. $|x - 4|$ ______ $|4 - x|$
6. $|x - 8|$ ______ $|8 - x|$
7. If $x < y$, then $|x - y|$ ______ $y - x$.
8. If $x < y$, then $|x - y|$ ______ $x - y$.
9. If $|x| \leq 5$, then -5 ______ x ______ 5.
10. If $|3x| \leq 1$, then -1 ______ $3x$ ______ 1.
11. If $|x| \geq 4$, then x ______ 4 or x ______ -4.
12. If $|x + 2| \geq 3$, then $x + 2$ ______ 3 or $x + 2$ ______ -3.
13. $|3| \cdot |-5|$ ______ $|3(-5)|$
14. $|3| \cdot |x|$ ______ $|3x|$
15. $|4 + 3|$ ______ $|4| + |3|$
16. $|5 - 1|$ ______ $|5| + |-1|$
17. $|-2 + 3|$ ______ $|-2| + |3|$
18. $|5 - 1|$ ______ $|5| - |1|$
19. $|3 - 2|$ ______ $|3| - |2|$
20. $|3 - 4|$ ______ $|3| - |4|$
21. $\frac{|7|}{|-1|}$ ______ $\left|\frac{7}{-1}\right|$
22. $\frac{|-1|}{|-3|}$ ______ $\left|\frac{-1}{-3}\right|$

Graph the following intervals on the number line.

23. $x \geq 5$
24. $-2 < x < 4$
25. $x + 2 \leq 8$ and $x \geq 3$
26. $x - 1 > 0$ and $x < 0$
27. $x > 3$ or $x < 1$
28. $2x > -2$ or $x + 2 < 4$
29. $x \geq 2$ or $x > 5$

Solve.

30. $|x| = 4$
31. $|x| = -2$
32. $|-x| = -1$
33. $|-x| = 1$
34. $|x - 2| = 5$
35. $|2x + 5| = 3$
36. $\left|x - \frac{1}{2}\right| = 3$
37. $\left|\frac{x}{3} + 1\right| = 2$

ALGEBRAIC EXPRESSIONS AND EXPONENTS

An **algebraic expression** is any expression of variables or constants obtained by performing a finite number of the basic operations of addition, subtraction, multiplication, division (except by zero), or extraction of roots. The simplest such algebraic expression is a polynomial, which we shall discuss in the first section.

2.1 Polynomials

If n is any natural number, and a is any real number, then we define the symbol a^n by writing

▶ $$a^n = a \cdot a \cdot a \cdots a,$$

where a appears as a factor n times. Here n is called the **exponent,** or **power,** a is called the **base,** and the expression a^n is called an **exponential.** Thus

$$4^3 = 4 \cdot 4 \cdot 4 = 64, \qquad \text{and} \qquad (-6)^2 = 36.$$

Any product of a constant (called the **coefficient**) and variables raised to powers is called a **term.** A finite sum of terms with whole number exponents on the variables, such as $-2x^3 + 4x^2 - 5$, or $2m^3n^2y - 3$, or $3p$, is called a **polynomial.**

If every term in a polynomial contains only one variable, such as x, the polynomial is called a **polynomial in *x*.** If every coefficient in a polynomial is a real number, the polynomial is called a **polynomial over the reals;** a **polynomial over the rationals** and a **polynomial over the integers** are defined in a similar way. Thus, $6x^3 - 4x^2 + 3x - 5$ is a polynomial in x over the integers, and $\sqrt{3}m^4 - 5m^3y + 3y^4$ is a polynomial over the reals.

The highest exponent in a polynomial in one variable is called the **degree** of the polynomial. A constant is said to have degree zero. By the definition above, $3x^4 - 5x^2 + 2x + 3$ is a polynomial of degree 4. A term containing more than one variable is said to have degree equal to the sum of all exponents appearing on the variables in the term. For example, $-3x^4y^3z^5$ is of degree $4 + 3 + 5 = 12$. The degree of a polynomial in more than one variable is equal to the highest degree of any term appearing in the polynomial. Thus, $2x^4y^3 - 3x^5y + x^6y^2$ is of degree 8. A polynomial containing exactly three terms is called a **trinomial,** one containing exactly two terms is a **binomial,** while a single term polynomial is called a **monomial.**

Capital letters are often used to name polynomials; we shall use the notation $P(x)$ to represent a polynomial P containing x as the only variable. If $P(x) = 3x^2 - 4x + 5$, we shall use $P(2)$ to represent the value of the polynomial P when $x = 2$. Here,

$$P(x) = 3x^2 - 4x + 5,$$

and so

$$\begin{aligned} P(2) &= 3 \cdot 2^2 - 4 \cdot 2 + 5 \\ &= 3 \cdot 4 - 4 \cdot 2 + 5 \\ &= 12 - 8 + 5 \\ &= 9. \end{aligned}$$

In the same way, $P(-1) = 12$ and $P(\frac{1}{2}) = \frac{15}{4}$.

Sums and differences of polynomials can be found by using the distributive property. Thus

$$\begin{aligned} 3m^5 - 8m^5 &= (3 - 8)m^5 \\ &= -5m^5, \end{aligned}$$

and

$$\begin{aligned} (2y^4 - 3y^2 + y) + (4y^4 + 7y^2 + 6y) &= (2 + 4)y^4 + (-3 + 7)y^2 + (1 + 6)y \\ &= 6y^4 + 4y^2 + 7y. \end{aligned}$$

Before we can find products of polynomials we need to develop a few properties of exponents. By definition, a^m (m a natural number, a any real number) means $a \cdot a \cdots a$, where a appears as a factor m times. In the same way, a^n means $a \cdot a \cdots a$, where a appears as a factor n times. Thus in the product $a^m \cdot a^n$, a appears as a factor a total of $m + n$ times. Hence, for any natural numbers m and n and any real number a,

$$a^m \cdot a^n = a^{m+n}.$$

Similar proofs can be given for the other properties of exponents summarized in the following theorem.

THEOREM 2.1 For any natural numbers m and n, and for any real numbers a and b:

(a) $a^m \cdot a^n = a^{m+n}$

(b) $\dfrac{a^m}{a^n} = \begin{cases} a^{m-n} \text{ if } m > n \\ 1 \text{ if } m = n \\ \dfrac{1}{a^{n-m}} \text{ if } m < n \end{cases} \quad (a \neq 0)$

(c) $(a^m)^n = a^{mn}$

(d) $(ab)^m = a^m b^m$

(e) $\left(\dfrac{a}{b}\right)^m = \dfrac{a^m}{b^m}$, $(b \neq 0)$.

Example 1 Each of the following examples is justified by one or more parts of Theorem 2.1, and the properties of multiplication.

(a) $(4x^3)(-3x^5) = (4)(-3)x^3 \cdot x^5 = -12x^8$

(b) $-3^4(-3)^5 = -(3)^4 \cdot [-(3)^5] = 3^9$

(c) $-3^5(-3)^4 = -3^9$

(d) $\dfrac{k^{r+1}}{k^r} = k^{(r+1)-r} = k$, for any natural number r, $k \neq 0$

We can also use the distributive property to find the product of two polynomials when one of them is not a monomial. For example, if we treat $3x - 4$ as a single quantity, we can use the distributive property to write

$$(3x - 4)(2x^2 - 3x + 5) = (3x - 4)(2x^2) + (3x - 4)(-3x) + (3x - 4)(5).$$

If we use the distributive property again, we have

$$\begin{aligned}(3x - 4)&(2x^2 - 3x + 5) \\ &= (3x)(2x^2) - 4(2x^2) + (3x)(-3x) - 4(-3x) + (3x)5 - 4(5) \\ &= 6x^3 - 8x^2 - 9x^2 + 12x + 15x - 20 \\ &= 6x^3 - 17x^2 + 27x - 20.\end{aligned}$$

It is sometimes more convenient to write such a product as shown below.

$$\begin{array}{rrrr} & 2x^2 - & 3x + & 5 \\ & & 3x - & 4 \\ \hline & -\ 8x^2 + & 12x - & 20 \\ 6x^3 - & 9x^2 + & 15x & \\ \hline 6x^3 - & 17x^2 + & 27x - & 20 \end{array}$$

Example 2

$$\begin{aligned}(6m + 1)(4m - 3) &= (6m)(4m) + (6m)(-3) + 1(4m) + 1(-3) \\ &= 24m^2 - 14m - 3\end{aligned}$$

By Theorem 2.1 (b), we know that if a is any nonzero real number, and m and n are natural numbers, then

$$\frac{a^m}{a^n} = \begin{cases} a^{m-n} \text{ if } m > n \\ 1 \text{ if } m = n \\ \dfrac{1}{a^{n-m}} \text{ if } m < n. \end{cases}$$

What happens if we try to generalize this a little? For example,

$$1 = \frac{6^4}{6^4},$$

but if we try to subtract exponents, as in the theorem, we have

$$1 = \frac{6^4}{6^4} = 6^{4-4} = 6^0.$$

We shall use this example to make the following definition plausible. If a is any nonzero real number,

▶ $$a^0 = 1.$$

The symbol 0^0 is meaningless.

Example 3

(a) $3^0 = 1$ (b) $(-4)^0 = 1$
(c) $-4^0 = -1$ (d) $-(-4)^0 = -1$

It can be shown that the conditions of Theorem 2.1 can be broadened to include the cases when m or n equal 0; parts of the proof of this statement are included in the exercises.

2.1 Exercises

Let $P(x) = 2x^2 - 4x$ and $Q(x) = 2x + 1$. Find each of the following.

1. $P(-3)$
2. $P(4)$
3. $Q(1)$
4. $Q(-3)$
5. $P(\frac{1}{2})$
6. $Q(\frac{1}{4})$
7. $Q(m)$
8. $P(r)$
9. $P(-m)$
10. $Q(1 - r)$
11. $P(7 + h)$
12. $P(-3 + 2r)$
13. $P(-2x + 3)$
14. $Q\left(\dfrac{x-1}{2}\right)$
15. $P(x + r)$
16. $Q(x - r)$
17. $Q[P(2)]$
18. $P[Q(-1)]$
19. $P[P(-2)]$
20. $Q[Q(0)]$

Perform each of the following indicated operations.

21. $(3x^2 - 4x + 5) + (-2x^2 + 3x - 2)$
22. $(4m^3 - 3m^2 + 5) + (-3m^3 - m^2 + 5)$

23. $(12y^2 - 8y + 6) - (3y^2 - 4y + 2)$
24. $(8p^2 - 5p) - (3p^2 - 2p + 4)$
25. $-3(4q^2 - 3q + 2) + 2(-q^2 + q - 4)$
26. $2(3r^2 - 4r + 2) - 3(-r^2 + 4r - 5)$
27. $(4r - 1)(7r + 2)$
28. $(5m - 6)(3m + 4)$
29. $(6p + 5)(3p - 7)$
30. $(2z + 1)(3z - 4)$
31. $(5r + 2)(5r - 2)$
32. $(6z + 5)(6z - 5)$
33. $(2k - 3)^2$
34. $(3p + 5)^2$
35. $(k + 2)(12k^3 - 3k^2 + k + 1)$
36. $(2p - 1)(3p^2 - 4p + 5)$
37. $(2z - 1)^3$
38. $(3p - 1)(9p^2 + 3p + 1)$
39. $(2m + 1)(4m^2 - 2m + 1)$
40. $(x - 1)(x^2 - 1)$

Simplify each of the following. Assume all variables appearing as exponents represent natural numbers.

41. $(3m)^2(-2m)^3(-3m^4)$
42. $(-p^3)(2p^2)^3(-2p^4)$
43. $\dfrac{(8m^2)^3(12m^4)^2(-6m^3)}{(-4m^3)^4(3m^5)^2(2m^2)}$
44. $\dfrac{(3^3)^3(-2z^4)^5(32z^3)}{(2z^2)^3(-3z^5)^2(-2z^2)^3}$
45. $y^m \cdot y^{1+m}$
46. $z^{2r}z^{r+3}$
47. $\dfrac{(2m^n)^2(-4m^{2+n})}{8m^{4n}}$, $(n > 2)$
48. $\dfrac{(3 - y)^k(3 - y)^{2k}}{(3 - y)^2}$, $(k > 1)$

Prove each of the following statements for $n = 0$ and any natural number m.

49. $a^m a^n = a^{m+n}$
50. $(a^m)^n = a^{mn}$

2.2 Quotients of Polynomials

We shall define the **quotient** of the polynomials $P(x)$ and $Q(x)$ as

$$\frac{P(x)}{Q(x)} = K \qquad \text{if and only if} \qquad P(x) = K \cdot Q(x),$$

for some algebraic expression K. A quotient of polynomials is not defined if any of the variables assume a value for which the denominator is zero. If the expression K is a polynomial, then $P(x)$ is said to be exactly (or evenly) divisible by $Q(x)$; if K is not a polynomial, then $P(x)$ is not exactly divisible by $Q(x)$. To find the quotient of a polynomial $P(x)$ and a monomial $Q(x)$, we use the definition of division and find the product

$$P(x) \cdot \frac{1}{Q(x)}, \qquad (Q(x) \neq 0).$$

Example 4

(a) $$\frac{2m^5 - 6m^3}{2m^3} = (2m^5 - 6m^3) \cdot \frac{1}{2m^3} = m^2 - 3$$

This is because $2m^5 - 6m^3 = 2m^3(m^2 - 3)$. Since the quotient $m^2 - 3$ is a polynomial, we say that $2m^5 - 6m^3$ is exactly divisible by $2m^3$.

(b) $$\frac{3y^6x^3 - 6y^3x^6 + 8y^5x}{3y^3x^3} = \frac{3y^6x^3}{3y^3x^3} - \frac{6y^3x^6}{3y^3x^3} + \frac{8y^5x}{3y^3x^3} = y^3 - 2x^3 + \frac{8y^2}{3x^2}$$

Here the quotient is not a polynomial, and so $3y^6x^3 - 6y^3x^6 + 8y^5x$ is not exactly divisible by $3y^3x^3$.

To divide one polynomial by another, it is often convenient to use a **division algorithm.** (An algorithm is an orderly procedure for performing an operation.) For example, we can use a division algorithm here very similar to that used for whole numbers.

$$\begin{array}{r|l}
 & 2m^2 - 3m + \tfrac{1}{2} \\ \hline
2m - 1 & 4m^3 - 8m^2 + 4m + 6 \\
 & \underline{4m^3 - 2m^2} \\
 & \quad - 6m^2 + 4m \\
 & \quad \underline{- 6m^2 + 3m} \\
 & \qquad\qquad m + 6 \\
 & \qquad\qquad \underline{m - \tfrac{1}{2}} \\
 & \qquad\qquad\quad \tfrac{13}{2}
\end{array}$$

Hence, $$\frac{4m^3 - 8m^2 + 4m + 6}{2m - 1} = 2m^2 - 3m + \tfrac{1}{2} + \frac{\tfrac{13}{2}}{2m - 1}.$$

We shall often need to divide a polynomial by a first degree binomial of the form $x + k$, where the coefficient of x is 1. Thus it will prove beneficial to develop a shortcut for this process. To begin, let us consider an example.

$$\begin{array}{r|l}
 & 3x^2 + 10x + 40 \\ \hline
x - 4 & 3x^3 - 2x^2 \qquad\qquad - 150 \\
 & \underline{3x^3 - 12x^2} \\
 & \qquad 10x^2 \\
 & \qquad \underline{10x^2 - 40x} \\
 & \qquad\qquad\quad 40x - 150 \\
 & \qquad\qquad\quad \underline{40x - 160} \\
 & \qquad\qquad\qquad\qquad 10
\end{array}$$

To simplify this process, we can omit all variables and write only coefficients, using 0 to represent the coefficient of any missing variables. Since the coefficient of x in the division is always 1 in problems of this type, we can omit it too. These omissions simplify the problem as follows.

$$
\begin{array}{r|rrrr}
 & 3 & 10 & 40 & \\ \cline{2-5}
-4 & 3 & -\ 2 & 0 & -\ 150 \\
 & \mathbf{3} & -\ 12 & & \\ \cline{2-3}
 & & 10 & & \\
 & & \mathbf{10} & -\ 40 & \\ \cline{3-4}
 & & & 40 & -\ 150 \\
 & & & \mathbf{40} & -\ 160 \\ \cline{4-5}
 & & & & 10
\end{array}
$$

The boldface numbers are repetitions of the numbers directly above, and can be omitted.

$$
\begin{array}{r|rrrr}
 & 3 & 10 & 40 & \\ \cline{2-5}
-4 & 3 & -\ 2 & 0 & -\ 150 \\
 & & -\ 12 & & \\ \cline{2-3}
 & & 10 & & \\
 & & & -\ 40 & \\ \cline{3-4}
 & & & 40 & -\ 150 \\
 & & & & -\ 160 \\ \cline{4-5}
 & & & & 10
\end{array}
$$

The entire problem can now be compacted, with the top row of numbers omitted, since it merely duplicates the bottom row.

$$
\begin{array}{r|rrrr}
\cline{2-5}
-4 & 3 & -\ 2 & 0 & -\ 150 \\
 & & -\ 12 & -\ 40 & -\ 160 \\ \cline{2-5}
 & 3 & 10 & 40 & 10
\end{array}
$$

We obtained the bottom row by subtracting -12, -40, and -160 from the corresponding terms above. For reasons that will become clear later, we change the -4 at the left to 4, which also changes the sign of the numbers in the second row, and then add. Doing this, we have

$$
\begin{array}{r|rrrr}
\cline{2-5}
4 & 3 & -\ 2 & 0 & -\ 150 \\
 & & 12 & 40 & 160 \\ \cline{2-5}
 & 3 & 10 & 40 & 10
\end{array}
$$

This abbreviated process is called **synthetic division.**

Example 5 Divide $5m^3 - 6m^2 - 28m - 2$ by $m + 2$.

Using synthetic division, we begin by writing

$$
\begin{array}{r|rrrr}
\cline{2-5}
-2 & 5 & -\ 6 & -\ 28 & -\ 2
\end{array}
$$

First, we bring down the 5.

$$\begin{array}{r|rrrr} -2 & 5 & -6 & -28 & -2 \\ & & & & \\ \hline & 5 & & & \end{array}$$

Now, multiply (-2) by (5) to get -10, and add. The result is -16.

$$\begin{array}{r|rrrr} -2 & 5 & -6 & -28 & -2 \\ & & -10 & & \\ \hline & 5 & -16 & & \end{array}$$

Here, $(-2)(-16) = 32$. Again we add.

$$\begin{array}{r|rrrr} -2 & 5 & -6 & -28 & -2 \\ & & -10 & 32 & \\ \hline & 5 & -16 & 4 & \end{array}$$

Since $(-2)(4) = -8$, we have

$$\begin{array}{r|rrrr} -2 & 5 & -6 & -28 & 2 \\ & & -10 & 32 & -8 \\ \hline & 5 & -16 & 4 & -10 \end{array}$$

The quotient is read directly from the bottom row.

$$\frac{5m^3 - 6m^2 - 28m - 2}{m + 2} = 5m^2 - 16m + 4 + \frac{-10}{m + 2}$$

The degree of the quotient will always be one less than the degree of the polynomial to be divided.

2.2 Exercises

Perform each of the following divisions.

1. $\dfrac{4m^3 - 8m^2 + 16m}{2m}$
2. $\dfrac{30k^5 - 12k^3 + 18k^2}{6k^2}$
3. $\dfrac{6y^2 + y - 2}{2y - 1}$
4. $\dfrac{16r^2 + 2r - 3}{2r + 1}$
5. $\dfrac{20p^2 + 11p - 3}{4p + 3}$
6. $\dfrac{32m^2 + 20m - 3}{8m - 1}$
7. $\dfrac{12x^2 - 7x - 12}{4x - 5}$
8. $\dfrac{12p^2 - 13p - 16}{3p - 2}$
9. $\dfrac{8z^2 + 14z - 20}{2z + 5}$
10. $\dfrac{6a^2 - 13a - 18}{2a + 1}$
11. $\dfrac{2x^3 - 11x^2 + 19x - 10}{2x - 5}$
12. $\dfrac{3p^3 - 11p^2 + 5p + 3}{3p + 1}$

13. $\dfrac{15x^3 + 11x^2 + 20}{3x + 4}$

14. $\dfrac{4r^3 + 2r^2 - 14r + 15}{2r + 5}$

15. $\dfrac{4z^5 - 4z^2 - 5z + 3}{2z^2 + z + 1}$

16. $\dfrac{12z^4 - 25z^3 + 35z^2 - 26z + 10}{4z^2 - 3z + 5}$

Use synthetic division for each of the following.

17. $\dfrac{m^4 - 3m^3 - 4m^2 + 12m}{m - 2}$

18. $\dfrac{p^4 - 3p^3 - 5p^2 + 2p - 16}{p + 2}$

19. $\dfrac{3x^3 - 11x^2 - 20x + 3}{x - 5}$

20. $\dfrac{4p^3 + 8p^2 - 16p - 9}{p + 3}$

21. $\dfrac{4m^3 - 3m - 2}{m + 1}$

22. $\dfrac{3q^3 - 4q + 2}{q - 1}$

23. $\dfrac{x^5 + 3x^4 + 2x^3 + 2x^2 + 3x + 1}{x + 2}$

24. $\dfrac{m^6 - 3m^4 + 2m^3 - 6m^2 - 5m + 3}{m + 2}$

25. $\dfrac{\frac{1}{3}x^3 - \frac{2}{9}x^2 + \frac{1}{27}x + 1}{x - \frac{1}{3}}$

26. $\dfrac{x^3 + x^2 + \frac{1}{2}x + \frac{1}{8}}{x + \frac{1}{2}}$

27. $\dfrac{x^4 - 1}{x - 1}$

28. $\dfrac{x^7 + 1}{x + 1}$

29. $\dfrac{x^4 + 1}{x - 1}$

30. $\dfrac{x^7 - 1}{x + 1}$

2.3 Special Products and Factoring

The products mentioned in the following theorem occur so often in practice that they are referred to as special products.

THEOREM 2.2 For all real numbers a, b, c, d, and for all variables x and y:

(a) $x^2 + 2xy + y^2 = (x + y)^2$
(perfect square trinomial)

(b) $x^3 + 3x^2y + 3xy^2 + y^3 = (x + y)^3$

(c) $x^2 - y^2 = (x + y)(x - y)$
(difference of two squares)

(d) $x^2 + (a + b)x + ab = (x + a)(x + b)$

(e) $acx^2 + (ad + bc)x + bd = (ax + b)(cx + d)$

(f) $x^3 + y^3 = (x + y)(x^2 - xy + y^2)$
(sum of two cubes)

(g) $x^3 - y^3 = (x - y)(x^2 + xy + y^2)$
(difference of two cubes).

This result can be proved by the methods of multiplication of polynomials presented in Section 2.2.

The process of finding polynomials whose product equals a given polynomial is called **factoring** the given polynomial. For example, by (c) above,

$$4x^2 - 9 = (2x + 3)(2x - 3).$$

Example 6 Factor $m^2 + 2m - 24$.

We need to find two real numbers a and b such that $(m + a)(m + b) = m^2 + 2m - 24$. By Theorem 2.2(d) we see that the coefficient for m must be $(a + b)$ and the constant term must equal ab. Thus, $a + b = 2$ and $ab = -24$. That is, we need two numbers whose sum is 2 and whose product is -24. Since -4 and 6 satisfy these requirements, we can write

$$m^2 + 2m - 24 = (m - 4)(m + 6).$$

Example 7 Factor $6p^2 - 7pq - 5q^2$.

We use Theorem 2.5(e). We need real numbers a, b, c, and d such that $ac = 6$, $ad + bc = -7$, and $bd = -5$. In general, we use the method of trial and error to find these numbers. By inspection, we can find that

$$6p^2 - 7pq - 5q^2 = (3p - 5q)(2p + q).$$

Although it is true that $x^2 - 5 = (x + \sqrt{5})(x - \sqrt{5})$, we shall in general be interested only in factors having integer coefficients. For this reason, we say that $x^2 - 5$ cannot be factored over the integers. Note that $x^2 - 5$ can be factored over the reals, but not over the integers. Unless otherwise specified, factor shall mean to factor over the integers.

Example 8 Factor $m^3 - 64n^3$.

In Theorem 2.2(g), if we replace x with m and y with $4n$, we have

$$\begin{aligned} m^3 - 64n^3 &= m^3 - (4n)^3 \\ &= (m - 4n)[m^2 + m(4n) + (4n)^2] \\ &= (m - 4n)(m^2 + 4mn + 16n^2). \end{aligned}$$

Example 9 Factor $(a - 1)^3 + 1$.

We can use Theorem 2.2(f) with $x = a - 1$ and $y = 1$. We have

$$\begin{aligned} (a - 1)^3 + 1 &= [(a - 1) + 1][(a - 1)^2 - (a - 1) + 1] \\ &= a(a^2 - 2a + 1 - a + 1 + 1) \\ &= a(a^2 - 3a + 3). \end{aligned}$$

Example 10 Factor $p^6 - q^6$.

Use Theorem 2.2(g) with $x = p^2$ and $y = q^2$.

$$p^6 - q^6 = (p^2 - q^2)(p^4 + p^2q^2 + q^4)$$

Since $p^2 - q^2$ can be further factored, as $(p + q)(p - q)$, the result is

$$p^6 - q^6 = (p + q)(p - q)(p^4 + p^2q^2 + q^4).$$

Example 11 Factor $p^6 - q^6$ using Theorem 2.2(c).

We have
$$\begin{aligned} p^6 - q^6 &= (p^3)^2 - (q^3)^2 \\ &= (p^3 - q^3)(p^3 + q^3) \\ &= (p - q)(p^2 + pq + q^2)(p + q)(p^2 - pq + q^2). \end{aligned}$$

From this result, we see that we did not factor completely in Example 10 above. For this reason, the approach of Example 11 is better for this type of problem.

2.3 Exercises

Factor the following. Assume all variables used as exponents are real numbers.

1. $12mn - 8m$
2. $3pq - 18pqr$
3. $6px^2 - 8px^3 - 12px$
4. $9m^2n^3 - 18m^3n^2 + 27m^2n^4$
5. $4p^2 + 3p - 1$
6. $6x^2 + 7x - 3$
7. $2z^2 + 7z - 30$
8. $4m^2 + m - 3$
9. $6q^2 - q - 12$
10. $10b^2 - 19b - 15$
11. $12r^2 + 24r - 15$
12. $12p^2 + p - 20$
13. $18r^2 - 3rs - 10s^2$
14. $12m^2 + 16mn - 35n^2$
15. $64p^2q^2 - 121m^2n^2$
16. $169r^2 - 81s^2$
17. $81m^4 - 16n^8$
18. $256p^4 - 81q^4$
19. $9x^2 - 6x^3 + x^4$
20. $m^{16} - 1$
21. $m^6 - 1$
22. $a^3 + b^6$

23. $8m^6 - 27n^3$
24. $p^6 - q^6$
25. $x^2 - (x - y)^2$
26. $16m^2 - 25(m - n)^2$
27. $8y^3 + 27(x - y)^3$
28. $64p^6 - 81(p^2 - q)^3$
29. $(x + y)^2 + 2(x + y)z - 15z^2$
30. $(m + n)^2 + 3(m + n)p - 10p^2$
31. $6(a + b)^2 + (a + b)c - 40c^2$
32. $12(g + h)^2 - 5(g + h)j - 25j^2$
33. $(p + q)^2 - (p - q)^2$
34. $(p - q)^2 - (p + q)^2$
35. $m^{2n} - 16$
36. $p^{4n} - 49$
37. $x^{3n} - y^{6n}$
38. $a^9 - b^9$
39. $2x^{2n} - 23x^n y^n - 39y^{2n}$

2.4 Rational Expressions

Any expression which can be stated as the quotient of two polynomials (with nonzero denominator) is called a **rational expression.** Since a rational expression is analogous to a rational number (which is a quotient of integers), many of the results we proved in Chapter 1 for rational numbers generalize quickly to rational expressions. In particular, based on Theorem 1.6, we have the following result.

THEOREM 2.3 For all polynomials $P(x)$, $Q(x) \neq 0$, $R(x)$, and $S(x) \neq 0$, we have:

(a) $\dfrac{P(x)}{Q(x)} = \dfrac{R(x)}{S(x)}$ if and only if $P(x) \cdot S(x) = Q(x) \cdot R(x)$.

(b) $\dfrac{P(x)}{Q(x)} = \dfrac{P(x) \cdot S(x)}{Q(x) \cdot S(x)}$.
(fundamental theorem of rational expressions)

(c) $\dfrac{P(x)}{Q(x)} \cdot \dfrac{R(x)}{S(x)} = \dfrac{P(x) \cdot R(x)}{Q(x) \cdot S(x)}$.
(multiplication of rational expressions)

(d) $\dfrac{P(x)}{Q(x)} + \dfrac{R(x)}{Q(x)} = \dfrac{P(x) + R(x)}{Q(x)}$.
(addition of rational expressions)

(e) $\dfrac{P(x)}{Q(x)} - \dfrac{R(x)}{Q(x)} = \dfrac{P(x) - R(x)}{Q(x)}$.
(subtraction of rational expressions)

(f) $\dfrac{P(x)}{Q(x)} \div \dfrac{R(x)}{S(x)} = \dfrac{P(x) \cdot S(x)}{Q(x) \cdot R(x)}$, $(R(x) \neq 0)$.
(division of rational expressions)

We must be particularly careful to indicate the domain of the variable when using part (b) of Theorem 2.3. For example, -2 is not in the domain of the rational expression

$$\frac{x+6}{x+2},$$

since $x = -2$ makes the denominator zero. In the same way, -2 and -4 are not in the domain of

$$\frac{(x+6)(x+4)}{(x+2)(x+4)}.$$

Although $$\frac{x+6}{x+2} = \frac{(x+6)(x+4)}{(x+2)(x+4)}$$

by part (b) of the theorem, we should remember that this holds only when

$$x \neq -2, \; x \neq -4.$$

Example 12 $$\frac{3m^2-2m-8}{3m^2+14m+8} \cdot \frac{3m+2}{3m+4} = \frac{(m-2)(3m+4)}{(m+4)(3m+2)} \cdot \frac{3m+2}{3m+4}$$
$$= \frac{(m-2)(3m+4)(3m+2)}{(m+4)(3m+2)(3m+4)}$$
$$= \frac{m-2}{m+4} \qquad \text{for all } m \neq -4, \; m \neq -\tfrac{2}{3}, \; m \neq -\tfrac{4}{3}$$

From now on, we shall not state (unless specifically required) all restrictions on the variable. But keep in mind that no statement involving rational expressions is meaningful for values of the variable which make any denominator zero.

Example 13

$$\frac{6p^2-7pq-20q^2}{6p^2-25pq+25q^2} \div \frac{6p^2+11pq+4q^2}{2p^2+9pq+4q^2}$$
$$= \frac{(2p-5q)(3p+4q)(2p+q)(p+4q)}{(2p-5q)(3p-5q)(2p+q)(3p+4q)}$$
$$= \frac{p+4q}{3p-5q}$$

Example 14 $$\frac{y+2}{y-1} + \frac{y}{y+1} = \frac{(y+2)(y+1)}{(y-1)(y+1)} + \frac{y(y-1)}{(y-1)(y+1)}$$

(Here we use the fact that $(y-1)(y+1)$ is common denominator.)

$$= \frac{y^2+3y+2}{(y-1)(y+1)} + \frac{y^2-y}{(y-1)(y+1)}$$
$$= \frac{y^2+3y+2+y^2-y}{(y-1)(y+1)}$$
$$= \frac{2y^2+2y+2}{(y-1)(y+1)}$$
$$= \frac{2(y^2+y+1)}{(y-1)(y+1)}$$

We usually leave the denominator in factored form in a problem of this type.

Any quotient of two rational expressions in fractional form is called a **complex fraction.** Complex fractions can often be simplified by using Theorem 2.3, as shown in the following example.

Example 15 Simplify $$\frac{\dfrac{a}{a+1}+\dfrac{1}{a}}{\dfrac{1}{a}+\dfrac{1}{a+1}}.$$

We can first simplify numerator and denominator respectively.

$$\frac{\dfrac{a}{a+1}+\dfrac{1}{a}}{\dfrac{1}{a}+\dfrac{1}{a+1}} = \frac{\dfrac{a^2+1(a+1)}{a(a+1)}}{\dfrac{1(a+1)+1(a)}{a(a+1)}}$$

$$= \frac{\dfrac{a^2+a+1}{a(a+1)}}{\dfrac{2a+1}{a(a+1)}}$$

$$= \frac{a^2+a+1}{a(a+1)} \cdot \frac{a(a+1)}{2a+1}$$

$$= \frac{a^2+a+1}{2a+1}$$

We can also simplify the complex fraction by multiplying both numerator and denominator by the common denominator of all the denominators, in this case $a(a+1)$. We have

$$\frac{\dfrac{a}{a+1}+\dfrac{1}{a}}{\dfrac{1}{a}+\dfrac{1}{a+1}} = \frac{\left(\dfrac{a}{a+1}+\dfrac{1}{a}\right)a(a+1)}{\left(\dfrac{1}{a}+\dfrac{1}{a+1}\right)a(a+1)}$$

$$= \frac{a^2+(a+1)}{(a+1)+a}$$

$$= \frac{a^2+a+1}{2a+1}.$$

2.4 Exercises

Perform each of the following operations.

1. $\dfrac{2x-2}{3} \cdot \dfrac{6x-6}{(x-1)^3}$

2. $\dfrac{3m-15}{4m-20} \cdot \dfrac{m^2-10m+25}{12m-60}$

3. $\dfrac{a^2-a-6}{a+2} \div \dfrac{a-3}{a+2}$

4. $\dfrac{m^2+11m+30}{m+6} \div \dfrac{m+5}{m+6}$

5. $\dfrac{p^2 - p - 12}{p^2 - 2p - 15} \cdot \dfrac{p^2 - 9p + 20}{p^2 - 8p + 16}$

6. $\dfrac{x^2 + 2x - 15}{x^2 + 11x + 30} \cdot \dfrac{x^2 + 2x - 24}{x^2 - 8x + 15}$

7. $\dfrac{2m^2 - 5m - 12}{2m^2 + 5m - 12} \div \dfrac{2m^2 + 9m + 9}{2m^2 + 3m - 9}$

8. $\dfrac{3z^2 + z - 2}{4z^2 - z - 5} \div \dfrac{3z^2 + 11z + 6}{4z^2 + 7z - 15}$

9. $\left(1 + \dfrac{1}{x}\right)\left(1 - \dfrac{1}{x}\right)$

10. $\left(3 + \dfrac{2}{y}\right)\left(3 - \dfrac{2}{y}\right)$

11. $\dfrac{x^3 + y^3}{x^2 - y^2} \cdot \dfrac{x + y}{x^2 - xy + y^2}$

12. $\dfrac{4y^3 - 125}{4y^2 - 20y + 25} \cdot \dfrac{2y - 5}{y}$

13. $\dfrac{x^3 + y^3}{x^3 - y^3} \cdot \dfrac{x^2 - y^2}{x^2 + 2xy + y^2}$

14. $\dfrac{x^2 - y^2}{(x - y)^2} \cdot \dfrac{x^2 - xy + y^2}{x^2 - 2xy + y^2} \div \dfrac{x^3 + y^3}{(x - y)^4}$

15. $\dfrac{m}{m + n} + \dfrac{n}{m + n}$

16. $\dfrac{3}{y} + \dfrac{4}{y}$

17. $\dfrac{1}{y} + \dfrac{1}{y + 1}$

18. $\dfrac{2}{3(x - 1)} + \dfrac{1}{4(x - 1)}$

19. $\dfrac{2}{a + b} - \dfrac{1}{2(a + b)}$

20. $\dfrac{3}{m} - \dfrac{1}{m - 1}$

21. $\dfrac{1}{a + 1} - \dfrac{1}{a - 1}$

22. $\dfrac{1}{x + z} + \dfrac{1}{x - z}$

23. $\dfrac{m + 1}{m - 1} + \dfrac{m - 1}{m + 1}$

24. $\dfrac{2}{x - 1} + \dfrac{1}{1 - x}$

25. $\dfrac{3}{a - 2} - \dfrac{1}{2 - a}$

26. $\dfrac{3}{m-n} - \dfrac{m}{m+n}$

27. $\dfrac{1}{a^2-5a+6} - \dfrac{1}{a^2-4}$

28. $\dfrac{-3}{m^2-m-2} - \dfrac{1}{m^2+3m+2}$

29. $\dfrac{1}{x^2+x-12} - \dfrac{1}{x^2-7x+12} + \dfrac{1}{x^2-16}$

30. $\dfrac{2}{2p^2-9p-5} + \dfrac{p}{3p^2-17p+10} - \dfrac{2p}{6p^2-p-2}$

31. $\left(\dfrac{3}{p-1} - \dfrac{2}{p+1}\right)\left(\dfrac{p-1}{p}\right)$

32. $\left(\dfrac{y}{y^2-1} - \dfrac{y}{y^2-2y+1}\right)\left(\dfrac{y-1}{y+1}\right)$

33. $\dfrac{\dfrac{1}{x+h} - \dfrac{1}{x}}{h}$

34. $\dfrac{1}{h}\left(\dfrac{1}{(x+h)^2+9} - \dfrac{1}{x^2+9}\right)$

35. $\dfrac{1+\dfrac{1}{x}}{1-\dfrac{1}{x}}$

36. $\dfrac{2-\dfrac{2}{y}}{2+\dfrac{2}{y}}$

37. $\dfrac{1+\dfrac{1}{1-b}}{1-\dfrac{1}{1+b}}$

38. $m - \dfrac{m}{m+\dfrac{1}{2}}$

39. $1 + \dfrac{1}{1+\dfrac{1}{x}}$

40. $1 - \dfrac{1}{1-\dfrac{1}{x}}$

41. $1 - \dfrac{1}{1+\dfrac{1}{1+\dfrac{1}{1+\dfrac{1}{1+1}}}}$

42. $1 - \dfrac{1}{1+\dfrac{1}{1-\dfrac{1}{1+\dfrac{1}{1-\dfrac{1}{1+1}}}}}$

2.5 Integer Exponents

We have discussed natural number and zero exponents, and now we need to generalize this work to include all rational numbers as exponents. In this section we shall begin by discussing integer exponents (negative as well as nonnegative.) It would certainly be desirable to define integer exponents in such a way that all past results are still valid. For example, we know that for any nonnegative integers m and n, and any nonzero real number a,

$$\frac{a^m}{a^n} = a^{m-n}, \qquad \text{if } m > n.$$

To generalize this result, we let $m = 0$, and then we have

$$\frac{a^0}{a^n} = a^{0-n} = a^{-n}.$$

Since $a^0 = 1$, we have

$$a^{-n} = \frac{a^0}{a^n} = \frac{1}{a^n}.$$

Hence if negative integer exponents are to satisfy the same rules as nonnegative integer exponents, we must define a^{-n} as follows: for all integers n and all real numbers $a \neq 0$,

▶ $$a^{-n} = \frac{1}{a^n}.$$

For example, $3^{-1} = \frac{1}{3}$, $2^{-3} = \frac{1}{2^3} = \frac{1}{8}$, and $-4^{-2} = \frac{-1}{4^2} = \frac{-1}{16}$.

It can be shown that Theorem 2.1 can be generalized to include all integer exponents, and not just nonnegative integers. The result is given below.

THEOREM 2.4 For all integers m and n, and all real numbers $a \neq 0$ and $b \neq 0$, we have:

(a) $a^m \cdot a^n = a^{m+n}$

(b) $\dfrac{a^m}{a^n} = a^{m-n}$

(c) $(a^m)^n = a^{mn}$

(d) $(ab)^m = a^m b^m$

(e) $\left(\dfrac{a}{b}\right)^n = \dfrac{a^n}{b^n}$.

Note that our definition of integer exponents permits us to simplify part (b) of Theorem 2.1. Thus, instead of writing as we did before

$$\frac{m^5}{m^7} = \frac{1}{m^2}, \qquad \text{(since } 5 < 7\text{)}$$

we can now simply write

$$\frac{m^5}{m^7} = m^{5-7} = m^{-2}.$$

Example 16

(a) $2^{-4} \cdot 2^5 = 2^{-4+5} = 2^1 = 2$ Theorem 2.4 (a)

(b) $3x^{-2}(4^{-1}x^{-5})^2 = 3x^{-2}(4^{-2}x^{-10})$ Theorem 2.4 (d) and (c)

$= 3 \cdot 4^{-2}x^{-12}$ Theorem 2.4 (a)

$= \dfrac{3}{16x^{12}}$ definition of a negative exponent

Example 17 Simplify $\dfrac{(x+y)^{-1}}{x^{-1}+y^{-1}}$.

Using the definition of negative integer exponent, we can write

$$\frac{(x+y)^{-1}}{x^{-1}+y^{-1}} = \frac{\dfrac{1}{x+y}}{\dfrac{1}{x}+\dfrac{1}{y}} = \frac{\dfrac{1}{x+y}}{\dfrac{x+y}{xy}} = \frac{xy}{(x+y)^2}.$$

Example 18 Simplify $\dfrac{x^{-1}}{x^{-1}-y^{-1}}$.

Here we have

$$\frac{x^{-1}}{x^{-1}-y^{-1}} = \frac{\dfrac{1}{x}}{\dfrac{1}{x}-\dfrac{1}{y}} = \frac{\dfrac{1}{x}}{\dfrac{y-x}{xy}} = \frac{y}{y-x}.$$

2.5 Exercises

Simplify each of the following.

1. 2^{-3}
2. 3^{-2}
3. $(-4)^{-3}$
4. $(-5)^{-2}$
5. $(\frac{1}{2})^{-3}$
6. $(\frac{2}{3})^{-2}$
7. $2^{-3} - 2^{-2}$
8. $3^{-3} + 3^{-1}$
9. $(2^{-3})^{-2}$
10. $[(3^{-2})^2]^{-1}$

Write each of the following with no negative exponents. Assume all variables are nonzero real numbers.

11. $\dfrac{2^{-1}x^3y^{-3}}{xy^{-2}}$

12. $\dfrac{5^{-2}m^2y^{-1}}{5^2m^{-1}y^{-2}}$

13. $\dfrac{p^3r^2}{p^{-1}r^{-3}}$

14. $\dfrac{a^3b^{-2}}{a^{-1}b^4}$

15. $(-2x^{-3}y^2)^{-2}$

16. $(-3p^2q^{-2}s^{-1})^{12}$

17. $\dfrac{(3^{-1}m^{-2}n^2)^{-2}}{(mn)^{-1}}$

18. $\dfrac{(-4x^3y^{-2})^{-2}}{(4x^5y^4)^{-1}}$

19. $\dfrac{a^{-1}+b^{-1}}{(ab)^{-1}}$

20. $\dfrac{p^{-1}-q^{-1}}{(pq)^{-1}}$

21. $\dfrac{r^{-1}+q^{-1}}{r^{-1}-q^{-1}} \cdot \dfrac{r-q}{r+q}$

22. $\dfrac{xy^{-1}+yx^{-1}}{x^2+y^2}$

23. $(a+b)^{-1}(a^{-1}+b^{-1})$

24. $(m^{-1}+n^{-1})^{-1}$

25. $\dfrac{x-9y^{-1}}{(x-3y^{-1})(x+3y^{-1})}$

26. $\dfrac{(m+n)^{-1}}{m^{-2}-n^{-2}}$

2.6 Rational Exponents

We have discussed natural number exponents, and in the previous section we found that all results for natural number exponents apply also to integer exponents, provided we define integer exponents as we did in the last section. In this section we shall discuss rational number exponents, and find that rational exponents properly defined also obey the same rules as natural number exponents.

To begin, let us define the symbol $\sqrt[n]{a}$ as follows. If a is a nonnegative real number, and n is any natural number, or if a is negative, and n is an odd natural number, then

$$\sqrt[n]{a} = b \qquad \text{if and only if} \qquad b^n = a.$$

We call $\sqrt[n]{a}$ an **nth root of *a*.** When $a \geq 0$ and n is an even natural number, then $\sqrt[n]{a}$ represents the *positive* nth root of a. The general proof that such nth roots exist depends upon the completeness axiom (see Chapter 1) and is left for the reader to complete in Exercise 41. It is customary to write $\sqrt[2]{a}$ as $\sqrt{a}$ and to read both symbols as "square root of a", while $\sqrt[3]{a}$ is read "cube root of a." Note that $\sqrt[5]{32} = 2$ since $2^5 = 32$. It is important to note the restrictions stated in the definition. As long as a is a nonnegative real number, and n is any natural number, then $\sqrt[n]{a}$ is always a real number. However, if a is negative, then $\sqrt[n]{a}$ is a real number only for odd values of the natural number n. Thus, $\sqrt[3]{-27} = -3$, since $(-3)^3 = -27$, but $\sqrt[4]{-16}$ is not a real number, since there is no real number whose fourth power is -16.

The symbol $\sqrt[n]{a}$ is called a **radical,** while a itself is called a **radicand,** and n is called the **index** of the radical. Note that if a is positive and n is even, there are two different nth roots of a. One of these roots is positive, written $\sqrt[n]{a}$, while the other is negative and is written $-\sqrt[n]{a}$. Thus both -3 and 3 are 4th roots of 81, since $(-3)^4 = 81$ and also $3^4 = 81$. If n is odd, then for any real value of a there is exactly one nth root. The only cube root of 125 is 5, since 5 is the only real number with the property that $5^3 = 125$. If a is negative, and n is even, there are no real nth roots of a.

The basic properties of nth roots are summarized in the following results.

THEOREM 2.5 For any real numbers a and b and any natural number n for which the following roots are real, we have:

(a) $(\sqrt[n]{a})^n = a$

(b) $\sqrt[n]{a^n} = a$ if either $\begin{cases} a > 0 \\ a < 0 \text{ and } n \text{ is odd} \end{cases}$

(c) $\sqrt[n]{a^n} = |a|$ for $a < 0$, n even

(d) $\sqrt[n]{a} \cdot \sqrt[n]{b} = \sqrt[n]{ab}$

(e) $\dfrac{\sqrt[n]{a}}{\sqrt[n]{b}} = \sqrt[n]{\dfrac{a}{b}}$

(f) $\sqrt[m]{\sqrt[n]{a}} = \sqrt[mn]{a}$.

We shall prove only part of this result, with the remainder left as exercises. To prove (a), let $x = \sqrt[n]{a}$. Then, by the definition of nth root, $x^n = a$. By substitution, $(\sqrt[n]{a})^n = x^n = a$, which is what we wanted to show.

To verify (d), let $x = \sqrt[n]{a}$, and $y = \sqrt[n]{b}$. Then, by the definition of nth root, $x^n = a$ and $y^n = b$. Hence, $x^n \cdot y^n = ab$. However, by our previous work with exponents, we can write $x^n \cdot y^n = ab$ as

$$(xy)^n = ab,$$

which means xy is an nth root of ab, or

$$xy = \sqrt[n]{ab}.$$

Substituting the original values of x and y gives

$$\sqrt[n]{a} \cdot \sqrt[n]{b} = \sqrt[n]{ab}.$$

Let us now define the symbol $a^{1/n}$. We want to do this so as to be consistent with all earlier work on exponents. Thus, we must define $a^{1/n}$ so as to preserve the rule $(b^m)^n = b^{mn}$. If we replace b^m with $a^{1/n}$, we have

$$(a^{1/n})^n = a^{n/n} = a^1 = a.$$

Thus, to be consistent with past work we must define $a^{1/n}$ so that it is an nth root of a. That is, if n is a natural number and if $\sqrt[n]{a}$ exists, then

$$a^{1/n} = \sqrt[n]{a}.$$

What about rational exponents in general? Before we can define a symbol such as $a^{m/n}$, we might first note that in order for past results to be valid, we will need to have

$$(a^{1/n})^m = a^{m/n}$$

and

$$(a^m)^{1/n} = a^{m/n}.$$

Therefore, before we define $a^{m/n}$ we must first prove the equality of $(a^{1/n})^m$ and $(a^m)^{1/n}$.

THEOREM 2.6 For all integers m and n and all real numbers a for which all the indicated roots exist,

$$(a^m)^{1/n} = (a^{1/n})^m.$$

To prove this, we can use some of the previous work of this chapter.

$$[(a^{1/n})^m]^n = (a^{1/n})^{mn} = [(a^{1/n})^n]^m = a^m$$

Hence, $(a^{1/n})^m$ is an nth root of a^m, or $(a^{1/n})^m = (a^m)^{1/n}$.

Since $(a^{1/n})^m = (a^m)^{1/n}$ we shall define $a^{m/n}$ as follows. For all integers m and n and all real numbers a for which all the following roots exist,

$$a^{m/n} = (a^m)^{1/n} = (a^{1/n})^m$$

or, equivalently, $a^{m/n} = \sqrt[n]{a^m} = (\sqrt[n]{a})^m$.

It can now be shown that all the results of Theorem 2.4 concerning integer exponents also apply to rational exponents. We shall not prove this, but we shall assume it from now on.

Example 19

(a) $(-27)^{4/3} = [(-27)^{1/3}]^4 = (-3)^4 = 81$
Also, $(-27)^{4/3} = [(-27)^4]^{1/3} = 81$. (Note that $(-27)^4$ is positive.)

(b) $8^{4/3} = (2^3)^{4/3} = 2^4 = 16$

(c) $(-81)^{5/4} = \sqrt[4]{-81^5}$ which is not a real number.

(d) $81^{5/4} \cdot 4^{-3/2} = (3^4)^{5/4} \cdot (2^2)^{-3/2} = 3^5 \cdot 2^{-3} = \frac{3^5}{8}$

(e) $\left(\frac{3m^{5/6}}{y^{3/4}}\right)^2 \cdot \left(\frac{8y^3}{m^6}\right)^{2/3} = \frac{9m^{5/3}}{y^{3/2}} \cdot \frac{4y^2}{m^4} = \frac{36y^{1/2}}{m^{7/3}}$

2.6 Exercises

Simplify each of the following that exist. Assume all variables represent nonnegative real numbers.

1. $4^{1/2}$
2. $25^{1/2}$
3. $8^{2/3}$
4. $81^{3/4}$
5. $27^{-2/3}$
6. $32^{-4/5}$
7. $(\frac{4}{9})^{-3/2}$
8. $(\frac{1}{8})^{-5/3}$
9. $125^{-2/3}$
10. $64^{5/6}$
11. $(16p^4)^{1/2}$
12. $(36r^6)^{1/2}$
13. $(27x^6)^{2/3}$
14. $(64a^{12})^{5/6}$

Write each of the following with only positive exponents. Assume all variables represent positive real numbers, and variables used as exponents represent rational numbers.

15. $\left(\frac{x^4y^3z}{16x^{-6}yz^5}\right)^{-1/2}$
16. $\left(\frac{p^3r^9}{27p^{-3}r^{-6}}\right)^{-1/3}$
17. $(r^{1/2} - r^{-1/2})^2$
18. $(p^{1/2} - p^{-1/2})(p^{1/2} + p^{-1/2})$
19. $\frac{k^{3/2} \cdot k^{-1/2}}{k^{1/4} \cdot k^{3/4}}$
20. $\frac{m^{2/5} \cdot m^{3/5} \cdot m^{-4/5}}{m^{1/5} \cdot m^{-6/5}}$
21. $\frac{x^{-2/3} \cdot x^{4/3}}{x^{1/2} \cdot x^{-3/4}}$
22. $\frac{-4a^{1/2} \cdot a^{2/3}}{a^{-5/6}}$
23. $\frac{8y^{2/3}y^{-1}}{2^{-1}y^{3/4} \cdot y^{-1/6}}$
24. $\frac{9 \cdot k^{1/3} \cdot k^{-1/2} \cdot k^{-1/6}}{k^{-2/3}}$
25. $\frac{m^{1-a} \cdot m^a}{m^{-1/2}}$
26. $\frac{y^{3-b}y^{2b-1}}{y^{1/2}}$
27. $(r^{3/p})^{2p} \cdot (r^{1/p})^{p^2}$
28. $(m^{2/x})^{x/3} \cdot (m^{x/4})^{2/x}$
29. $[(x^{1/2} - x^{-1/2})^2 + 4]^{1/2}$
30. $[(x^{1/2} + x^{-1/2})^2 - 4]^{1/2}$

Write each of the following using rational exponents instead of radical signs. Assume all variables represent positive real numbers.

31. $\sqrt{12}$
32. $\sqrt{40}$
33. $\sqrt[3]{x^2}$
34. $p\sqrt[3]{p^4}$
35. $\sqrt[3]{y^4}$
36. $\sqrt[3]{z^5}$
37. $y^3\sqrt[4]{y^6}$
38. $p^2\sqrt[3]{\sqrt[4]{p^8}}$
39. $m^{-2/3}\sqrt[4]{m^6}$
40. $y^{-3/2}\sqrt[4]{\sqrt{y^6}}$
41. We can use the completeness axiom to verify that a number such as $\sqrt{2}$ actually exists. (Recall the completeness axiom: a nonempty set of real numbers with an upper bound has a least upper bound.) Let

$$A = \{x | x^2 < 2\}.$$

(a) Verify that A is not empty.
(b) Verify that $x = 2$ is an upper bound for set A.
(c) We showed in (b) that set A has an upper bound. By the completeness axiom, what conclusion can we draw about set A?
(d) What is a good name for the least upper bound of set A?

2.7 Simplifying Radicals

We shall consider an expression containing radicals **simplified** when all the following conditions are satisfied.

▶ (1) All radicals are removed from any denominators (a process called **rationalizing the denominator**).

▶ (2) All possible factors have been removed from under radical signs.

▶ (3) The index on the radical is as small as possible.

Using this definition, we write, for example,

$$\sqrt[3]{81x^5y^7z^6} = \sqrt[3]{27 \cdot 3x^3 \cdot x^2y^6 \cdot y \cdot z^6} = 3xy^2z^2\sqrt[3]{3x^2y}.$$

Also, $\sqrt[3]{\frac{3}{4}} = \sqrt[3]{\frac{3 \cdot 4 \cdot 4}{4 \cdot 4 \cdot 4}} = \sqrt[3]{\frac{48}{64}} = \frac{\sqrt[3]{48}}{\sqrt[3]{64}} = \frac{\sqrt[3]{48}}{4} = \frac{2\sqrt[3]{6}}{4} = \frac{\sqrt[3]{6}}{2}.$

Although we can write $\sqrt[6]{3^2}$ as $3^{2/6} = 3^{1/3} = \sqrt[3]{3}$, extreme care must be used in simplifying radicals such as $\sqrt[6]{x^2}$, since x can be positive or negative. Note that

$$\sqrt[6]{(-8)^2} = (-8)^{2/6} = [(-8)^2]^{1/6} = 64^{1/6} = 2;$$

while, even though $2/6 = 1/3$, $(-8)^{1/3} = -2$. Hence, in this case

$$\sqrt[6]{x^2} \neq \sqrt[3]{x}.$$

If x is nonnegative, then it is always true that $x^{m/n} = x^{mp/np}$; however it is not necessarily true if x is negative (although, it may be since $(-8)^{1/3} = (-8)^{3/9} = -2$). Reducing rational exponents on negative bases must be considered case by case.

The process of rationalizing the denominator, mentioned in (1) above, is perhaps best illustrated with a few examples.

Example 20 Rationalize the denominator of $\frac{1}{1 - \sqrt{2}}$.

The best approach to a problem such as this is to multiply both numerator and denominator by the conjugate of the denominator, in this case $1 + \sqrt{2}$. Doing this, we obtain

$$\frac{1}{1 - \sqrt{2}} = \frac{1 \cdot (1 + \sqrt{2})}{(1 - \sqrt{2})(1 + \sqrt{2})} = \frac{1 + \sqrt{2}}{1 - 2} = -1 - \sqrt{2}.$$

Example 21 Find a decimal approximation to $\frac{3}{2 + \sqrt{5}}$.

We could find a decimal approximation for $\sqrt{5}$ in the Appendix:

$$\sqrt{5} \approx 2.236.$$

Then we could write

$$\frac{3}{2+\sqrt{5}} \approx \frac{3}{2+2.236} = \frac{3}{4.236}.$$

which requires a lengthy calculation. It would be better to first rationalize the denominator.

$$\frac{3}{2+\sqrt{5}} = \frac{3(2-\sqrt{5})}{(2+\sqrt{5})(2-\sqrt{5})} = \frac{6-3\sqrt{5}}{4-5} = -6+3\sqrt{5}$$

Now we have

$$\frac{3}{2+\sqrt{5}} = -6+3\sqrt{5} \approx -6+3(2.236) = -6+6.708 = .708.$$

Example 22 Rationalize the denominator of $\dfrac{1}{x^{1/3}+y}$.

We use a process similar to, but more complicated than, the process used above for square roots. Recall that $a^3+b^3 = (a+b)(a^2-ab+b^2)$. Here we can let $a = x^{1/3}$ and $b = y$, and then multiply top and bottom by

$$a^2-ab+b^2 = x^{2/3}-x^{1/3}y+y^2.$$

Doing this, we have

$$\frac{1}{x^{1/3}+y} = \frac{1(x^{2/3}-x^{1/3}y+y^2)}{(x^{1/3}+y)(x^{2/3}-x^{1/3}y+y^2)} = \frac{x^{2/3}-x^{1/3}y+y^2}{x+y^3}.$$

Example 23 Simplify $\sqrt[3]{64m^4n^5} - \sqrt[3]{-27m^{10}n^{14}}$.

We have

$$\begin{aligned}\sqrt[3]{64m^4n^5} - \sqrt[3]{-27m^{10}n^{14}} &= 4mn\sqrt[3]{mn^2} - (-3)m^3n^4\sqrt[3]{mn^2}\\ &= 4mn\sqrt[3]{mn^2} + 3m^3n^4\sqrt[3]{mn^2}\\ &= (4+3m^2n^3)\,mn\sqrt[3]{mn^2}.\end{aligned}$$

Example 24 Write $\sqrt[3]{\sqrt{6}}$ using rational exponents.

Here $\sqrt[3]{\sqrt{6}} = \sqrt[3]{6^{1/2}} = (6^{1/2})^{1/3} = 6^{1/6}$.

2.7 Exercises

Simplify each of the following. Assume all variables represent positive real numbers.

1. $\sqrt[3]{81}$
2. $\sqrt{50}$
3. $\sqrt[4]{\dfrac{32}{81}}$
4. $\sqrt[3]{\dfrac{81}{16}}$
5. $\sqrt{24 \cdot 3^3 \cdot 2^4}$
6. $\sqrt[3]{16 \cdot (-2)^4 \cdot 2^8}$
7. $\sqrt{5}+\sqrt{20}$
8. $\sqrt{6}+\sqrt{54}$

9. $4\sqrt{3} - 5\sqrt{12} + 3\sqrt{75}$

10. $2\sqrt{5} - 3\sqrt{20} + 2\sqrt{45}$

11. $\sqrt{50} - 8\sqrt{8} + 4\sqrt{18}$

12. $6\sqrt{27} - 3\sqrt{12} + 5\sqrt{48}$

13. $3\sqrt[3]{16} - 4\sqrt[3]{2}$

14. $\sqrt[3]{2} - \sqrt[3]{16} + 2\sqrt[3]{54}$

15. $\dfrac{1}{\sqrt{3}} - \dfrac{2}{\sqrt{12}} + 2\sqrt{3}$

16. $\dfrac{1}{\sqrt{2}} + \dfrac{3}{\sqrt{8}} + \dfrac{1}{\sqrt{32}}$

17. $\sqrt{98x^5z^8}$

18. $\sqrt[3]{16z^5x^8y^4}$

19. $\sqrt[3]{81p^5x^2y^6}$

20. $\sqrt{242a^4b^3}$

21. $\sqrt[3]{x^{2/3}y^3}$

22. $\sqrt[4]{m^5 \cdot n^{3/2}}$

23. $\sqrt{\dfrac{2}{3x}}$

24. $\sqrt{\dfrac{5}{3p}}$

25. $\sqrt{\dfrac{9x}{2xy}}$

26. $\sqrt{\dfrac{8y}{9yz}}$

27. $\sqrt[3]{\dfrac{8}{x^2}}$

28. $\sqrt[3]{\dfrac{9}{16p^4}}$

29. $\sqrt{\dfrac{x^5y^3}{z^2}}$

30. $\sqrt[3]{\dfrac{k^5m^3 \cdot r^2}{r^8}}$

31. $\sqrt[4]{\dfrac{a^8b^5}{c^3}}$

32. $\sqrt{\dfrac{g^3h^5}{r^3}}$

33. $\sqrt[3]{\dfrac{9x^5y^6}{z^5w^2}}$

34. $\sqrt[4]{\dfrac{32m^5p^8}{q^3}}$

35. $\sqrt[4]{\dfrac{r^5 \cdot s^3}{t^2}}$

36. $\sqrt[3]{\dfrac{a^8 \cdot b^5}{c^2 \cdot d^5}}$

37. $\sqrt[4]{\dfrac{x^2}{y^5}}$

38. $\sqrt[4]{\dfrac{32}{8y^3}}$

39. $\dfrac{\sqrt[3]{mn} \cdot \sqrt[3]{m^2}}{\sqrt[3]{n^2}}$

40. $\dfrac{\sqrt[3]{8m^2n^3} \cdot \sqrt[3]{2m^2}}{\sqrt[3]{32m^4n^3}}$

41. $\sqrt{a^3b^5} - 2\sqrt{a^7b^3} + \sqrt{a^3b^9}$

42. $\sqrt{p^7q^3} - \sqrt{p^5q^9} + \sqrt{p^9q}$

43. $\dfrac{x}{\sqrt{x+1}}$

44. $\dfrac{\sqrt{p}}{1 - \sqrt{p}}$

45. $\dfrac{2}{\sqrt{1+r}}$

46. $\dfrac{y}{\sqrt{y+4}}$

47. $\dfrac{\sqrt{x} + \sqrt{x+1}}{\sqrt{x} - \sqrt{x+1}}$

48. $\dfrac{\sqrt{p} + \sqrt{p^2-1}}{\sqrt{p} - \sqrt{p^2-1}}$

Rationalize the numerator of each of the following.

49. $\dfrac{\sqrt{3}}{2}$ 50. $\dfrac{\sqrt{6}}{5}$ 51. $\dfrac{1+\sqrt{2}}{2}$ 52. $\dfrac{1-\sqrt{3}}{3}$

53. $\dfrac{\sqrt{x}}{1+\sqrt{x}}$

54. $\dfrac{\sqrt{p}}{1-\sqrt{p}}$

55. $\dfrac{\sqrt{x}+\sqrt{x+1}}{\sqrt{x}-\sqrt{x+1}}$

56. $\dfrac{\sqrt{p}+\sqrt{p^2-1}}{\sqrt{p}-\sqrt{p^2-1}}$

3

FIRST DEGREE RELATIONS AND FUNCTIONS

Relations and functions are a central idea of modern mathematics. In fact, much of the rest of this text is devoted to a discussion of some of the many types of relations and functions: quadratic, such as $y = x^2$; cubic, such as $y = (x + 1)^3$; exponential, such as $y = 2^x$; and so on. In this chapter, we discuss first degree relations and functions, such as $y = 2x - 4$ or $y = x - 2$. To begin, we will first define the term relation, and then the term function.

3.1 Relations and Functions

An **ordered pair** of numbers consists of two numbers written in parentheses, in which the sequence of numbers is important, such as (2, 4). Here 2 is called the **first component** and 4 the **second component.** A **relation** is any set of ordered pairs. Thus, $\{(2, -1), (4, -2)\}$ is a relation, as is

$$R = \{(x, y) \mid x < 4 \text{ and } y > 0\}.$$

(This is read "the set of all ordered pairs (x, y) such that $x < 4$ and $y > 0$.) The set of all first components of the ordered pairs of a relation is called the **domain** of the relation, while the set of all second components is called the **range** of the relation. In relation R above, the domain can be expressed as $\{x \mid x < 4\}$, while the range is $\{y \mid y > 0\}$.

Example 1 Find the domain and range of each of the following.

(a) $\{(x, y) \mid y = x^2\}$

Here x can take on any value, while y is always nonnegative, since $y = x^2$. The domain can be written $\{x \mid x \text{ is a real number}\}$, while the range is written $\{y \mid y \geq 0\}$.

(b) $\{(x, y) \mid y = \sqrt{x}\}$

Since $\sqrt{x}$ is defined only for nonnegative x, the domain here is $\{x \mid x \geq 0\}$. The symbol $\sqrt{x}$ indicates the nonnegative square root of x, and so the range is $\{y \mid y \geq 0\}$.

A **function** is a special type of relation with the property that for each domain element (x-value) there is exactly one range element (y-value). By this definition, $\{(x, y) \mid x = y^2\}$ is not a function (although it is a relation) since the domain element $x = 16$ leads to two range numbers, $y = 4$ and $y = -4$. On the other hand, the relations $\{(x, y) \mid y = x^2\}$ and $\{(x, y) \mid y = 2x - 5\}$ are functions since both have exactly one value of y for each value of x.

Letters such as f, g, and h are often used to name functions. Thus, we can write

$$f = \{(x, y) \mid y = 3x - 4\}.$$

For this function f, if $x = 5$, then $y = 3 \cdot 5 - 4 = 11$. Hence, $(5, 11) \in f$. This is often abbreviated by saying $f(5) = 11$ (read "f of 5 equals 11.") In the same way, $f(-3) = 3(-3) - 4 = -13$, $f(0) = -4$, and $f(x) = 3x - 4$. Using this new notation we shall often write

$$f(x) = 3x - 4$$

or

$$y = 3x - 4$$

instead of

$$f = \{(x, y) \mid y = 3x - 4\}.$$

Example 2 Let $f(x) = 3x - 4$ and $g(x) = 5x + 6$.

(a) Find $g(-6)$.
$g(-6) = 5(-6) + 6 = -30 + 6 = -24$.

(b) Find $f(4) + g(4)$.
$f(4) + g(4) = (3 \cdot 4 - 4) + (5 \cdot 4 + 6) = 8 + 26 = 34$.

(c) Find $f[g(2)]$.
Since $g(2) = 5 \cdot 2 + 6 = 16$, we have $f[g(2)] = f(16) = 44$.

(d) Find $g[g(-1)]$.
Since $g(-1) = 1$, we have $g[g(-1)] = g(1) = 11$.

An expression such as $f[g(x)]$ is called a **composite function.**

Example 3 Let $f(x) = 5x - 1$ and $g(x) = x^2$. Find $f[g(x)]$ and $g[f(x)]$.

We have $\quad f[g(x)] = f(x^2) = 5x^2 - 1,$

and $\quad g[f(x)] = g(5x - 1) = (5x - 1)^2.$

It is often helpful to draw a graph of the ordered pairs belonging to a relation. To do this we normally use the crossed number lines of a Cartesian coordinate system (named for its inventor, the philosopher-mathematician René Descartes). We shall let the number lines cross at right angles at their respective zero points, with positive numbers measured to the right on the horizontal number line, called the **x-axis,** and up along the vertical number line, called the **y-axis.** Thus in Figure 3.1, A represents the point whose coordinates are given by the ordered pair (3, 4), while B represents (−5, 6), C represents (−2, −4), D represents (4, −3) and E represents (−3, 0). Point A is said to be in the first **quadrant,** B is in the second quadrant, C is in the third quadrant, and D is in the fourth quadrant. The points of the axes themselves, such as E, belong to no quadrant.

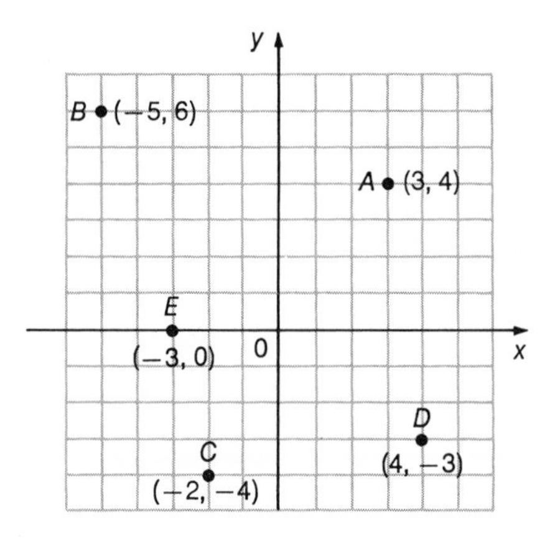

Figure 3.1

We can derive a useful graphical test for determining when a relation is also a function. By the definition of function, for each value of x in the domain there must be exactly one value of y in the range. Thus for a relation to be a function, as shown in Figure 3.2, a vertical line must never cut the graph of

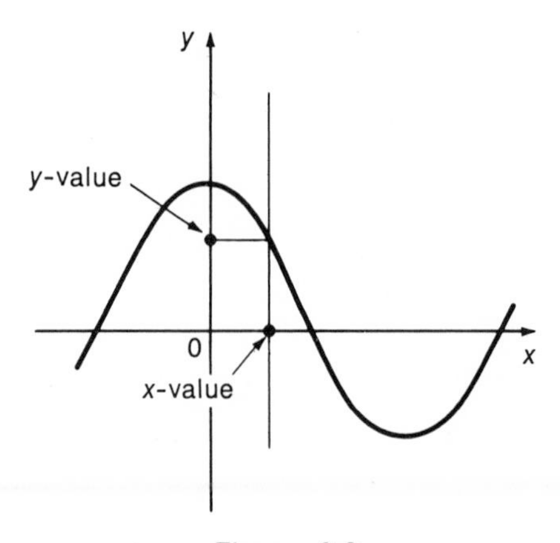

Figure 3.2

the relation in more than one point. Hence, Figure 3.3 represents a function, while Figure 3.4 and Figure 3.5 do not.

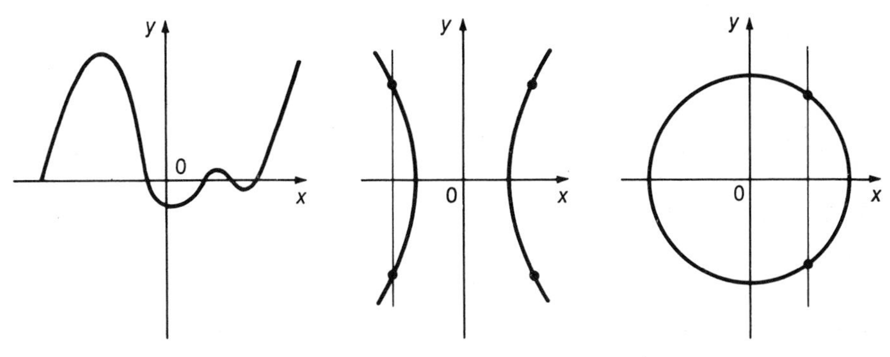

Figure 3.3 Figure 3.4 Figure 3.5

3.1 Exercises

Give the domain and range of each of the following relations. Identify those which are functions.

1. $\{(x,y) \mid y = 2x\}$
2. $\{(x,y) \mid y - 2 = x\}$
3. $\{(x,y) \mid y < x + 1\}$
4. $\{(x,y) \mid 3x - 1 < y\}$
5. $\{(x,y) \mid y = x^2\}$
6. $\{(x,y) \mid y = (x - 2)^2\}$
7. $\{(x,y) \mid y = \sqrt{16 - x^2}\}$
8. $\{(x,y) \mid y = |x|\}$
9. $\{(x,y) \mid y = -2\}$
10. $\{(x,y) \mid x = 3\}$
11. $\left\{(x, y) \mid y = \dfrac{1}{x}\right\}$
12. $\left\{(x, y) \mid y = \dfrac{2}{x - 1}\right\}$

Let $f(x) = 3x - 1$ and $g(x) = x^2$. Find each of the following.

13. $f(0)$
14. $f(-1)$
15. $f(-3)$
16. $f(4)$
17. $g(2)$
18. $g(0)$
19. $f(a)$
20. $g(b)$
21. $f[g(2)]$
22. $g[f(2)]$
23. $f[f(1)]$
24. $g[g(-2)]$

For all positive integers x, let $f(x) = 2^x$. Find each of the following.

25. $f(1)$
26. $f(2)$
27. $f(4)$
28. $f(m)$

Let $g = \{(x, y) \mid y = 3^{-x}$ for all integers $x\}$. Find each of the following.

29. $g(0)$
30. $g(2)$
31. $g(-1)$
32. $g(-2)$

A function f with the property that $f(-x) = f(x)$ for all values of x is called an *even function. A function f such that $f(-x) = -f(x)$ is called an odd function. Identify each of the following as odd, even, or neither.*

33. $f(x) = x^2$
34. $f(x) = x^3$
35. $f(x) = x^4 + x^2 + 5$
36. $f(x) = x^3 - x + 1$
37. $f(x) = 2x + 3$
38. $f(x) = |x|$

3.2 Linear Functions

As mentioned in the introduction to this chapter, there are many special kinds of functions. In this section we discuss an important first degree function (all terms having degree no greater than 1), the linear function. First, we define a **linear relation** as any relation of the form

$$\{(x, y) \mid ax + by = c;\ a,\ b \text{ not both } 0\}.$$

Here a, b, and c are real numbers, a restriction which will be assumed in the future. Any linear relation which is also a function is called a **linear function.** Hence, $\{(x, y) \mid 2x = y\}$ is a linear function since it is equivalent to

$$\{(x, y) \mid 2x - y = 0\},$$

where $a = 2$, $b = -1$, and $c = 0$. From now on, we shall abbreviate relations such as $\{(x, y) \mid 2x = y\}$ by merely writing the associated equation, in this case $2x = y$.

Which (if any) linear relations are not also linear functions? A relation is not a function if for some value of x, there is more than one value of y. If we let x equal some given real number, say $x = x_1$, then the definition of linear relation yields the equation $ax_1 + by = c$. For fixed real number values of a, x_1, and c, the only way there can be more than one value of y which satisfies $ax_1 + by = c$ is if $b = 0$. Hence, we have the following result.

THEOREM 3.1 If $b \neq 0$, then $ax + by = c$ is a linear function.

By Theorem 3.1, the only equations leading to linear relations that are not also functions are those of the form $x = k$, for some real number k.

Every linear relation has a graph which is a straight line. To find this straight line graph, it is often useful to find the x-intercept and y-intercept. An **x-intercept** is an x-value (if any) at which the graph of the relation crosses the x-axis, while a **y-intercept** is a y-value (if any) at which the graph crosses the y-axis. Note that the graph of Figure 3.6 has x-intercept x_1 and y-intercept y_1.

To find the x-intercept of a line $ax + by = c$, we can let $y = 0$ (see Figure 3.6; note that when the graph crosses the x-axis we have $y = 0$). Doing this, we have

$$\begin{aligned} ax + b \cdot 0 &= c \\ ax &= c \\ x &= c/a, \end{aligned}$$

where $a \neq 0$. In the same way, to find the y-intercept we let $x = 0$. Doing so, we have

$$\begin{aligned} a \cdot 0 + by &= c \\ by &= c \\ y &= c/b, \end{aligned}$$

where $b \neq 0$. Hence, we have proved the following result.

THEOREM 3.2 For any linear function $ax + by = c$, with $a \neq 0$ and $b \neq 0$, the x-intercept is c/a and the y-intercept is c/b.

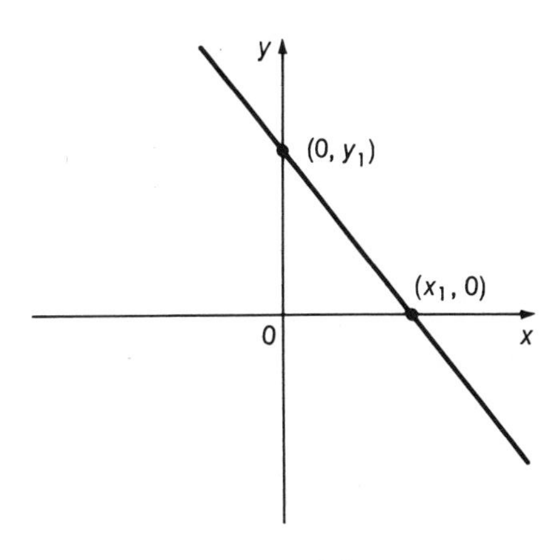

Figure 3.6

Example 4 Graph $3x + 2y = 6$.

Here $a = 3$, $b = 2$ and $c = 6$. Hence, the x-intercept is given by $c/a = 6/3 = 2$ and the y-intercept by $c/b = 6/2 = 3$. The graph is shown in Figure 3.7 on page 56.

Example 5 Graph $y = -3$.

To have an x-intercept there must be a value of x to make $y = 0$. However, here $y = -3 \neq 0$. Hence the line has no x-intercept. This happens only if the line is parallel to the x-axis, as shown in Figure 3.8 on page 56.

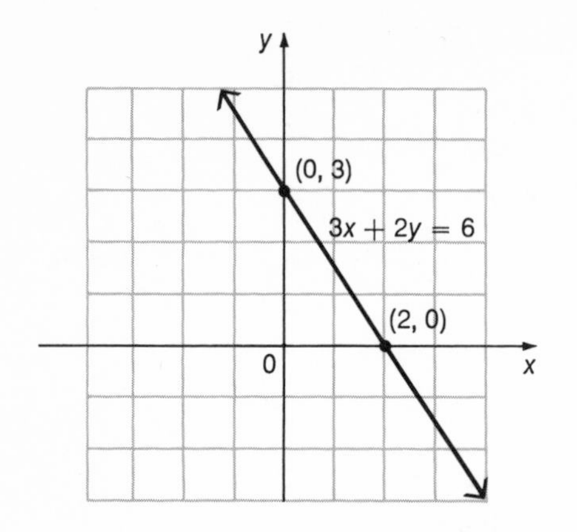

Figure 3.7

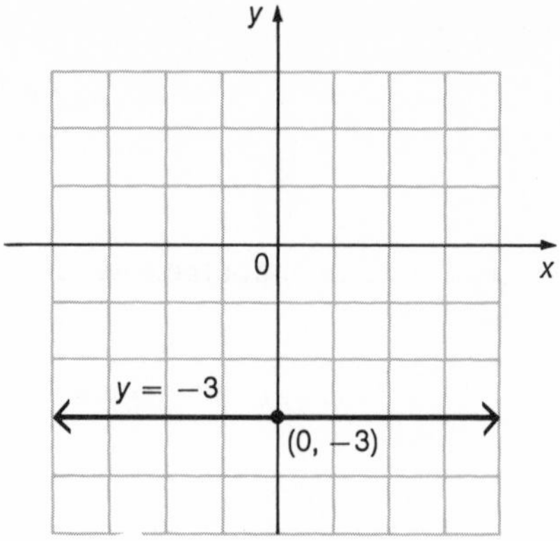

Figure 3.8

Example 6 Find the simultaneous solution set (the set of all ordered pairs satisfying both equations) of $2x - y = 7$ and $x - 2y = 8$.

One method we can use to find the solution set is to graph both lines and from the graph estimate the point (or points) of intersection (if any), as was done for this example in Figure 3.9. From Figure 3.9 we see that $\{(2, -3)\}$ is the simultaneous solution set.

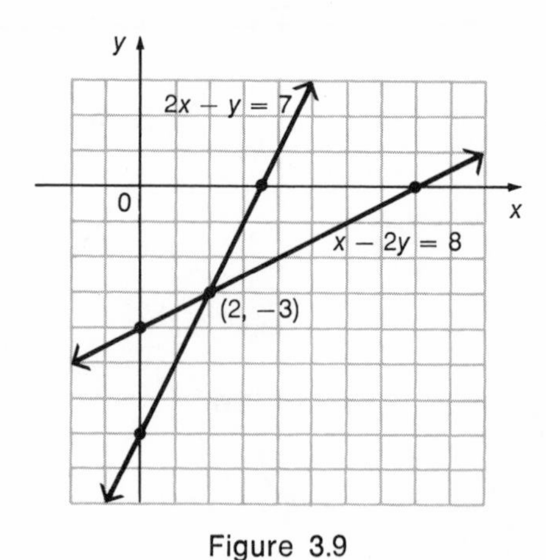

Figure 3.9

3.2 Exercises

Graph each of the following.

1. $y = 3x$
2. $x + y = 4$
3. $3x - y = 6$
4. $2x + 3y = 6$
5. $2x + 5y = 10$
6. $3x + 4y = 12$
7. $4x - 3y = 9$
8. $5x + 3y = 15$
9. $x = 2$
10. $y = -3$
11. $y = -5$
12. $x = 6$

Find the point of intersection (if any) of the following pairs of lines by graphing.

13. $x + y = 6$
    $x - y = 2$

14. $x + 2y = 9$
    $x - y = 0$

15. $3x + 4y = 7$
    $2x + 5y = 0$

16. $4x + y + 1 = 0$
    $2x + 3y = 7$

17. $3x + 2y = -2$
    $3x = 6 - 2y$

18. $4x + 5y = 10$
    $10y = 5 - 8x$

3.3 Straight Lines

As mentioned in the previous section, linear relations of the form $ax + by = c$ have straight lines as graphs. An important characteristic of a straight line is its slope, a numerical measure of the steepness of the line. To find this measure, let us consider the line going through the two distinct points (x_1, y_1) and (x_2, y_2), as shown in Figure 3.10. The difference

$$x_2 - x_1$$

is called the **change in *x*,** and is denoted Δx (read "delta x",) where Δ is the Greek letter delta. In the same way, the **change in *y*** can be written

$$\Delta y = y_2 - y_1.$$

The **slope** of a line (usually denoted m) is defined to be the change in y divided by the change in x, or

▶ $$m = \frac{\Delta y}{\Delta x} = \frac{y_2 - y_1}{x_2 - x_1} \qquad (\Delta x \neq 0).$$

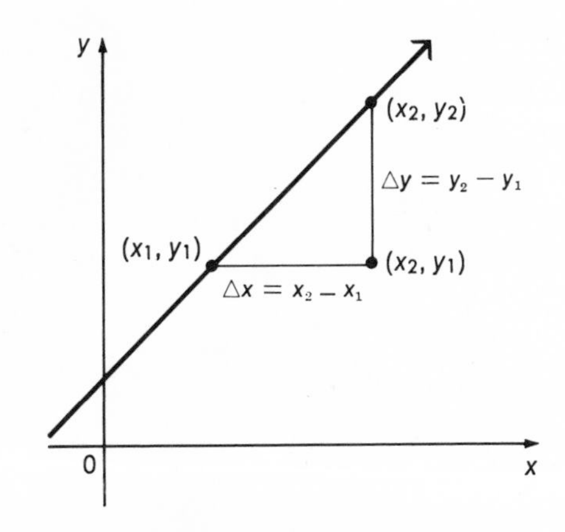

Figure 3.10

It can be shown (using theorems for similar triangles) that the slope is independent of the choice of points on the line. That is, the slope of a line remains the same no matter which pair of distinct points of the line determines the slope.

Example 7 Find the slope of the line through $(-4, 8)$ and $(2, -3)$.

We can choose $x_1 = -4$, $y_1 = 8$, $x_2 = 2$ and $y_2 = -3$. Hence, $\Delta y = -3 - 8 = -11$ and $\Delta x = 2 - (-4) = 6$. Thus the slope is given by

$$m = \frac{\Delta y}{\Delta x} = \frac{-11}{6}.$$

Example 8 Find the slope of the line $3x - 4y = 12$.

We first find any two points that lie on the line, for example, $(0, -3)$ and $(4, 0)$. Using these points and the definition of slope, we have

$$m = \frac{\Delta y}{\Delta x} = \frac{0 - (-3)}{4 - 0} = \frac{3}{4}.$$

Example 9 Find the slope of the line $y = -3$.

Again, we select any two points that lie on the line. Here we might select $(5, -3)$ and $(-2, -3)$. We can then use the definition of slope.

$$m = \frac{\Delta y}{\Delta x} = \frac{-3 - (-3)}{-2 - 5} = 0$$

Recall that the graph of $y = -3$ is a horizontal line. In general, as suggested by this example, the slope of a horizontal line is 0.

Example 10 Find the slope of $x = 2$.

Here $(2, 1)$ and $(2, -4)$ lie on the line. The slope is given by

$$m = \frac{\Delta y}{\Delta x} = \frac{-4 - 1}{2 - 2} = -\frac{5}{0},$$

which is not a real number. The graph of $x = 2$ is a vertical line; by generalizing, a vertical line has no slope.

The next theorem provides a convenient method to determine the slope and y-intercept of a line. We shall then use these numbers as an aid in drawing the graph.

THEOREM 3.3 SLOPE-INTERCEPT FORM
If a linear function is given by $\boldsymbol{y = mx + b}$,
then m is the slope of the line and b is the y-intercept.

The proof that b is the y-intercept will be left as an exercise. To prove that m is the slope we must consider several cases; the proof of the case where $m \neq 0$, $b \neq 0$ is as follows.

As above, to find the slope we find two distinct points that lie on the line. Here (as can be seen by substitution) the points $(0, b)$ and $(-b/m, 0)$ lie on the line. Since $b \neq 0$, these points are distinct. Using these two points and the definition of slope we can find the slope of the line.

$$\text{slope} = \frac{\Delta y}{\Delta x} = \frac{0 - b}{-\frac{b}{m} - 0} = \frac{-b}{-\frac{b}{m}} = m$$

Hence, the slope is given by m, which is what we wanted to show. To complete the proof, we need to consider the case where $m \neq 0$, and $b = 0$, and the case where $m = 0$ and $b \neq 0$.

Example 11 Find the slope and y-intercept of $3x - y = 2$. Graph the line.

We first write $3x - y = 2$ in the form $y = mx + b$. Here, $y = 3x - 2$. From this we see that the slope is $m = 3$ and the y-intercept is $b = -2$. To draw the graph we first locate the y-intercept. (See Figure 3.11.) By the definition of slope, $m = \frac{\Delta y}{\Delta x}$, if $\Delta x = 1$, then $\Delta y = m$. Hence, if x changes by 1 unit, y changes by m units. Here we know $m = 3 = 3/1$. Thus, if x changes 1 unit, y changes 3 units. We can use this fact to find another point of the graph, as shown in Figure 3.11.

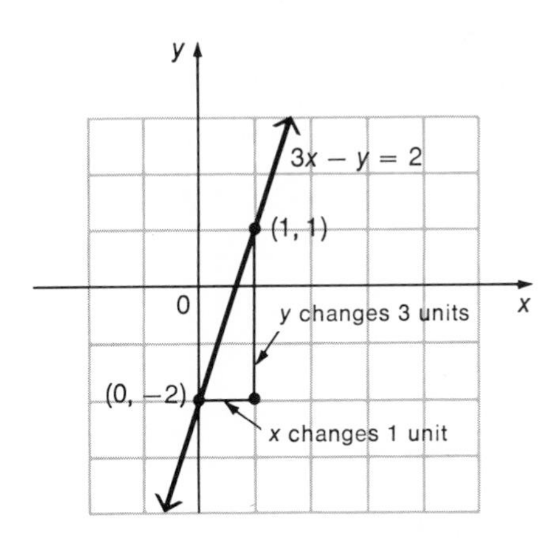

Figure 3.11

Sometimes we know the slope of a line together with one point that the line passes through. In such a case it is often helpful to use the point-slope form of the equation of a line given in the next theorem.

THEOREM 3.4 POINT-SLOPE FORM

If a line has slope m and passes through (x_1, y_1), then the equation of the line is given by

$$y - y_1 = m(x - x_1).$$

We leave it to the reader to show that the point (x_1, y_1) satisfies the given equation and that m is the slope of the line $y - y_1 = m(x - x_1)$. (Hint: Solve the equation for y and then use Theorem 3.3.)

Example 12 Write and simplify the equation of the line going through $(-4, 1)$ with slope $m = -3$.

Here, $x_1 = -4$, $y_1 = 1$ and $m = -3$. Using the point-slope form of the equation of a line, we have

$$\begin{aligned} y - 1 &= -3[x - (-4)] \\ y - 1 &= -3(x + 4) \\ y - 1 &= -3x - 12 \\ y &= -3x - 11. \end{aligned}$$

Example 13 Find and simplify the equation of the line through $(-3, 2)$ and $(2, -4)$.

Here $m = \dfrac{-4 - 2}{2 - (-3)} = \dfrac{-6}{5}$. If we let $x_1 = -3$ and $y_1 = 2$, we have

$$\begin{aligned} y - 2 &= -\frac{6}{5}[x - (-3)] \\ 5(y - 2) &= -6(x + 3) \\ 5y - 10 &= -6x - 18 \\ 5y + 6x + 8 &= 0. \end{aligned}$$

Verify that we get the same equation if we use $(2, -4)$ instead of $(-3, 2)$ in the point-slope form.

3.3 Exercises

Find the slope of each of the following lines.

1. $y = 2x$
2. $y = 3x - 2$
3. $y = 4x - 5$
4. $y = -2x + 1$
5. $3x + 4y = 6$
6. $2x + y = 8$
7. through $(-2, 1)$ and $(3, 2)$
8. through $(-2, 3)$ and $(-1, 2)$
9. through $(0, 4)$ and $(-1, -3)$
10. through $(-4, -3)$ and $(5, 0)$
11. $y = 4$
12. $x = -6$
13. the x-axis
14. the y-axis

Find an equation for each of the following lines.

15. through $(-1, 3)$, $m = 2$
16. through $(2, 4)$, $m = -1$
17. through $(-5, 4)$, $m = -\frac{3}{2}$
18. through $(-4, 3)$, $m = \frac{3}{4}$
19. through $(2, 0)$, $m = -\frac{3}{4}$
20. through $(0, -1)$, $m = \frac{3}{5}$
21. through $(-3, 2)$, $m = 0$
22. through $(6, 1)$, $m = 0$
23. through $(-8, 1)$, no slope
24. through $(0, 4)$, no slope
25. through $(4, 2)$ and $(1, 3)$
26. through $(8, -1)$ and $(4, 3)$
27. through $(-1, 3)$ and $(3, 4)$
28. through $(6, 0)$ and $(3, 2)$
29. through $(0, 3)$ and $(4, 0)$
30. through $(-3, 0)$ and $(0, -5)$
31. x-intercept 3, y-intercept -2
32. x-intercept -2, y-intercept 4
33. x-intercept -5, no y-intercept
34. y-intercept 3, no x-intercept

Graph each of the following lines.

35. through $(-1, 2)$, $m = \frac{3}{2}$
36. through $(-2, 8)$, $m = -1$
37. through $(3, -4)$, $m = -\frac{1}{3}$
38. through $(-2, -3)$, $m = -\frac{3}{4}$
39. through $(-3, -4)$, $m = -\frac{3}{5}$
40. through $(3, 4)$, $m = \frac{2}{3}$

Two lines having the same slope are parallel, while two lines whose slopes have a product of -1 are perpendicular. Write equations for each of the following lines.

41. through $(-1, 4)$, parallel to $x + 3y = 5$
42. through $(2, -5)$, parallel to $y - 4 = 2x$
43. through $(3, -4)$, perpendicular to $x + y = 4$
44. through $(-2, 6)$, perpendicular to $2x - 3y = 5$
45. Do the points $(4, 3)$, $(2, 0)$, and $(-18, -12)$ lie on the same line? (Hint: find the equation of the line through two of the points.)
46. Do the points $(4, -5)$, $(3, -5/2)$, and $(-6, 18)$ lie on a line?
47. Prove that b is the y-intercept in Theorem 3.3.

3.4 Solving Equations and Inequalities

In this section we shall discuss methods of finding the solution sets of linear equations and inequalities. First, we shall say that any two equations having the same domain and the same solution set are **equivalent.** Hence, $2x = 16$ and $3x - 5 = 19$ are equivalent, since they both have the same solution set, $\{8\}$. The next theorem, which follows from the addition and multiplication properties of equality, is useful in finding the solution set of linear equations.

THEOREM 3.5 If $P(x)$, $Q(x)$, and $R(x)$ are algebraic expressions, then:

(a) $P(x) = Q(x)$ and $P(x) + R(x) = Q(x) + R(x)$ are equivalent;

(b) $P(x) = Q(x)$ and $P(x) \cdot R(x) = Q(x) \cdot R(x)$ are equivalent, if $R(x) \neq 0$.

To solve an equation such as

$$\frac{3p-1}{3} - \frac{2p}{p-1} = p,$$

we first obtain a (simpler) equivalent equation by multiplying both sides by $3(p-1)$, where we must assume $p \neq 1$, (Why?) From this, we have,

$$\begin{aligned} (3p-1)(p-1) - 3(2p) &= 3p(p-1) \\ 3p^2 - 4p + 1 - 6p &= 3p^2 - 3p. \end{aligned}$$

We can obtain an even simpler equivalent equation by combining terms and adding $-3p^2$ to both sides.

$$-10p + 1 = -3p$$

If we now add $10p$ to both sides, we obtain the equivalent equation

$$1 = 7p.$$

Multiplying both sides by $\frac{1}{7}$ gives the equivalent equation

$$\tfrac{1}{7} = p,$$

so that the solution set of the given equation is $\{\frac{1}{7}\}$.

To solve

$$\frac{x}{x-2} = \frac{2}{x-2} + 2,$$

we multiply both sides of the equation by $x - 2$, assuming that $x - 2 \neq 0$. This gives

$$\begin{aligned} x &= 2 + 2x - 4 \\ x &= 2. \end{aligned}$$

But we had to assume that $x - 2 \neq 0$ in order to use Theorem 3.5(b). Since $x = 2$, $x - 2 = 0$, and the theorem does not apply. The solution set is $\varnothing$.

Example 14 Solve $J\left(\frac{x}{k} + a\right) = x$ for x.

If we first multiply both sides by k, we have

$$\begin{aligned} kJ\left(\frac{x}{k} + a\right) &= kx \\ kJ\left(\frac{x}{k}\right) + kJa &= kx \\ Jx + kJa &= kx. \end{aligned}$$

If we add $-Jx$ to both sides, we have

$$kJa = kx - Jx$$
$$kJa = x(k - J).$$

If we assume $k \neq J$, we can multiply both sides by $\frac{1}{k - J}$ to find the solution

$$\frac{kJa}{k - J} = x.$$

Example 15 Melinda and Hortense win equal amounts of cash in a contest. Melinda will spend her money in 8 years, while Hortense will require 10. If they both work together on only one of the prizes, how long will it take them to spend it?

Let x represent the number of years it will take the two women working together to spend a single prize. In one year, Melinda will spend $\frac{1}{8}$ of the amount while Hortense will spend $\frac{1}{10}$. Hence,

$$\frac{1}{8} + \frac{1}{10} = \frac{1}{x}$$
$$5x + 4x = 40$$
$$x = \frac{40}{9} \text{ years.}$$

Working together, the women can spend one prize in $\frac{40}{9}$ years, or about 4 years, 5 months.

To solve a linear inequality, such as $3x - 5 < 7$, we use the following properties of inequalities which are consequences of Theorem 1.8.

THEOREM 3.6 If $P(x)$, $Q(x)$, and $R(x)$ are algebraic expressions, then:
(a) $P(x) < Q(x)$ and $P(x) + R(x) < Q(x) + R(x)$ are equivalent (have the same solution set);
(b) $P(x) < Q(x)$ and $P(x) \cdot R(x) < Q(x) \cdot R(x)$ are equivalent, if $R(x) > 0$;
(c) $P(x) < Q(x)$ and $P(x) \cdot R(x) > Q(x) \cdot R(x)$ are equivalent, if $R(x) < 0$.

To use the theorem to solve the inequality $3x - 5 < 7$, we write the following series of equivalent inequalities.

$$3x - 5 + 5 < 7 + 5$$
$$3x < 12$$
$$x < 4$$

Thus, the solution set of $3x - 5 < 7$ is $\{x \mid x < 4\}$.

Example 16 Solve $4 - 3y \le 7 + 2y$.

We can write the following series of equivalent inequalities.

$$4 - 3y \le 7 + 2y$$
$$-5y \le 3$$
$$y \ge -\frac{3}{5}$$

From this last inequality, we see that the solution set is $\{y | y \ge -\frac{3}{5}\}$.

To solve an equation involving absolute values, such as $|2x + 5| = 9$, we use the definition of absolute value to write

$$2x + 5 = 9 \quad \text{or} \quad 2x + 5 = -9.$$

The solution to this compound sentence is

$$x = 2 \quad \text{or} \quad x = -7.$$

A similar technique is used for inequalities, as shown in the next two examples.

Example 17 Solve $|3p - 1| \le 5$.

By Theorem 1.9, $|x| \le b$ for nonnegative b implies $-b \le x \le b$. Hence, for $|3p - 1| \le 5$ we may write

$$-5 \le 3p - 1 \le 5$$
$$-4 \le 3p \le 6$$
$$-\frac{4}{3} \le p \le 2.$$

The solution set is $\{p | -\frac{4}{3} \le p \le 2\}$.

Example 18 Solve $|4m - 3| \ge 13$.

Again by Theorem 1.9, for all nonnegative b, $|x| \ge b$ implies $x \ge b$ or $x \le -b$. Hence, we have

$$4m - 3 \ge 13 \quad \text{or} \quad 4m - 3 \le -13$$
$$4m \ge 16 \quad \text{or} \quad 4m \le -10$$
$$m \ge 4 \quad \text{or} \quad m \le -\frac{5}{2}.$$

The solution set can be written $\{m | m \ge 4 \text{ or } m \le -\frac{5}{2}\}$.

3.4 Exercises

Solve each of the following.

1. $4x - 1 = 15$
2. $-3y + 2 = 5$

3. $-2[m - (4 + 2m) + 3] = 2m + 2$
4. $4[2p - (3 - p) + 5] = -7p - 2$
5. $\frac{3x - 2}{7} = \frac{x + 2}{5}$
6. $\frac{2p + 5}{5} = \frac{p + 2}{3}$
7. $\frac{x}{3} - 7 = 6 - \frac{3x}{4}$
8. $\frac{4p - 2}{7} = \frac{3p - 2}{4}$
9. $\frac{m}{2} - \frac{1}{m} = \frac{6m + 5}{12}$
10. $\frac{6z + 2}{5} - \frac{3z + 5}{3} = \frac{z + 1}{15}$
11. $\frac{2r}{r - 1} = 5 + \frac{2}{r - 1}$
12. $\frac{5}{2a + 3} + \frac{1}{a - 6} = 0$
13. $\frac{x + 4}{x - 3} - \frac{8x}{2x + 5} + \frac{3x}{x - 3} = 0$
14. $\frac{5}{2p + 3} - \frac{3}{p - 2} = \frac{4}{2p + 3}$
15. $2x + 1 \leq 9$
16. $3y - 2 < 10$
17. $-3p - 2 \leq 1$
18. $-5r + 3 \leq -2$
19. $|a - 2| = 1$
20. $|3 - x| = 2$
21. $|3m - 1| = 2$
22. $|4p + 2| = 5$
23. $|5 - 3x| = 3$
24. $|-2 + 5a| = 3$
25. $|a - 4| < 6$
26. $|z + 5| \leq 2$
27. $|3m - 2| > 1$
28. $|4z + 6| \geq 6$
29. $|8b + 5| \geq 7$
30. $|2m - 5| > 3$

Solve each of the following for the letter indicated.

31. $PV = k$ for V
32. $s = \frac{1}{2}gt^2$ for g
33. $V = V_0 + gt$ for g
34. $C = \frac{5}{9}(F - 32)$ for F
35. $\frac{1}{R} = \frac{1}{r_1} + \frac{1}{r_2}$ for R, (all denominators not equal to 0)
36. $A = \frac{1}{2}h(b + B)$ for B
37. Harry can do a job in 9 hours, and Mert in 6. How long will it take them if they work together on the same job?
38. An inlet pipe can fill Tom's pool in 5 hours, while an outlet pipe can empty it in 8 hours. In his haste to watch television, Tom left both pipes open. How long would it then take to fill the pool?
39. Suppose Tom discovered his error (see Exercise 38) after an hour long program. If he then closed the outlet pipe, how much longer would be needed to fill the pool?

3.5 Graphing Linear Inequalities

A line divides a plane into three sets of points–the points of the line itself, and the points belonging to the two regions determined by the line. Each of these

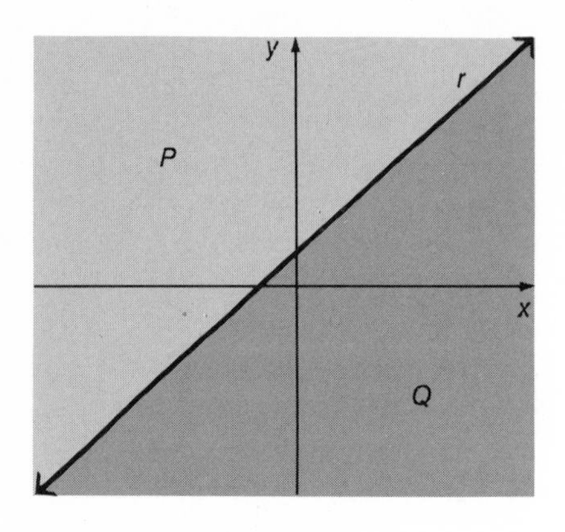

Figure 3.12

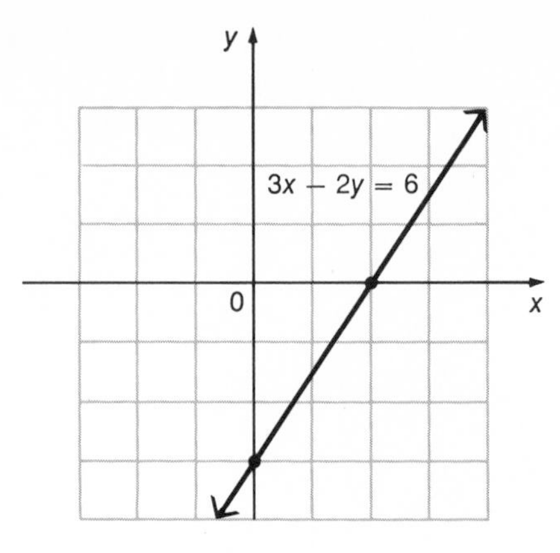

Figure 3.13

two regions is called a **half-plane.** In Figure 3.12, line r divides the plane into three different sets of points — line r, half-plane P, and half-plane Q. Note that r belongs neither to P nor to Q.

We shall use the idea of a half-plane as an aid in graphing linear inequalities, as described in the next theorem which we state without proof.

THEOREM 3.7 The graph of the linear inequality $ax + by < c$ is the half-plane, determined by the line $ax + by = c$, whose points satisfy the inequality $ax + by < c$. Also, the graph of the linear inequality $ax + by \leq c$ is the union of the appropriate half-plane determined by the line $ax + by = c$ and the points of the line itself.

By Theorem 3.7, the graph of $3x - 2y \leq 6$ is a half-plane and its boundary line; to find the graph we first draw the line $3x - 2y = 6$, as shown in Figure 3.13. To determine which half-plane belongs to the graph, we select any point not on the line. If and only if this point satisfies the inequality, then the half-plane from which the point was chosen belongs to the graph. Often, $(0, 0)$ is a useful point for this test. If we select $(0, 0)$ here, we have $3(0) - 2(0) \leq 6$. Since this statement is true, the half-plane including $(0, 0)$ belongs to the graph of $3x - 2y \leq 6$. The final graph is shown in Figure 3.14.

Example 19 Graph $x + 4y > 4$.

Here the graph consists only of a half-plane. It is customary to indicate this by making the boundary line dashed, as was done in Figure 3.15.

Example 20 Graph $y < -1$.

Recall that $y = -1$ is a horizontal line. Using this fact we can complete the graph, as shown in Figure 3.16.

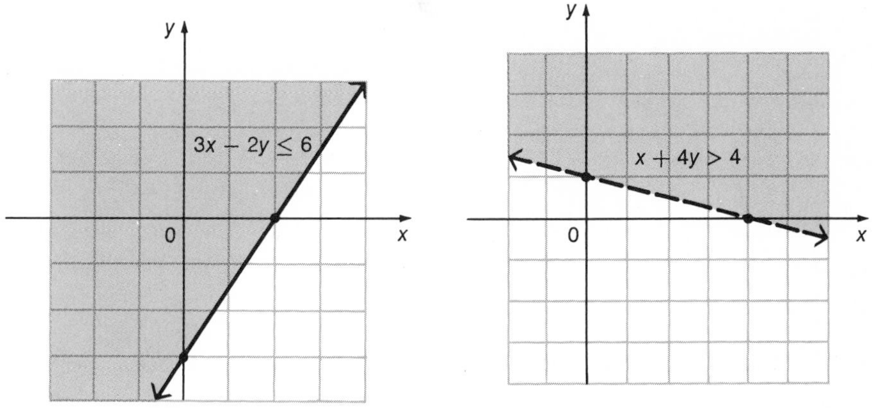

Figure 3.14

Figure 3.15

Example 21 Graph the intersection of $2x - y \leq 4$ and $x < 3$.

Figure 3.17 shows the graphs of both relations. The heavily shaded area represents the intersection of the two regions. Note that the point of intersection of the boundary lines, (3, 2), belongs to the first relation, but not the second, and hence not to the intersection. To show this, we use an open dot on the graph at (3, 2).

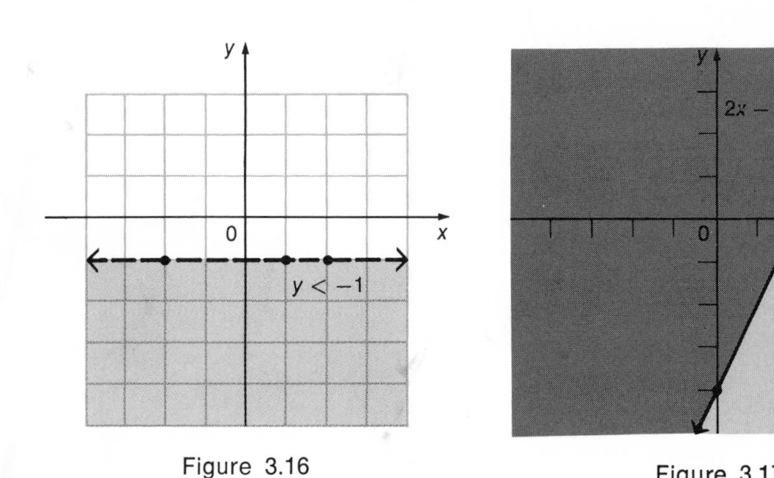

Figure 3.16

Figure 3.17

3.5 Exercises

Graph each of the following.

1. $x \leq 3$
2. $y \leq -2$
3. $x + 2y \leq 6$
4. $x - y \geq 2$
5. $2x + 3y \geq 4$
6. $4y - 3x < 5$

7. $3x - 5y > 6$
8. $x < 3 + 2y$
9. $5x \leq 4y - 2$
10. $2x > 3 - 4y$

Graph the intersection of the following relations.

11. $x - 3y < 4$ and $x \leq 0$
12. $2x + y > 5$, and $y \geq 0$
13. $3x + 2y \geq 6$ and $y \leq 2$
14. $4x + 3y \leq 6$ and $x \leq -2$
15. $x + 2y < 4$ and $3x - y > 5$
16. $4x + 1 < 2y$ and $-x > y - 3$
17. $2x + 3y < 6$ and $x - 5y \geq 10$
18. $3x - 4y > 6$ and $2x + 3y > 4$

3.6 Graphing Nonlinear First Degree Relations

Many first degree relations which we shall use in later work are **nonlinear.** Thus, the equations defining these relations *cannot* be written in the form $ax + by = c$. In this section we shall graph several such nonlinear relations. For example, to graph $y = |x|$, let us recall that $|x| \geq 0$ for all values of x. As long as $x \geq 0$, then $|x| = x$. On the other hand, if $x < 0$ then $|x| = -x$. Using these facts, and selecting some ordered pairs belonging to the relation, we obtain the graph shown in Figure 3.18. Note that the graph is the graph of a function.

Example 22 Graph $y < |x - 2|$.

As in earlier work with linear inequalities, we begin by graphing the boundary, $y = |x - 2|$. We select some points that satisfy the boundary relation, and then sketch the graph, as shown in Figure 3.19. To determine which region belongs to the relation we select any point off the boundary line, such as $(5, -3)$. Since $-3 < |5 - 2|$, we shade the side of the graph including $(5, -3)$, as shown in Figure 3.19.

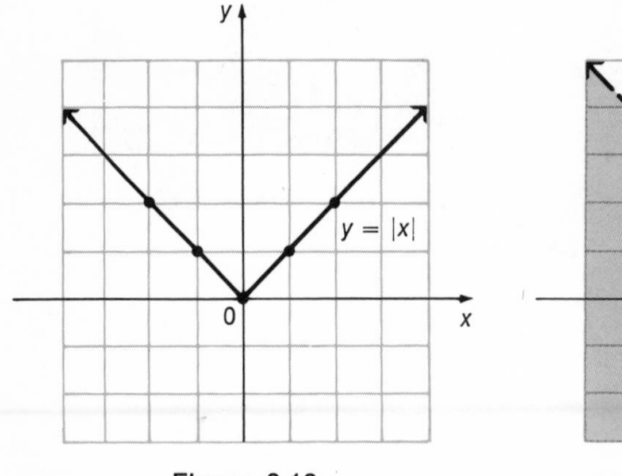

Figure 3.18

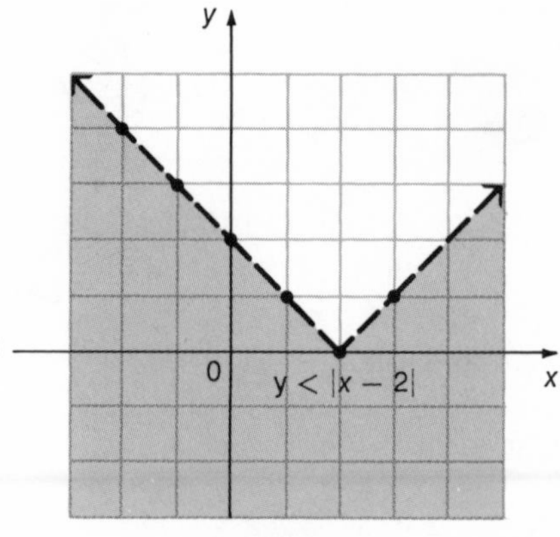

Figure 3.19

Example 23 Graph $y \geq |3x - 4|$.

Again, we graph the boundary $y = |3x - 4|$. Since (0, 0) does not satisfy the relation, we shade the region that does not include (0, 0). The graph is shown in Figure 3.20.

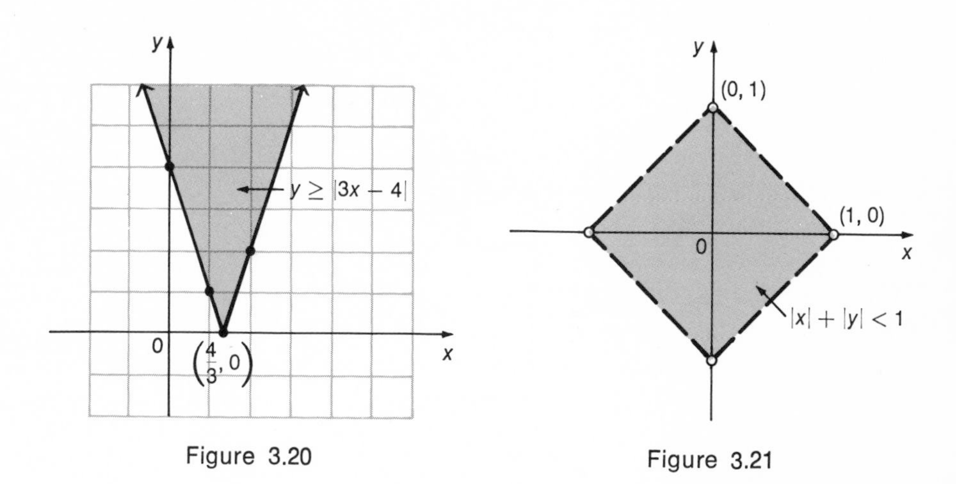

Figure 3.20

Figure 3.21

Example 24 Graph $|x| + |y| < 1$.

First, we graph the boundary, $|x| + |y| = 1$. We can do this by considering the following cases. First, if $x \geq 0$ and $y \geq 0$, then $|x| + |y| = 1$ becomes $x + y = 1$. Hence, in quadrant I we graph $x + y = 1$. Next, in quadrant II, we have $x < 0$ and $y > 0$. Hence, $|x| + |y| = 1$ becomes $-x + y = 1$. In quadrant III we have $-x - y = 1$, while we have $x - y = 1$ in quadrant IV. We now graph these dashed lines in the appropriate quadrants. Note that the four points (0, 1), (1, 0), (−1, 0), (0, −1) are not part of the graph. Since (0, 0) satisfies $|x| + |y| < 1$, we have the graph of Figure 3.21.

The **greatest integer function,** written $y = [x]$ is defined by saying that $[x]$ represents the greatest integer less than or equal to x. Hence, $[8] = 8$, $[-5] = -5$, $[\pi] = 3$, $[-3.4] = -4$, and so on. To graph this function, we might note that for $0 \leq x < 1$, we have $[x] = 0$. Also, for $1 \leq x < 2$, we have $[x] = 1$. Thus, the graph, as shown in Figure 3.22, is composed of a series of line segments. Note that in each case the left endpoint of the segment is included, and the right endpoint is excluded. The greatest integer function is an example of a **step function.**

Example 25 Graph the **nearest integer function,** $y = \langle x \rangle$, defined by saying $\langle x \rangle$ is the distance from x to the nearest integer.

Note here that $\langle \frac{3}{4} \rangle = \frac{1}{4}$, since the distance from $\frac{3}{4}$ to the nearest integer, 1, is $\frac{1}{4}$. Also, $\langle 12.3 \rangle = .3$, $\langle -4.1 \rangle = .1$, and $\langle \frac{3}{2} \rangle = \frac{1}{2}$. Using these points we can sketch the graph, as shown in Figure 3.23. The nearest integer function is an example of a **periodic function.**

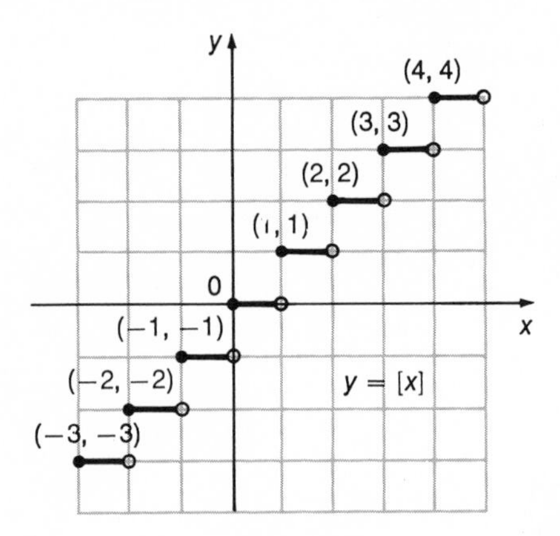

Figure 3.22

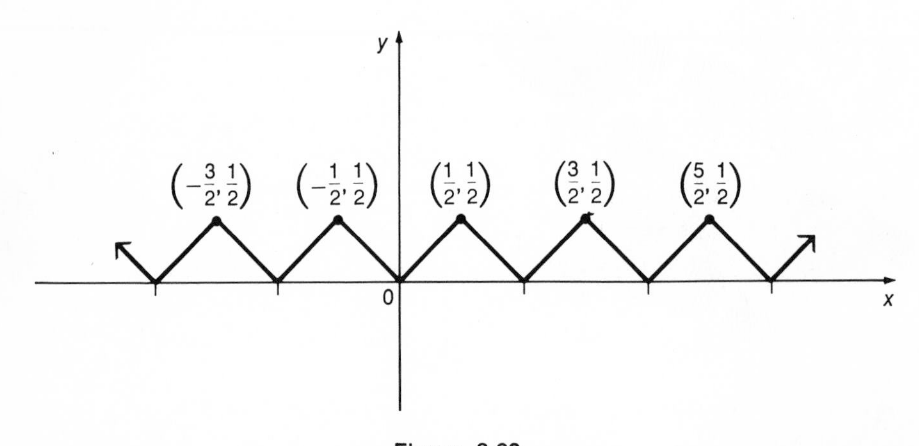

Figure 3.23

Example 26 Graph the relation

$$y = \begin{cases} x + 1 \text{ if } x \leq 2 \\ -2x + 7 \text{ if } x > 2. \end{cases}$$

For $x \leq 2$, we graph $y = x + 1$. For $x > 2$, we graph $y = -2x + 7$, as shown in Figure 3.24.

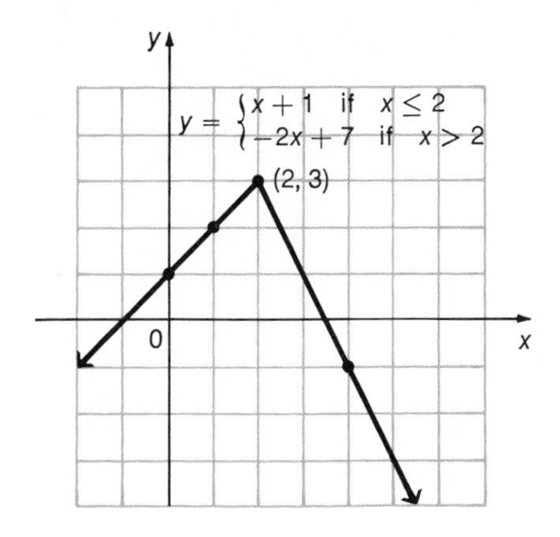

Figure 3.24

3.6 Exercises

Graph each of the following relations.

1. $y \le |x|$
2. $x \ge |y+1|$
3. $x \le |y-3|$
4. $y \ge |x+2|$
5. $2x > |3y-1|$
6. $x - 1 < |3y+2|$
7. $y = |x| - x$
8. $x = |y| + y$
9. $y < |x| - x$
10. $x < |y| + y$
11. $y = [x + \frac{1}{2}]$ (Hint: this is similar to the greatest integer function.)
12. $y = [2x-1]$
13. $y = [3x-2]$
14. $y = [3x] + 1$
15. $y = \langle 2x \rangle$
16. $y = \langle 3x \rangle - 1$
17. $x + |y| = 1$
18. $|x| - y = 1$
19. $|x+y| = 1$
20. $|x+3y| = 2$
21. $y = \begin{cases} x - 1 \text{ if } x \le 3 \\ 2 \text{ if } x > 3 \end{cases}$
22. $y = \begin{cases} 6 - x \text{ if } x \le 3 \\ 3x - 6 \text{ if } x > 3 \end{cases}$
23. $y = \begin{cases} 4 - x \text{ if } x < 2 \\ 1 + 2x \text{ if } x \ge 2 \end{cases}$
24. $y = \begin{cases} -2 \text{ if } x \ge 1 \\ 2 \text{ if } x \le 1 \end{cases}$

3.7 Arithmetic Progressions

In this section we shall investigate a special type of first degree function–an arithmetic sequence, or arithmetic progression. First, let us say that a **sequence** is a function whose domain is a subset of the set of positive integers. The range elements of this function are called the **terms** of the sequence. Since the domain of every sequence is the same, we can exhibit a given sequence merely by listing its terms. In fact, we often refer to the list of terms as the sequence itself. For example, in using the terminology "the sequence 2, 4, 6, 8, . . .," we actually mean "the sequence whose terms (range elements), listed in order, are 2, 4, 6, 8, ..." The terms of the sequence

$$a = \{(x, a(x)) \mid a(x) = 2x + 1, x \text{ is a positive integer}\}$$

are $a(1) = 3$, $a(2) = 5$, $a(3) = 7, \ldots, a(n) = 2n + 1, \ldots$. Note that the terms of a sequence have an order automatically determined by the order of the positive integers. Hence it makes sense to speak of the first term of a sequence, the second term, the third term, and so on.

It is customary to write the term $a(1)$ as a_1, $a(2)$ as a_2, and in general, $a(n)$ as a_n. Using this notation, we see that for the sequence given above, $a_5 = 2(5) + 1 = 11$, $a_9 = 19$, and $a_k = 2k + 1$.

A sequence in which each term after the first is obtained by adding a fixed number (called the **common difference**) to the previous term is called an **arithmetic sequence** or **arithmetic progression.** Thus, the sequence

$$5, 9, 13, 17, 21, \ldots$$

is an arithmetic progression since each term after the first is obtained from the previous one by adding 4. That is,

$$\begin{aligned} 9 &= 5 + 4 \\ 13 &= 9 + 4 \\ 17 &= 13 + 4 \\ 21 &= 17 + 4, \end{aligned}$$

and so on. In general, if a_1 is the first term of an arithmetic sequence and d is the common difference, the sequence is given by

$$a_1, a_1 + d, a_1 + 2d, a_1 + 3d, \ldots, a_1 + (n - 1)d, \ldots.$$

As shown above, the ***n*th term of an arithmetic sequence** is given by

▶ $$a_n = a_1 + (n - 1)d.$$

Example 27 Find the 13th term of the sequence $-3, 1, 5, 9, \ldots$.

Here $a_1 = -3$ and $d = 4$. Hence, $a_{13} = -3 + (13 - 1)4 = 45$.

Example 28 Suppose $a_8 = -16$ and $a_{16} = -40$. Find a_1.

We know $a_8 = a_1 + (8 - 1)d$. Hence, $-16 = a_1 + 7d$ or $a_1 = -16 - 7d$. Also, $-40 = a_1 + 15d$ and $a_1 = -40 - 15d$. From these two equations, we have $-16 - 7d = -40 - 15d$, or $d = -3$. Since $-16 = a_1 + 7d$, we have $a_1 = 5$.

Given a_1 and d we can find the sum of the first n terms of an arithmetic progression. To begin, we write the sum of the first n terms of an arithmetic progression.

$$S_n = a_1 + [a_1 + d] + [a_1 + 2d] + \cdots + [a_1 + (n - 1)d]$$

Next, we write this same sum in reversed order.

$$S_n = [a_1 + (n - 1)d] + [a_1 + (n - 2)d] + \cdots + [a_1 + d] + a_1$$

Now we add the respective sides of these two equations.

$$S_n + S_n = (a_1 + [a_1 + (n - 1)d]) + ([a_1 + d] + [a_1 + (n - 2)d]) + \cdots + ([a_1 + (n - 1)d] + a_1)$$

From this, we have

$$2S_n = [2a_1 + (n - 1)d] + [2a_1 + (n - 1)d] + \cdots + [2a_1 + (n - 1)d].$$

Since there are n of the $[2a_1 + (n - 1)d]$ terms on the right, we have

$$2S_n = n[2a_1 + (n - 1)d]$$

▶ $$\mathbf{S_n = \frac{n}{2}[2a_1 + (n - 1)d].}$$

Since $a_n = a_1 + (n - 1)d$, we can also write

▶ $$\mathbf{S_n = \frac{n}{2}[a_1 + a_n].}$$

Example 29 Find the sum of the first 60 positive integers.
Here $n = 60$, $a_1 = 1$, and $a_{60} = 60$. We have

▶ $$S_{60} = \frac{60}{2}(1 + 60) = 30 \cdot 61 = 1830.$$

Example 30 The sum of the first 17 terms of an arithmetic progression is 187. If $a_{17} = -13$, find a_1 and d.
We can use the formula developed above. We have

$$\begin{aligned} 187 &= \frac{17}{2}(a_1 - 13) \\ 374 &= 17(a_1 - 13) \\ 22 &= a_1 - 13 \\ a_1 &= 35. \end{aligned}$$

We know $a_{17} = a_1 + (17 - 1)d$. Hence,

$$\begin{aligned} -13 &= 35 + 16d \\ -48 &= 16d \\ d &= -3. \end{aligned}$$

3.7 Exercises

Find the eighth term of each of the following arithmetic progressions.

1. $a_1 = 5,\ d = 2$
2. $a_1 = -3,\ d = -4$
3. $a_1 = -4,\ d = 2$
4. $a_1 = 0,\ d = 5$
5. $a_3 = 2,\ d = 1$
6. $a_4 = 5,\ d = -2$
7. $a_1 = 8,\ a_2 = 6$
8. $a_1 = 6,\ a_2 = 3$
9. $a_1 = 12,\ a_3 = 6$
10. $a_2 = 5,\ a_4 = 1$
11. $a_{10} = 6,\ a_{12} = 15$
12. $a_{15} = 8,\ a_{17} = 2$
13. $a_1 = x,\ a_2 = x + 3$
14. $a_2 = y + 1,\ d = -3$
15. $a_6 = 2m,\ a_7 = 3m$
16. $a_5 = 4p + 1,\ a_7 = 6p + 7$

Find the sum of the first 10 terms for each of the following arithmetic progressions.

17. $a_1 = 8,\ d = 3$
18. $a_1 = 2,\ d = 6$
19. $a_1 = 12,\ d = -2$
20. $a_1 = -9,\ d = 4$
21. $a_3 = 5,\ a_4 = 8$
22. $a_2 = 9,\ a_4 = 13$
23. $5, 9, 13, \ldots$
24. $8, 6, 4, \ldots$
25. $3\frac{1}{2}, 5, 6\frac{1}{2}, \ldots$
26. $2\frac{1}{2}, 3\frac{3}{4}, 5, \ldots$
27. $a_1 = 10,\ a_{10} = 5\frac{1}{2}$
28. $a_1 = -8,\ a_{10} = -\frac{5}{4}$

We may write $f(1) + f(2) + f(3) + \cdots + f(n)$ in abbreviated form as

$$\sum_{i=1}^{n} f(i).$$

($\sum$ is the Greek letter sigma which is used to represent sum.) For instance, if $f(i) = 5i - 3$ and $n = 6$, then

$$\begin{aligned}\sum_{i=1}^{6} (5i - 3) &= (5 \cdot 1 - 3) + (5 \cdot 2 - 3) \\ &\quad + (5 \cdot 3 - 3) + (5 \cdot 4 - 3) + (5 \cdot 5 - 3) + (5 \cdot 6 - 3) \\ &= 2 + 7 + 12 + 17 + 22 + 27 \\ &= 87.\end{aligned}$$

Evaluate each of the following sums.

29. $\sum_{i=1}^{3} i$
30. $\sum_{i=1}^{5} (2i + 1)$
31. $\sum_{i=1}^{6} (3i + 2)$
32. $\sum_{i=1}^{7} (2 - 4i)$
33. $\sum_{i=4}^{9} (5i + 6)$
34. $\sum_{i=4}^{12} (7i + 3)$
35. Find the sum of all integers between 50 and 72.
36. Find the sum of all integers between -9 and 31.
37. If a clock strikes the proper number of bongs each hour on the hour, how many bongs will it bong in a month of 30 days?
38. A stack of telephone poles has 30 in the first row, 29 in the second, and so on, with one pole in the last row. How many poles are in the stack?

SECOND DEGREE RELATIONS AND FUNCTIONS

Many important applications of mathematics require a knowledge of second degree relations. In this chapter, we discuss methods of graphing such relations, as well as methods of finding real number solutions of second degree equations.

4.1 The Quadratic Function

A function of the form $y = ax^2 + bx + c$, $(a \neq 0)$ a, b, and c real numbers, is called a **quadratic function.** (Without the restriction $a \neq 0$ we could have $y = 0x^2 + bx + c = bx + c$, which is a linear function.) Perhaps the simplest quadratic function is $y = x^2$, where $a = 1$, $b = 0$, and $c = 0$. To find some ordered pairs belonging to the graph of this function we can choose values for x and then find the corresponding values for y, as shown in the chart accompanying Figure 4.1. These pairs can then be graphed, and a smooth curve

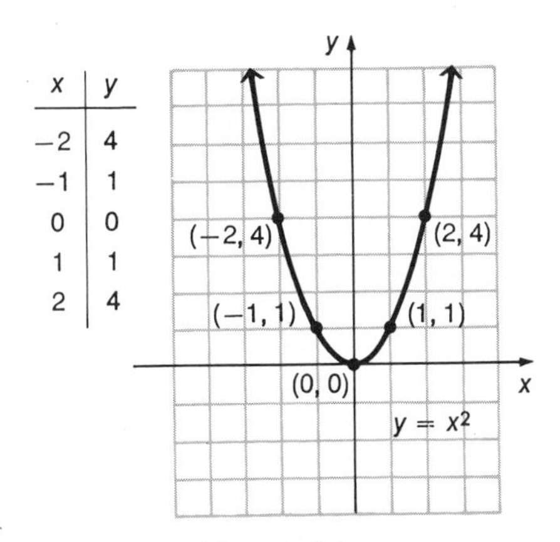

Figure 4.1

drawn through them, as in Figure 4.1. This curve is called a **parabola.** Since the coefficient of x^2 in the equation $y = x^2$ is positive, the parabola opens upward. On the other hand, the coefficient of x^2 in the equation $y = -x^2$ is negative, and the graph of this parabola would open downward.

Parabolas have many practical applications. For example, cross sections of spotlight reflectors, or radar dishes, form parabolas. Every quadratic function has a graph which is a parabola.

Example 1 Graph $y = 2x^2 + 8x + 5$.

To graph such a parabola, it is often convenient to rewrite the equation in a special form by **completing the square.** We want to write $y = 2x^2 + 8x + 5$ in the form $y = m(x + n)^2 + p$. To do so, we write

$$y = 2(x^2 + 4x \qquad) + 5.$$

The process of completing the square requires that we now convert the expression in parentheses, $x^2 + 4x$, into a perfect square by adding an appropriate number. To find this number, take half the coefficient of x ($\frac{1}{2} \cdot 4 = 2$) and square the result ($2^2 = 4$). Add the square, 4, to the quantity inside the parentheses, and subtract 8 outside the parentheses to compensate for adding $2 \cdot 4 = 8$. The result is

$$y = 2(x^2 + 4x + 4) + 5 - 8.$$

We can now write the expression as

$$y = 2(x + 2)^2 - 3.$$

Since $2(x + 2)^2 \geq 0$ for all real values of x, we see that the lowest possible value of y is $y = -3$, and that y has this value when $x = -2$. Thus, the lowest point on this graph, called the **vertex,** is $(-2, -3)$. By finding other ordered pairs belonging to the function, we get the graph shown in Figure 4.2.

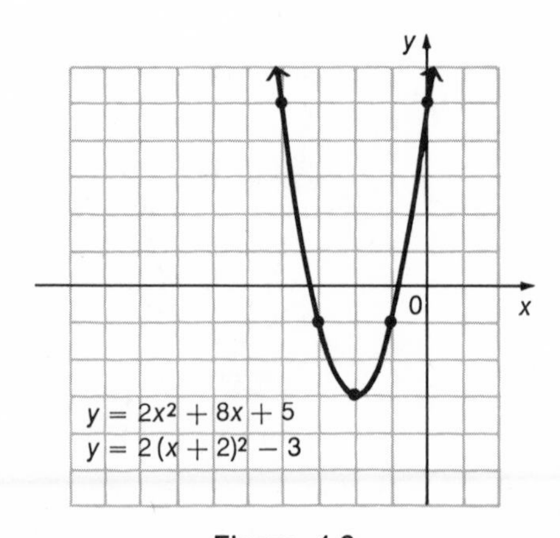

Figure 4.2

Example 2 Graph $y = -3x^2 - 2x + 1$.

Before completing the square here, we first rewrite the expression as

$$y = -3(x^2 + \tfrac{2}{3}x \qquad) + 1.$$

We then add $(\frac{1}{2} \cdot \frac{2}{3})^2 = \frac{1}{9}$ to the quantity inside the parentheses, and correspondingly, subtract $(-3)(\frac{1}{9}) = -\frac{1}{3}$ outside the parentheses.

$$y = -3(x^2 + \tfrac{2}{3}x + \tfrac{1}{9}) + 1 - (-\tfrac{1}{3})$$
$$y = -3(x + \tfrac{1}{3})^2 + \tfrac{4}{3}$$

Since $-3(x + \frac{1}{3})^2 \leq 0$ for all real values of x, the parabola will open downward, and hence the vertex, $(-\frac{1}{3}, \frac{4}{3})$, will be the *highest* point on this graph. By selecting some ordered pairs of the graph, we get the result of Figure 4.3.

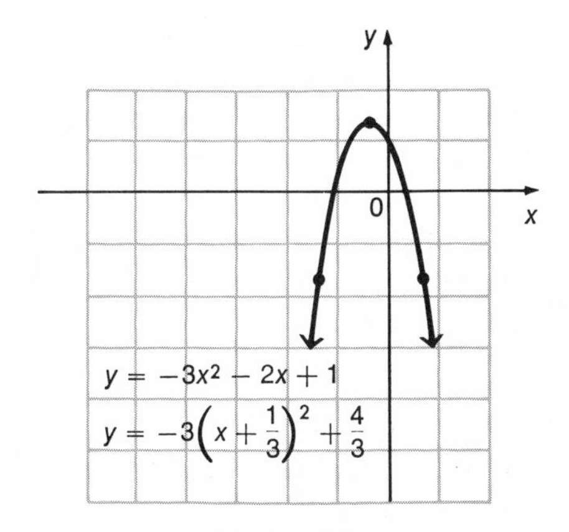

Figure 4.3

In the same way, we can complete the square on the general quadratic function

$$y = ax^2 + bx + c \qquad (a \neq 0).$$

Following the steps of the previous examples, we have

$$y = a\left(x^2 + \frac{b}{a}x \qquad\right) + c.$$

Now we take half of $\dfrac{b}{a}\left(\dfrac{1}{2} \cdot \dfrac{b}{a} = \dfrac{b}{2a}\right)$ and square this result.

This gives

$$y = a\left(x^2 + \frac{b}{a}x + \frac{b^2}{4a^2}\right) + c - \frac{b^2}{4a}$$
$$= a\left(x + \frac{b}{2a}\right)^2 + \frac{4ac - b^2}{4a}.$$

We use this last statement to prove the following result.

THEOREM 4.1 The vertex of the graph of the quadratic function $y = ax^2 + bx + c$, $(a \neq 0)$ is given by

$$\left(-\frac{b}{2a}, \frac{4ac - b^2}{4a}\right).$$

This vertex is the highest point on the graph if $a < 0$, and the lowest point on the graph if $a > 0$.

Example 3 Find the vertex of $y = -2x^2 + 3x - 5$.

Here $a = -2$, $b = 3$, and $c = -5$. Theorem 4.1 is a convenient way to find the vertex. Using the result of this theorem, we have

$$\left(-\frac{b}{2a}, \frac{4ac - b^2}{4a}\right) = \left(\frac{-3}{2(-2)}, \frac{4(-2)(-5) - 3^2}{4(-2)}\right)$$
$$= \left(\frac{3}{4}, -\frac{31}{8}\right).$$

Here the vertex is the highest point on the graph. The graph of this parabola could be obtained by plotting this vertex and some other ordered pairs that satisfy the equation.

Example 4 Graph $y < 2x^2 - 3$.

Using a result analagous to Theorem 3.7 for graphing linear inequalities, we first graph the boundary parabola, $y = 2x^2 - 3$, as shown in Figure 4.4. Then, we select any point not on the parabola, such as (0, 0). Since (0, 0) does not satisfy the original inequality, we shade the portion of the graph not including (0, 0), as in Figure 4.4.

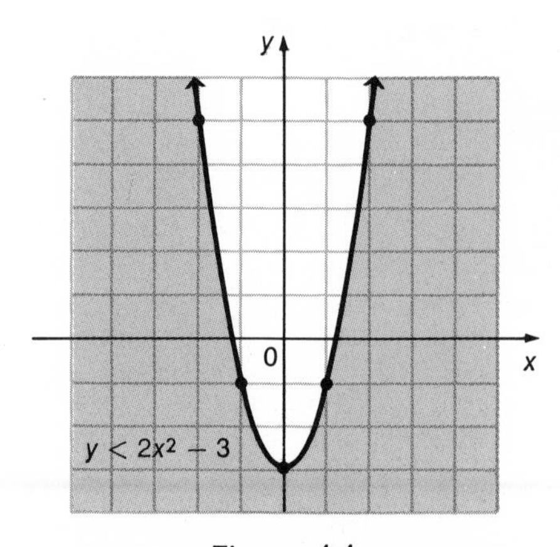

Figure 4.4

4.1 Exercises

Graph each of the following parabolas and identify the vertex.

1. $y = x^2 + 1$
2. $y = 2x^2$
3. $y = (x - 2)^2$
4. $y = (x + 4)^2$
5. $y = (x + 3)^2 - 4$
6. $y = (x - 5)^2 - 4$
7. $y = -2(x + 3)^2 + 2$
8. $y = -3(x - 2)^2 + 1$
9. $y = -\frac{1}{2}(x + 1)^2 - 3$
10. $y = \frac{2}{3}(x - 2)^2 - 1$
11. $y = x^2 - 2x + 3$
12. $y = x^2 + 6x + 5$
13. $y = x^2 + 8x + 13$
14. $y = x^2 - 12x + 30$
15. $y = -x^2 - 4x + 2$
16. $y = -x^2 + 6x - 6$
17. $y = 2x^2 - 4x + 5$
18. $y = -3x^2 + 24x - 46$

Graph each of the following relations.

19. $y < 3x^2 + 2$
20. $y \leq x^2 - 4$
21. $y > (x - 1)^2 + 2$
22. $y > 2(x + 3)^2 - 1$
23. Farmer Frieda wishes to fence a rectangular area which lies against a river, so that only three sides need to be fenced. She has 320 feet of fencing. What should the dimensions of the field be if the enclosed area is to be a maximum? (Hint: let x represent one side of the area, with $320 - 2x$ the other side. Graph the area parabola, $A = x(320 - 2x)$ and investigate the vertex.)
24. What would be the maximum area that could be enclosed by Farmer Frieda's 320 feet of fencing if she had to enclose all 4 sides of her field?
25. Find two numbers whose sum is 20 and whose product is a maximum.

4.2 Zeros of Quadratic Functions

A **zero** of a function $y = f(x)$ is a value of x that makes $f(x) = 0$. Thus, -2 is a zero of $y = 2x^2 + 3x - 2$, since $y = 2(-2)^2 + 3(-2) - 2 = 0$. In this section we shall discuss methods of finding real zeros of quadratic functions. The simplest method of finding such zeros, but one that is not always applicable, is by factoring. This method depends on the following result.

THEOREM 4.2 ZERO-FACTOR THEOREM
If a and b are real numbers with $ab = 0$,
then $a = 0$ or $b = 0$.

To prove this, let us assume that $ab = 0$ and $a \neq 0$. We must then show that $b = 0$. We have

$$\begin{aligned} a &\neq 0 \\ ab &= 0 \\ \frac{1}{a}(ab) &= \frac{1}{a}(0) \end{aligned}$$

(Why do we know that $\frac{1}{a}$ is a real number?)

$$\begin{aligned} \left(\frac{1}{a} \cdot a\right)b &= 0 \\ b &= 0. \end{aligned}$$

In a similar way, we can show that if $b \neq 0$, then $a = 0$.

Example 5 Find the zeros of $y = 6x^2 + 7x - 3$.

We want to find all the values of x that make $6x^2 + 7x - 3 = 0$. To use the zero-factor theorem, we factor $6x^2 + 7x - 3 = 0$ as $(3x - 1)(2x + 3) = 0$. Thus, we have

$$(3x - 1)(2x + 3) = 0,$$

and by the zero-factor theorem, $(3x - 1)(2x + 3) = 0$ only if

$$3x - 1 = 0 \qquad \text{or} \qquad 2x + 3 = 0.$$

The solution set of this compound sentence, $\{\frac{1}{3}, -\frac{3}{2}\}$, gives the zeros of the function.

Not all quadratic functions have zeros which can be found by factoring ($y = x^2 + x + 6$ is an example). Hence, we need a more general method for finding zeros. To find this more general method, we complete the square on the general quadratic equation

$$ax^2 + bx + c = 0 \qquad (a \neq 0).$$

We begin by writing

$$a\left(x^2 + \frac{b}{a}x \qquad \right) + c = 0.$$

We can now complete the square in a manner similar to that used when looking for the vertex of a parabola. First, we take half of b/a, and square the result, obtaining $b^2/4a^2$. We add this quantity inside the parentheses, and subtract $b^2/4a$, which compensates for adding $a(b^2/4a^2) = b^2/4a$.

$$a\left(x^2 + \frac{b}{a}x + \frac{b^2}{4a^2}\right) + c - \frac{b^2}{4a} = 0$$

Now we add $\frac{b^2}{4a} - c$ $\left(\text{which equals } \frac{b^2 - 4ac}{4a}\right)$ to both sides, and then simplify.

$$a\left(x^2 + \frac{b}{a}x + \frac{b^2}{4a^2}\right) = \frac{b^2 - 4ac}{4a}$$

We can now write the expression in parentheses on the left-hand side as the square of a binomial.

$$a\left(x + \frac{b}{2a}\right)^2 = \frac{b^2 - 4ac}{4a}$$

If we now divide both sides of the equation by a, we get

$$\left(x + \frac{b}{2a}\right)^2 = \frac{b^2 - 4ac}{4a^2}.$$

This last statement leads to the compound sentence

$$x + \frac{b}{2a} = \sqrt{\frac{b^2 - 4ac}{4a^2}} \quad \text{or} \quad x + \frac{b}{2a} = -\sqrt{\frac{b^2 - 4ac}{4a^2}},$$

which we can also write as

$$x = \frac{-b + \sqrt{b^2 - 4ac}}{2a} \quad \text{or} \quad x = \frac{-b - \sqrt{b^2 - 4ac}}{2a}.$$

A simplified form of this result is given in Theorem 4.3.

THEOREM 4.3 THE QUADRATIC FORMULA

The zeros of the quadratic function $y = ax^2 + bx + c$, $(a \neq 0)$ are given by

$$\boldsymbol{x = \frac{-b \pm \sqrt{b^2 - 4ac}}{2a}}.$$

Example 6 Find the zeros of $y = x^2 - 4x + 1$.

Here $a = 1$, $b = -4$, and $c = 1$. Using the quadratic formula, we have

$$x = \frac{-b \pm \sqrt{b^2 - 4ac}}{2a}$$

$$x = \frac{4 \pm \sqrt{16 - 4}}{2}$$

$$x = \frac{4 \pm 2\sqrt{3}}{2}$$

$$x = 2 \pm \sqrt{3}.$$

The zeros are $2 + \sqrt{3}$ and $2 - \sqrt{3}$.

The expression $b^2 - 4ac$ appearing in the quadratic formula is called the **discriminant.**

- If $b^2 - 4ac > 0$, the function has **two** distinct real zeros.
- If $b^2 - 4ac = 0$, the function has exactly **one** real zero.
- If $b^2 - 4ac < 0$, the function has **no** real zeros.

We shall discuss nonreal zeros of quadratic functions in a later chapter.

4.2 Exercises

Find all real solutions of the following equations.

1. $x(x - 2)(x + 3) = 0$
2. $(2m - 1)(3m + 2)(4m + 5) = 0$
3. $p^2 - 5p + 6 = 0$
4. $q^2 + 2q - 8 = 0$
5. $6z^2 - 5z - 50 = 0$
6. $6r^2 + 7r = 3$
7. $8k^2 + 14k + 3 = 0$
8. $18r^2 - 9r - 2 = 0$
9. $-m^2 + m + 1 = 0$
10. $y^2 - 3y + 2 = 0$
11. $2s^2 + 2s = 3$
12. $t^2 - t = 3$
13. $3a^2 + 7a - 2 = 0$
14. $4z^2 - 4z = 3$
15. $x^2 - 6x + 7 = 0$
16. $11p^2 - 7p + 1 = 0$
17. $12r^2 + 5r = 3$
18. $m^2 - 4m + 7 = 0$
19. $9p^2 - 30p = 25$
20. $4n^2 + 12n + 9 = 0$
21. $4 - \frac{11}{x} - \frac{3}{x^2} = 0$
22. $3 - \frac{4}{p} = \frac{2}{p^2}$
23. $2 - \frac{5}{r} + \frac{3}{r^2} = 0$
24. $2 - \frac{5}{k} + \frac{2}{k^2} = 0$
25. $m^2 - \sqrt{2}m - 1 = 0$
26. $z^2 - \sqrt{3}z - 2 = 0$
27. $\sqrt{2}p^2 - 3p + \sqrt{2} = 0$
28. $-\sqrt{6}k^2 - 2k + \sqrt{6} = 0$
29. $mx^2 + nx + 2 = 0$
30. $(p + q)r^2 - 2pr + p = 0$

Evaluate the discriminant $b^2 - 4ac$ for each of the following, and use it to predict the type of zeros of the function.

31. $y = x^2 + 8x + 16$
32. $y = x^2 - 5x + 4$
33. $y = 3x^2 - 5x + 2$
34. $y = 8x^2 - 14x + 3$
35. $y = 4x^2 - 6x - 3$
36. $y = 2x^2 - 4x + 1$
37. $y = 9x^2 + 11x + 4$
38. $y = 3x^2 - 8x + 5$

4.3 Equations Quadratic in Form

An equation that can be made into a quadratic equation, either by substituting variables or by raising both sides to a power, is described as **quadratic in form.** For example,

$$12m^4 - 11m^2 + 2 = 0$$

can be converted into a quadratic by letting $x = m^2$. Doing this, we have

$$12x^2 - 11x + 2 = 0,$$

which can be solved by factoring.

$$(3x - 2)(4x - 1) = 0$$
$$x = \tfrac{2}{3} \quad \text{or} \quad x = \tfrac{1}{4}.$$

Since $x = m^2$, we have

$$\begin{aligned} m^2 &= \tfrac{2}{3} && \text{or} && m^2 = \tfrac{1}{4} \\ m &= \pm\sqrt{\tfrac{2}{3}} && \text{or} && m = \pm\sqrt{\tfrac{1}{4}} \\ m &= \pm\frac{\sqrt{6}}{3} && \text{or} && m = \pm\frac{1}{2}. \end{aligned}$$

Example 7 Solve $(a - 2)^{2/3} + (a - 2)^{1/3} - 2 = 0$.

Here we can let $x = (a - 2)^{1/3}$. By substitution, we have

$$x^2 + x - 2 = 0,$$

which can be factored as follows.

$$(x + 2)(x - 1) = 0$$
$$x = -2 \quad \text{or} \quad x = 1.$$

Since $x = (a - 2)^{1/3}$, we have

$$\begin{aligned} (a - 2)^{1/3} &= -2 && \text{or} && (a - 2)^{1/3} = 1 \\ a - 2 &= -8 && \text{or} && a - 2 = 1 \\ a &= -6 && \text{or} && a = 3. \end{aligned}$$

These solutions can be verified by substituting in the original equation.

Example 8 Solve $\left(m - \frac{1}{m}\right)^2 + 2\left(m - \frac{1}{m}\right) = 8$.

Let $x = m - \frac{1}{m}$. Then we have

$$x^2 + 2x - 8 = 0,$$
$$(x + 4)(x - 2) = 0$$
$$x = -4 \quad \text{or} \quad x = 2.$$

Since $x = m - \frac{1}{m}$, we have

$$\begin{aligned} m - \frac{1}{m} &= -4 && \text{or} && m - \frac{1}{m} = 2 \\ m^2 - 1 &= -4m && \text{or} && m^2 - 1 = 2m \\ m^2 + 4m - 1 &= 0 && \text{or} && m^2 - 2m - 1 = 0. \end{aligned}$$

Using the quadratic formula, we can write

$$m = \frac{-4 \pm \sqrt{16 + 4}}{2} \qquad \text{or} \qquad m = \frac{2 \pm \sqrt{4 + 4}}{2}$$

$$m = -2 \pm \sqrt{5} \qquad \text{or} \qquad m = 1 \pm \sqrt{2}.$$

To solve an equation such as $x = \sqrt{15 - 2x}$ we can square both sides. That this procedure is valid is shown by the next theorem.

THEOREM 4.4 The solution set of an equation $P(x) = Q(x)$ is a subset of the solution set of the equation $[P(x)]^n = [Q(x)]^n$, for any positive integer value of n.

Note that the theorem only asserts that the solution set of the original equation is a *subset* of the solution set of the new equation. The new equation may have more solutions or **roots** than the original equation. Any such extra roots are called **extraneous roots.** To identify extraneous roots, all proposed roots must be checked in the *original* equation.

By Theorem 4.4, the equation $x = \sqrt{15 - 2x}$ can be solved by squaring both sides.

$$x^2 = (\sqrt{15 - 2x})^2$$
$$x^2 = 15 - 2x$$
$$x^2 + 2x - 15 = 0$$
$$(x + 5)(x - 3) = 0$$
$$x = -5 \qquad \text{or} \qquad x = 3.$$

We must now check the proposed roots in the original equation,

$$x = \sqrt{15 - 2x}.$$

If $x = -5$, does $x = \sqrt{15 - 2x}$?
$$-5 \neq \sqrt{15 + 10}$$
$$-5 \neq 5.$$

If $x = 3$, does $x = \sqrt{15 - 2x}$?
$$3 = \sqrt{15 - 6}$$
$$3 = 3.$$

By this check, only $x = 3$ is a root. The proposed root $x = -5$, which does not work in the original equation, is an extraneous root.

Example 9 Solve $\sqrt{2x + 3} - \sqrt{x + 1} = 1$.

We can write the equation as

$$\sqrt{2x + 3} = 1 + \sqrt{x + 1}.$$

If we now square both sides, we get

$$2x + 3 = 1 + 2\sqrt{x + 1} + x + 1$$
$$x + 1 = 2\sqrt{x + 1}.$$

One side of the equation still contains a radical, and to eliminate it, we need to square both sides again.

$$x^2 + 2x + 1 = 4(x + 1)$$
$$x^2 - 2x - 3 = 0$$
$$(x - 3)(x + 1) = 0$$
$$x = 3 \qquad \text{or} \qquad x = -1.$$

We must check these proposed roots in the original equation.

$$\text{If } x = 3, \text{ does } \sqrt{2x+3} - \sqrt{x+1} = 1?$$
$$\sqrt{9} - \sqrt{4} = 1$$
$$3 - 2 = 1.$$

$$\text{If } x = -1, \text{ does } \sqrt{2x+3} - \sqrt{x+1} = 1?$$
$$\sqrt{1} - \sqrt{0} = 1$$
$$1 - 0 = 1.$$

Here both proposed solutions, $x = 3$ and $x = -1$, belong to the solution set, $\{3, -1\}$.

4.3 Exercises

Solve each of the following equations.

1. $\sqrt{x} + 1 = 2$
2. $\sqrt{x} - 3 = 5$
3. $m^4 + 2m^2 - 15 = 0$
4. $p^4 - p^2 - 6 = 0$
5. $2z^4 + 5z^2 - 3 = 0$
6. $3k^4 + 7k^2 - 6 = 0$
7. $2r^4 - 7r^2 + 5 = 0$
8. $4x^4 - 8x^2 + 3 = 0$
9. $(g - 2)^2 - 6(g - 2) + 8 = 0$
10. $(p + 2)^2 - 2(p + 2) - 15 = 0$
11. $-(r + 1)^2 - 3(r + 1) + 3 = 0$
12. $2(z - 4)^2 - 2(z - 4) - 3 = 0$
13. $6(k + 2)^4 - 11(k + 2)^2 + 4 = 0$
14. $8(m - 4)^4 - 10(m - 4)^2 + 3 = 0$
15. $\left(m + \frac{1}{m}\right)^2 + 3\left(m + \frac{1}{m}\right) - 10 = 0$
16. $\left(p - \frac{1}{p}\right)^2 + \left(p - \frac{1}{p}\right) = 30$
17. $(r - 1)^{2/3} + (r - 1)^{1/3} = 12$
18. $(y + 3)^{2/3} - 2(y + 3)^{1/3} - 3 = 0$

19. $p = \sqrt{5p - 6}$

20. $m = \sqrt{8 - 2m}$

21. $x = -\sqrt{\dfrac{50 + 5x}{6}}$

22. $s = \sqrt{\dfrac{3 - 2s}{2}}$

23. $\sqrt{18 + \sqrt{x}} = \sqrt{2x + 3}$

24. $\sqrt{9 + \sqrt{x}} = \sqrt{x - 3}$

25. $\sqrt{2x + 5} + \sqrt{5x - 1} = 6$

26. $\sqrt{2x + 1} - \sqrt{x - 3} = 2$

4.4 Quadratic Inequalities

For what values of x is $x^2 - x - 12 < 0$? To find out, we might begin by graphing the parabola $y = x^2 - x - 12$. This can be done by completing the square

$$y = (x^2 - x + \tfrac{1}{4}) - 12 - \tfrac{1}{4}$$
$$y = (x - \tfrac{1}{2})^2 - \tfrac{49}{4}.$$

The graph of this parabola is shown in Figure 4.5. From the graph of the para-

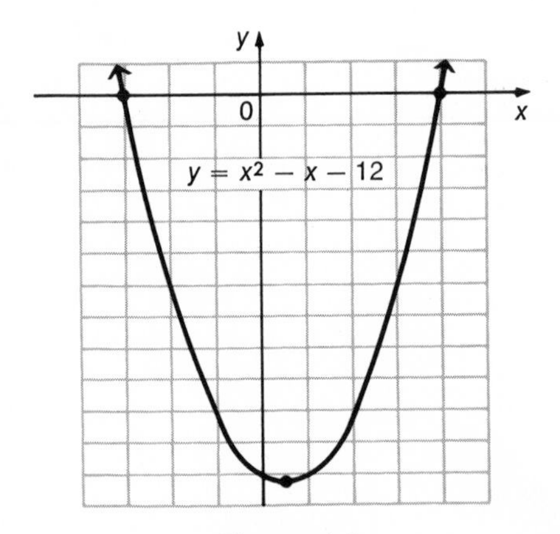

Figure 4.5

bola we can see that $y = x^2 - x - 12$ is negative for all values of x between -3 and 4. That is, the solution set of $x^2 - x - 12 < 0$ is $\{x \mid -3 < x < 4\}$. This solution set is graphed in Figure 4.6.

Figure 4.6

Since we want only the solution set, we can develop a shortcut. To solve $x^2 - x - 12 < 0$, we begin by finding the values of x which satisfy $x^2 - x - 12 = 0$.

$$x^2 - x - 12 = 0$$
$$(x + 3)(x - 4) = 0$$
$$x = -3 \qquad \text{or} \qquad x = 4.$$

Note that these roots of the equation are the points where the graph in Figure 4.5 crosses the x-axis. For all x-values between these points, the y-values will be negative since the graph there is below the x-axis. In the same way, the graph is above the x-axis when $x > 4$ or $x < -3$, so that the corresponding y-values are positive. Hence, all we need to do is determine one y-value in each of the three regions indicated in Figure 4.6. If one y-value is negative in a particular region, all y-values in that region will be negative and $x^2 - x - 12$ will be negative. By testing three points, using one x-value from each of the three regions, we can determine the solution.

In summary, to solve a quadratic inequality:

- ▶ (1) Find the roots of the corresponding quadratic equation.
- ▶ (2) Test one point in each of the regions determined by the roots to find the solution.

For example, to solve the inequality $2x^2 + 5x - 12 \geq 0$, we first find the roots of $2x^2 + 5x - 12 = 0$, namely, $x = \frac{3}{2}$ and $x = -4$. These two points, $x = \frac{3}{2}$ and $x = -4$, divide the number line into three regions, as shown in Figure 4.7. We select a point in each region and determine whether or not it

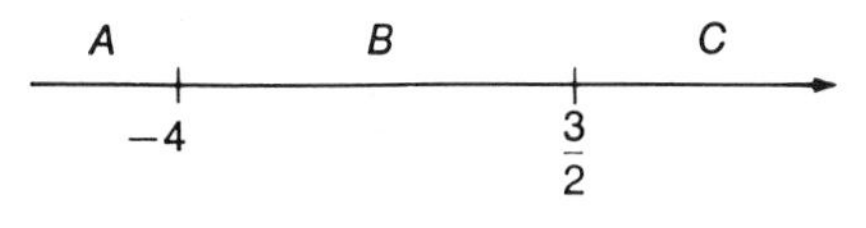

Figure 4.7

satisfies $2x^2 + 5x - 12 \geq 0$. If we try $x = -5$ from region A, we have

$$2(-5)^2 + 5(-5) - 12 = 13 \geq 0.$$

Since -5 satisfies the inequality, all the points of region A belong to the solution set. If we try $x = 0$, from region B, we have

$$2(0)^2 + 5(0) - 12 = -12 \ngeq 0.$$

This one point from region B did not satisfy the inequality, and thus no point from region B belongs to the solution set. For $x = 2$, from region C, we have

$$2(2)^2 + 5(2) - 12 = 6 \geq 0,$$

so that all points of C belong to the solution set. Hence, the solution set is $\{x | x \leq -4 \text{ or } x \geq \frac{3}{2}\}$, the graph of which is shown in Figure 4.8.

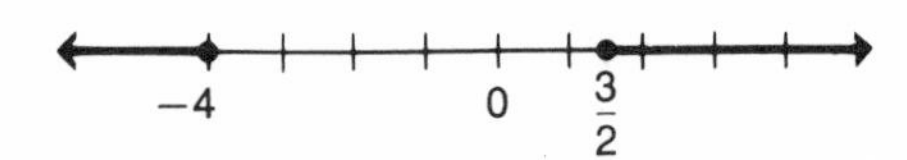

Figure 4.8

Example 10 Solve the inequality $1 \leq \dfrac{5}{x+4}$.

We could begin by multiplying both sides of the inequality by $x + 4$, but we would then have to consider whether $x + 4$ is positive or negative. To avoid this, we multiply both sides by $(x + 4)^2$, which is always nonnegative. Doing this, we have

$$\begin{aligned} (x+4)^2 &\leq 5(x+4) \\ x^2 + 8x + 16 &\leq 5x + 20 \\ x^2 + 3x - 4 &\leq 0. \end{aligned}$$

The roots of $x^2 + 3x - 4 = 0$ are $x = -4$ and $x = 1$. By testing a point from each of the three regions, we see that the solution set of $x^2 + 3x - 4 \leq 0$ is $\{x | -4 \leq x \leq 1\}$. However, the solution set of $1 \leq \dfrac{5}{x+4}$, is

$$\{x | -4 < x \leq 1\},$$

since $x = -4$ is meaningless in the original inequality.

We can also work this example in another way. We know that if we multiply both sides of

$$1 \leq \frac{5}{x+4}$$

by a positive number we do not change the direction of the inequality symbol. Thus, let us assume $x + 4 > 0$ and write:

CASE 1

$$\begin{aligned} x + 4 &> 0 \\ 1(x+4) &\leq \frac{5}{x+4}(x+4) \\ x + 4 &\leq 5 \\ x &\leq 1. \end{aligned}$$

Here we have $x + 4 > 0$, or $x > -4$, and $x \leq 1$. The solution set of $x > -4$ and $x \leq 1$ is given by

$$\begin{aligned} \{x | x > -4 \text{ and } x \leq 1\} &= \{x | x > -4\} \cap \{x | x \leq 1\} \\ &= \{x | -4 < x \leq 1\}. \end{aligned}$$

On the other hand, if we multiply both sides of the inequality by a negative number, we must change the direction of the inequality. Thus we have:

CASE 2

$$
\begin{aligned}
x + 4 &< 0 \\
1(x + 4) &\geq \frac{5}{x + 4}(x + 4) \\
x + 4 &\geq 5 \\
x &\geq 1.
\end{aligned}
$$

The solution set here is given by

$$\{x \mid x < -4\} \cap \{x \mid x \geq 1\} = \varnothing.$$

The solution set of the original inequality is the *union* of the solution sets from case 1 and case 2, or

$$\{x \mid -4 < x \leq 1\} \cup \varnothing = \{x \mid -4 < x \leq 1\}.$$

4.4 Exercises

Solve each of the following inequalities.

1. $x^2 \leq 9$
2. $p^2 > 16$
3. $y^2 - 10y + 25 < 25$
4. $m^2 + 6m + 9 < 9$
5. $r^2 + 4r + 4 \geq 3$
6. $z^2 + 6z + 9 < 8$
7. $x^2 - x \leq 6$
8. $r^2 + r < 12$
9. $2k^2 - 9k > -4$
10. $3n^2 < 10 - 13n$
11. $x^2 > 0$
12. $p^2 < -1$
13. $\frac{3}{x - 6} \leq 2$
14. $\frac{x + 1}{x + 2} \leq 3$
15. $\frac{1}{k - 2} < \frac{1}{3}$
16. $\frac{1}{m - 1} < 1$
17. $x^3 - 4x \leq 0$
18. $r^3 - 9r \geq 0$

4.5 Circles

In this section we shall discuss the circle. A **circle** is the set of points in a plane which lie a given distance from a given point. The given distance is called the **radius** and the given point is called the **center.** As an aid in developing the equation of a circle, we state the distance formula (see Exercise 27).

THEOREM 4.5 DISTANCE FORMULA
The distance between the points (x_1, y_1) and (x_2, y_2) is given by

$$\sqrt{(x_2 - x_1)^2 + (y_2 - y_1)^2}.$$

For example, the distance between (−5, 3) and (3, 18) is

$$\sqrt{[3-(-5)]^2+(18-3)^2} = \sqrt{8^2+15^2} = 17.$$

Figure 4.9 shows a circle of radius 3 with center at the origin. If we select any point (x, y) on the circle, and use the distance formula to find the distance from (x, y) to the center, we have

$$\begin{aligned} \sqrt{(x-0)^2+(y-0)^2} &= 3 \\ \sqrt{x^2+y^2} &= 3 \\ x^2+y^2 &= 9. \end{aligned}$$

Thus, $x^2 + y^2 = 9$ is the equation of a circle with center at the origin and radius 3. Note the characteristics of this equation: it contains both x^2 and y^2 and both have equal coefficients.

Example 11 Find the equation of a circle with center at (−2, 4) and radius 4 (see Figure 4.10).

We again use the distance formula.

$$\begin{aligned} \sqrt{(x+2)^2+(y-4)^2} &= 4 \\ (x+2)^2+(y-4)^2 &= 16 \end{aligned}$$

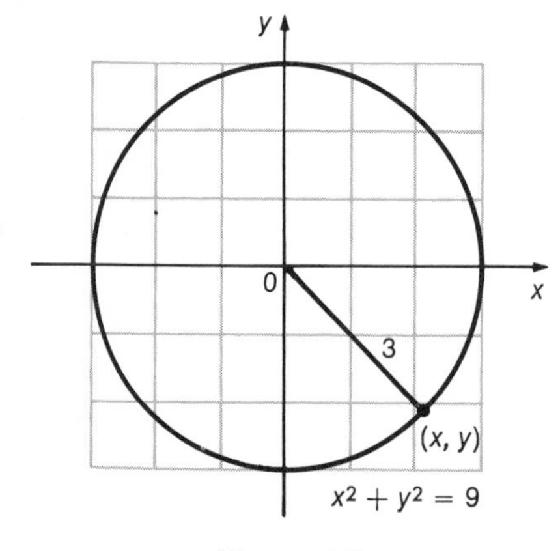

Figure 4.9

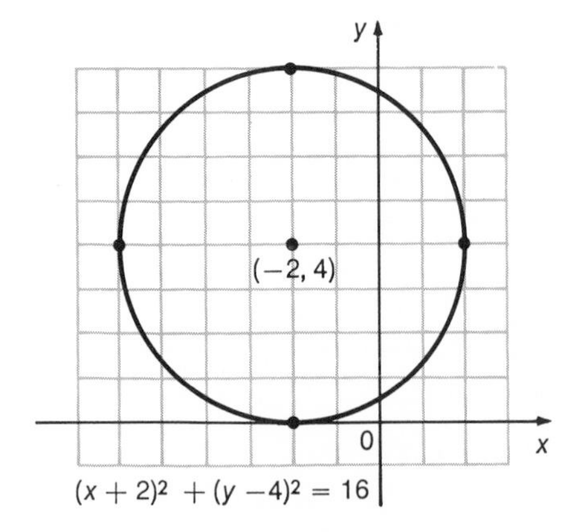

Figure 4.10

Generalizing from these examples, we see that the distance formula can be used to find the equation of a circle, as shown in the next theorem.

THEOREM 4.6 The equation of a circle having center (h, k) and radius r is

$$(x-h)^2 + (y-k)^2 = r^2.$$

Example 12 Graph $x^2 - 6x + y^2 + 10y = -25$.

We suspect that this equation represents a circle since the second degree terms have equal coefficients. To see if this equation really represents a circle, we complete the square on both x and y.

$$(x^2 - 6x + 9) + (y^2 + 10y + 25) = -25 + 9 + 25$$
$$(x - 3)^2 + (y + 5)^2 = 9$$

The equation is that of a circle having center at $(3, -5)$ and radius 3, with the graph as shown in Figure 4.11.

Example 13 Graph $x^2 + 10x + y^2 - 4y + 33 = 0$.

Completing the square gives

$$(x^2 + 10x + 25) + (y^2 - 4y + 4) = -33 + 25 + 4$$
$$(x + 5)^2 + (y - 2)^2 = -4.$$

There are no ordered pairs (x, y) satisfying this condition (why?).

Example 14 Graph $y = \sqrt{16 - x^2}$.

If we square both sides, we have $y^2 = 16 - x^2$, or $x^2 + y^2 = 16$, which is a circle with center at the origin and radius 4. However, in the original equation $y = \sqrt{16 - x^2}$, so that y can take on only nonnegative values. Hence, the graph is a semicircle, as shown in Figure 4.12. The domain of the function is $\{x \mid -4 \le x \le 4\}$, and the range is $\{y \mid 0 \le y \le 4\}$.

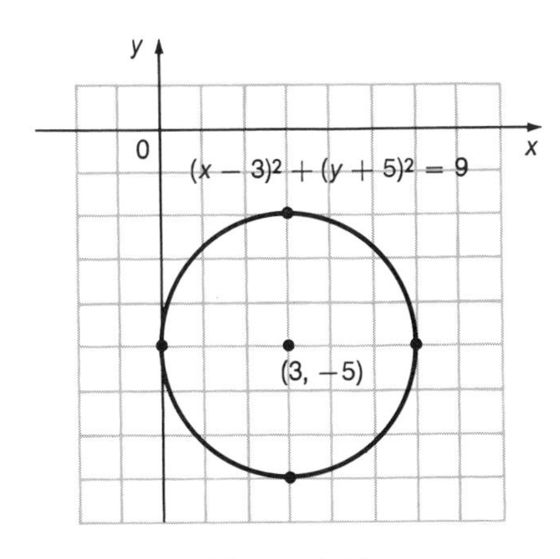

Figure 4.11

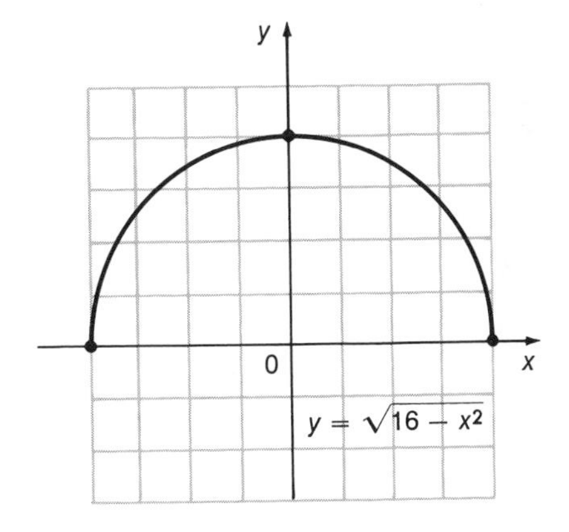

Figure 4.12

4.5 Exercises

Find the distance between each of the following pairs of points.

1. $(2, 4), (3, -1)$
2. $(0, 4), (-2, 3)$
3. $(-5, 1), (2, 3)$
4. $(-2, 1), (4, 2)$

5. Use the distance formula to verify that (1, 1), (2, -2), and (3, -5) lie on the same line. (Hint: if the three points do not lie on the same line, they will be the vertices of a triangle.)
6. Verify that (1, 1), (5, -11), and (4, 2) are the vertices of a right triangle. (Hint: if the sum of the squares of the length of two sides of a triangle equals the square of the third side, then the triangle is a right triangle.)

Graph each of the following.

7. $x^2 + y^2 = 36$
8. $x^2 + y^2 = 81$
9. $(x - 4)^2 + (y + 3)^2 = 4$
10. $(x + 3)^2 + (y - 2)^2 = 16$
11. $(x + 2)^2 + (y - 5)^2 = 12$
12. $(x - 4)^2 + (y - 3)^2 = 8$
13. $y = \sqrt{36 - x^2}$
14. $x = \sqrt{121 - y^2}$
15. $x = -\sqrt{8 - y^2}$
16. $y = -\sqrt{12 - x^2}$
17. $x^2 + (y + 3)^2 \leq 16$
18. $(x - 4)^2 + (y + 3)^2 \leq 9$
19. $y \leq \sqrt{49 - x^2}$, $y \geq 0$
20. $x > \sqrt{16 - y^2}$, $x \geq 0$

Find the center and radius of each of the following.

21. $x^2 + 6x + y^2 + 8y = -9$
22. $x^2 - 4x + y^2 + 12y = -4$
23. $x^2 - 12x + y^2 + 10y = -25$
24. $x^2 + 8x + y^2 - 6y = -16$
25. $x^2 + 8x + y^2 - 14y = -65$
26. $x^2 - 2x + y^2 = -1$
27. Prove the distance formula (Theorem 4.5).

4.6 Ellipses and Hyperbolas

An **ellipse** is the set of points in a plane the sum of whose distances from two distinct fixed points is constant. The two fixed points are called the **foci** (singular: **focus**) of the ellipse. See Figure 4.13. Using the distance formula and the definition of ellipse we have the following theorem (see Exercise 23).

THEOREM 4.7 The ellipse with x-intercepts $x = a$ and $x = -a$ and y-intercepts $y = b$ and $y = -b$ has the equation

$$\frac{x^2}{a^2} + \frac{y^2}{b^2} = 1.$$

Note that the coefficients of x^2 and y^2 are different, but both positive.

To graph an equation such as $4x^2 + 9y^2 = 36$, we first divide through by 36, obtaining

$$\frac{x^2}{9} + \frac{y^2}{4} = 1.$$

This equation is of the form shown in Theorem 4.7. Hence, the graph is an ellipse having x-intercepts $x = 3$ and $x = -3$, with y-intercepts given by

$y = 2$ and $y = -2$. Additional ordered pairs which satisfy the equation of the ellipse may be found to complete the graph. The graph of this ellipse is shown in Figure 4.14.

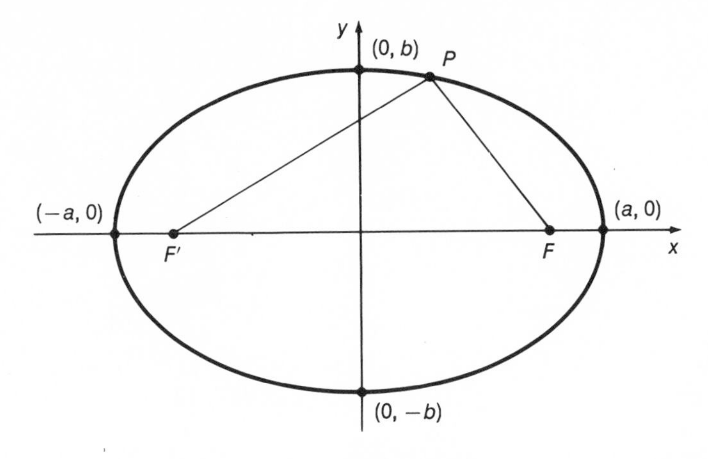

If F and F' are the foci of an ellipse, then for any point P on the ellipse, $PF' + PF = 2a$.

Figure 4.13

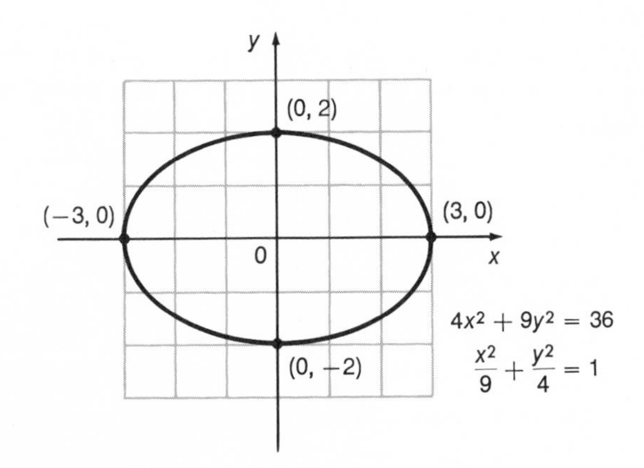

$4x^2 + 9y^2 = 36$

$\frac{x^2}{9} + \frac{y^2}{4} = 1$

Figure 4.14

Example 15 Graph $16x^2 + y^2 \leq 16$.

Dividing through by 16 gives the equation of the boundary, $x^2 + \frac{y^2}{16} = 1$. This equation represents an ellipse having x-intercepts of $x = 1$ and $x = -1$, with y-intercepts $y = 4$ and $y = -4$. Since the point (0, 0) satisfies $16x^2 + y^2 \leq 16$, the graph also includes the interior of the ellipse, as shown in Figure 4.15.

The graph in Figure 4.16 illustrates the idea of symmetry. For each point A of the graph, there is a corresponding point A', such that $AB = A'B$,

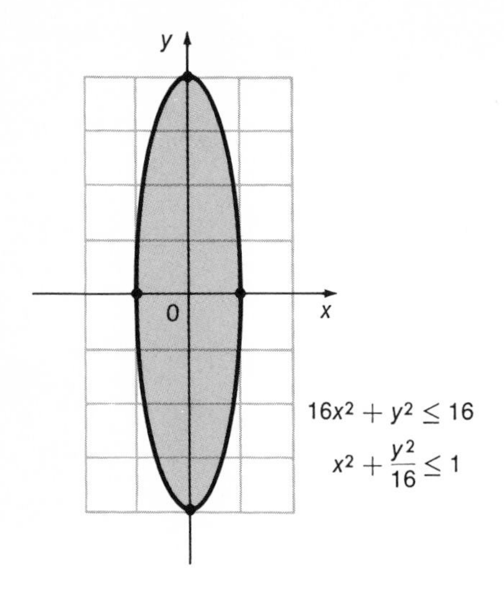

Figure 4.15

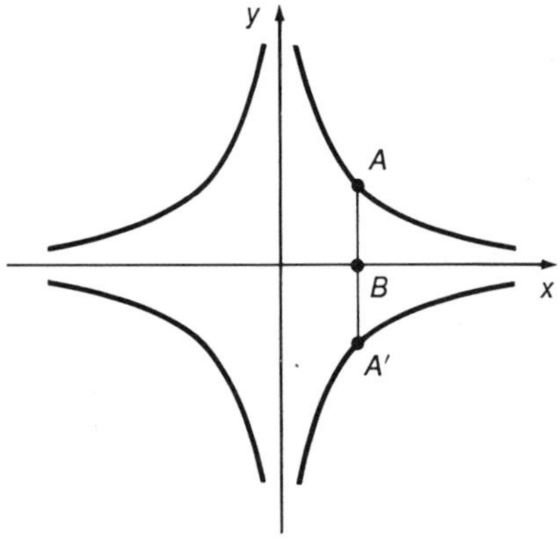

Figure 4.16

where AA' is perpendicular to the x-axis, and B is on the x-axis. Thus the graph is said to be **symmetric with respect to the x-axis.** Verify that the figure is also symmetric with respect to the y-axis.

A **hyperbola** is the set of points in a plane the difference of whose distances from two fixed points (called foci) is constant. The following theorem depends on the distance formula and the idea of symmetry.

THEOREM 4.8 A hyperbola having x-intercepts $x = a$ and $x = -a$, centered about the origin and symmetric to the x-axis, has an equation of the form

$$\frac{x^2}{a^2} - \frac{y^2}{b^2} = 1,$$

while a hyperbola having y-intercepts $y = a$ and $y = -a$, centered about the origin and symmetric to the y-axis, has an equation of the form

$$\frac{y^2}{a^2} - \frac{x^2}{b^2} = 1.$$

As an example, let us graph the hyperbola

$$\frac{x^2}{16} - \frac{y^2}{9} = 1.$$

By Theorem 4.8, the x-intercepts are $x = 4$ and $x = -4$. However, if $x = 0$, we have

$$-\frac{y^2}{9} = 1,$$

or

$$y^2 = -9,$$

which has no real solutions. Hence, the graph has no y-intercepts. To complete the graph we can find some other ordered pairs that belong to it. For example, if $x = 6$, we have

$$\frac{6^2}{16} - \frac{y^2}{9} = 1$$

$$-\frac{y^2}{9} = 1 - \frac{36}{16}$$

$$\frac{y^2}{9} = \frac{20}{16}$$

$$y^2 = \frac{180}{16} = \frac{45}{4}$$

$$y = \pm\frac{3\sqrt{5}}{2} \approx \pm 3.4.$$

Thus the graph includes the points (6, 3.4) and (6, −3.4). Also, if $x = -6$, we have $y \approx \pm 3.4$. Hence, the graph also includes (−6, 3.4) and (−6, −3.4), as shown in Figure 4.17.

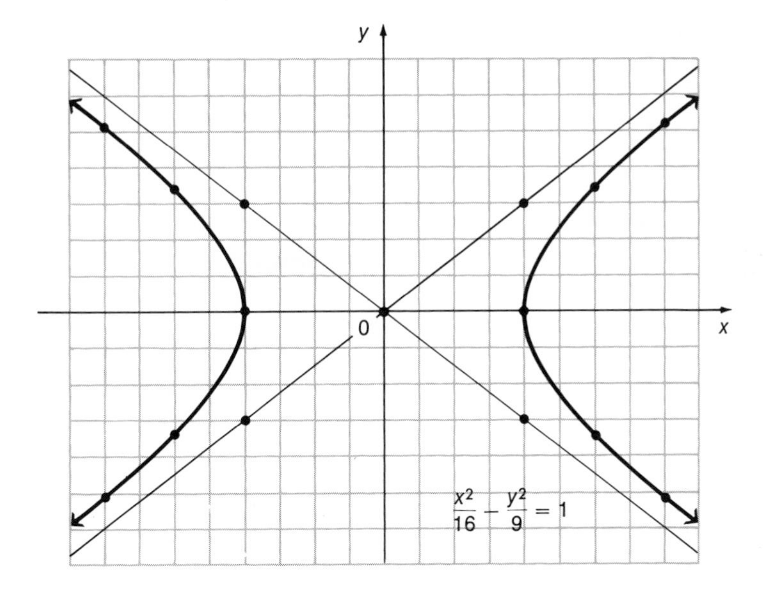

Figure 4.17

The straight lines approached by the graph of the hyperbola of Figure 4.17 are called the **asymptotes** of the hyperbola. A hyperbola of the form

$$\frac{x^2}{a^2} - \frac{y^2}{b^2} = 1$$

will have asymptotes whose equations are given by

$$\frac{x^2}{a^2} - \frac{y^2}{b^2} = 0.$$

This can be simplified by writing

$$\frac{x^2}{a^2} = \frac{y^2}{b^2},$$

from which

$$\frac{x}{a} = \frac{y}{b} \qquad \text{or} \qquad \frac{x}{a} = -\frac{y}{b},$$

so that

$$y = \frac{b}{a}x \qquad \text{or} \qquad y = -\frac{b}{a}x$$

are the equations of the asymptotes. In the same way,

$$\frac{y^2}{a^2} - \frac{x^2}{b^2} = 1$$

has asymptotes given by

$$y = \frac{a}{b}x \qquad \text{or} \qquad y = -\frac{a}{b}x$$

just as above. In general then, we have the following result.

THEOREM 4.9 The asymptotes of the hyperbola $\frac{x^2}{a^2} - \frac{y^2}{b^2} = 1$ are given by

$$\boldsymbol{y = \frac{b}{a}x \qquad \text{or} \qquad y = -\frac{b}{a}x,}$$

while the asymptotes of the hyperbola $\frac{y^2}{a^2} - \frac{x^2}{b^2} = 1$ are given by

$$\boldsymbol{y = \frac{a}{b}x \qquad \text{or} \qquad y = -\frac{a}{b}x.}$$

Example 16 Graph $4y^2 - x^2 = 16$.

If we divide both sides by 16, we have

$$\frac{y^2}{4} - \frac{x^2}{16} = 1.$$

Note that the y-intercepts are $y = 2$ and $y = -2$, and that there are no x-intercepts. We can use Theorem 4.9 to find the asymptotes. Here the asymptotes are given by

$$y = \frac{1}{2}x \qquad \text{or} \qquad y = -\frac{1}{2}x.$$

Using the intercepts and asymptotes, we get the graph shown in Figure 4.18.

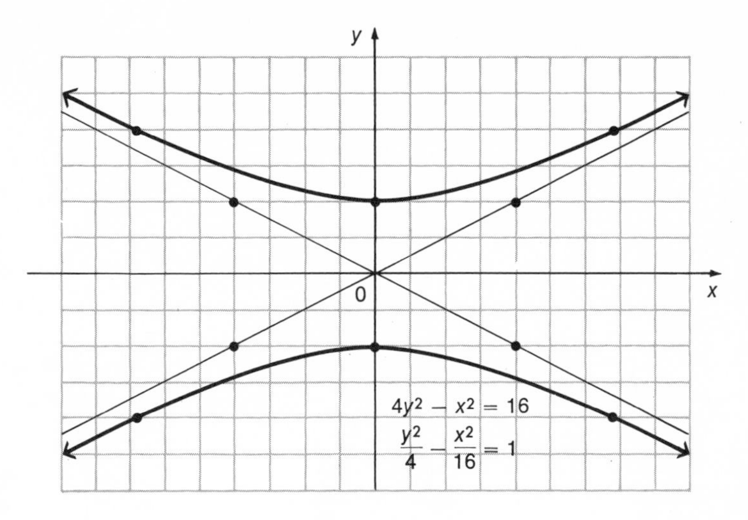

Figure 4.18

Example 17 Graph $y \leq \sqrt{1 + 4x^2}$, $y \geq 0$.

If we square both sides of $y = \sqrt{1 + 4x^2}$, we get

$$y^2 = 1 + 4x^2$$

or

$$y^2 - 4x^2 = 1,$$

which is a hyperbola with y-intercepts $y = 1$ and $y = -1$. Since $\sqrt{1 + 4x^2}$ is always nonnegative, we have only half a hyperbola, as shown in Figure 4.19. We then select any point not on the hyperbola and try it in the original inequality. Since (0, 0) satisfies this inequality we shade the side of the graph containing (0, 0), as shown in Figure 4.19.

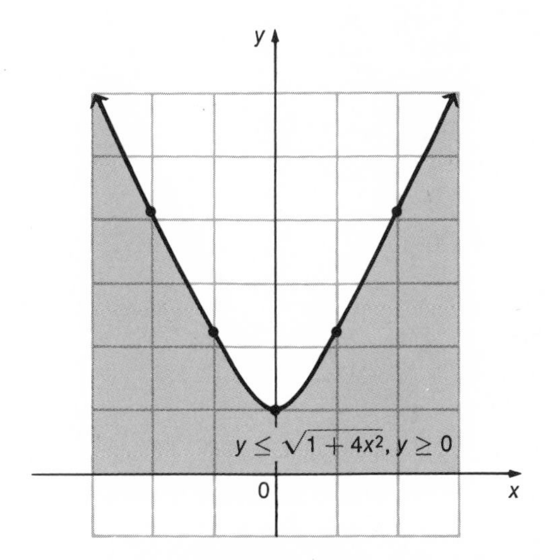

Figure 4.19

The graphs of the second degree relations we have studied, *parabolas, hyperbolas, ellipses* and *circles*, are called **conic sections** since each can be obtained by cutting a cone with a plane, as shown in Figure 4.20.

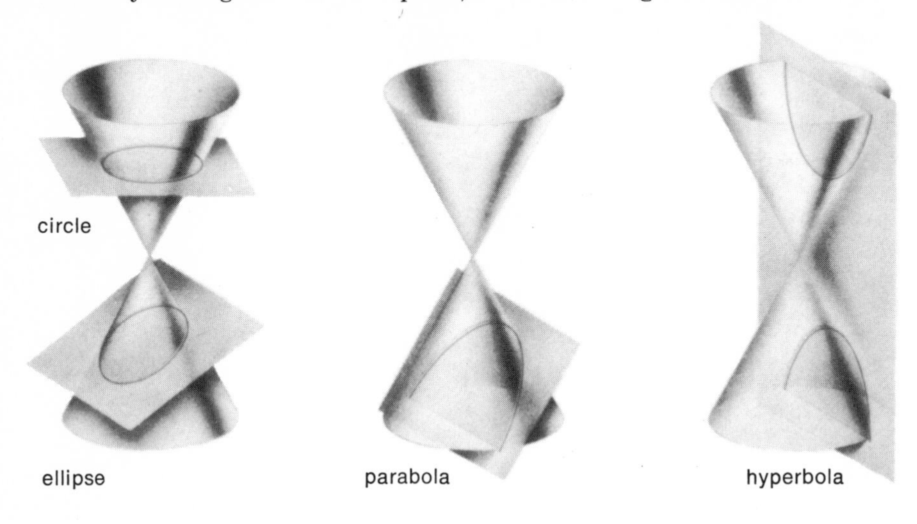

Figure 4.20

Note that the conic sections have second degree equations of the form

$$Ax^2 + Cy^2 + Dx + Ey + F = 0,$$

where either A or C must be nonzero. (The letter B is reserved for an xy term, which we have not discussed here.) The special characteristics of each of the relations we have discussed are summarized below.

- ▶ *parabola* either A or C equals 0, but not both
- ▶ *circle* $A = C \neq 0$
- ▶ *ellipse* $A \neq C$, $AC > 0$
- ▶ *hyperbola* $AC < 0$

4.6 Exercises

Sketch the graph of each of the following.

1. $\frac{x^2}{9} + \frac{y^2}{4} = 1$
2. $\frac{x^2}{16} + \frac{y^2}{36} = 1$
3. $\frac{x^2}{9} + y^2 = 1$
4. $\frac{y^2}{16} - \frac{x^2}{9} = 1$
5. $\frac{x^2}{6} + \frac{y^2}{9} = 1$
6. $\frac{x^2}{8} - \frac{y^2}{12} = 1$
7. $x^2 + 4y^2 = 16$
8. $25x^2 + 9y^2 = 225$
9. $x^2 = 9 + y^2$
10. $y^2 = 16 + x^2$
11. $\frac{x}{4} = \sqrt{1 - \frac{y^2}{9}}$

12. $\frac{y}{2} = \sqrt{1 - \frac{x^2}{25}}$

13. $\frac{y}{3} = \sqrt{1 + \frac{x^2}{16}}$

14. $x = \sqrt{1 + \frac{y^2}{36}}$

15. $x^2 \leq 9 - 4y^2$

16. $x^2 > 1 + 9y^2$

17. $\frac{x}{6} < \sqrt{1 - \frac{y^2}{121}}$, $x \geq 0$

18. $\frac{y}{4} \leq \sqrt{1 + \frac{x^2}{25}}$, $y \geq 0$

Find the equations of the asymptotes of each of the following hyperbolas.

19. $\frac{x^2}{9} - \frac{y^2}{25} = 1$

20. $\frac{y^2}{9} - \frac{x^2}{16} = 1$

21. $x^2 = 16 + 4y^2$

22. $y^2 = 25 + x^2$

23. Suppose $(c, 0)$ and $(-c, 0)$ are the foci of an ellipse. Suppose the sum of the distances from any point (x, y) of the ellipse to the two foci is $2a$. (See figure.)

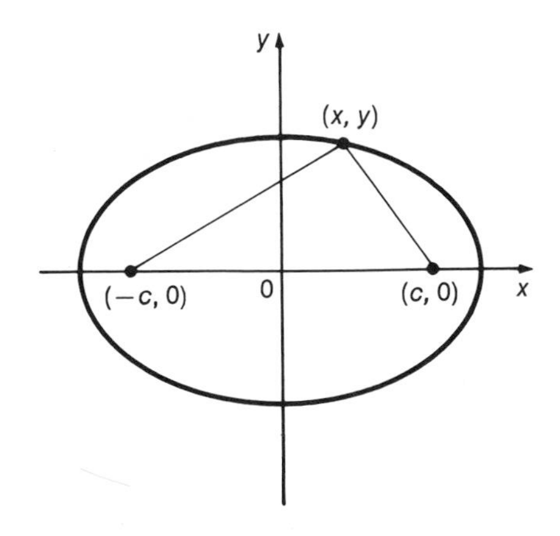

(a) Show that the equation of the resulting ellipse is

$$\frac{x^2}{a^2} + \frac{y^2}{a^2 - c^2} = 1.$$

(b) Show that a and $-a$ are the x-intercepts.

(c) Let $b^2 = a^2 - c^2$, and show that b and $-b$ are the y-intercepts.

4.7 Variation

Suppose Mert's salary is always three times Sam's salary. Then, if y represents Mert's salary, and x represents Sam's salary, we see that the two salaries satisfy a relationship of the form

$$y = kx \qquad (k \text{ a real number, } k \neq 0).$$

(In our example, $k = 3$.) Such a relationship is called a **direct variation;** here we can say that Mert's salary **varies directly** as Sam's salary. The real number k is sometimes called the **constant of variation.**

Example 18 Suppose the area of a certain rectangle varies directly as the width. If the area is 50 when the width is 10, find the area when the width is 25.

Since we know that area varies directly as width, we have

$$A = kw,$$

where A = area, w = width, and k is a nonzero constant which we must determine. When $A = 50$ and $w = 10$, we have

$$50 = 10k,$$

or $k = 5$. Thus, in this example, the relationship between area and width can be expressed as

$$A = 5w.$$

If $w = 25$, we have

$$A = 5(25) = 125.$$

If the variables satisfy a relationship of the form

$$y = \frac{k}{x} \qquad (k \text{ a real number, } k \neq 0),$$

then x and y are said to **vary inversely.** For example, suppose that in a certain manufacturing process the cost of producing a single item varies inversely as the number of items produced. Suppose further that if 100 items are produced, each costs \$2. To find the cost per item if 400 items are produced, we can let x represent the number of items produced and y the cost per item, and write

$$y = \frac{k}{x},$$

for some nonzero constant k. We know that $y = \$2$ when $x = 100$. Hence

$$2 = \frac{k}{100}$$

or

$$k = 200.$$

Thus, the relationship between x and y is given by

$$y = \frac{200}{x}.$$

When 400 items are produced, the cost per item is given by

$$y = \frac{200}{400} = \$.50.$$

Example 19 Suppose m varies directly as the square of p and inversely as q. Suppose also that $m = 8$ when $p = 2$ and $q = 6$. Find m if $p = 6$ and $q = 10$.

Here m depends on several variables. This is called **joint variation**. We can write

$$m = \frac{kp^2}{q}.$$

We know that $m = 8$ when $p = 2$ and $q = 6$. Thus,

$$8 = \frac{k \cdot 2^2}{6}$$

$$k = 12$$

and so,

$$m = \frac{12p^2}{q}.$$

If $p = 6$ and $q = 10$, we have

$$m = \frac{12 \cdot 6^2}{10}$$

$$m = \frac{216}{5}.$$

4.7 Exercises

1. Suppose m varies directly as z and p. If $m = 10$ when $z = 3$ and $p = 5$, find m when $z = 5$ and $p = 7$.
2. Suppose r varies directly as the square of m, and inversely as s. If $r = 12$ when $m = 6$ and $s = 4$, find r when $m = 4$ and $s = 10$.
3. The distance a body falls from rest varies directly as the square of the time it falls (disregarding air resistance). If an object falls 1024 feet in 8 seconds, how far will it fall in 12 seconds?
4. Hooke's law for an elastic spring states that the distance a spring stretches varies directly as the force applied. If a force of 15 pounds stretches a certain spring 8 inches, how much will a force of 30 pounds stretch the spring?
5. In electric current flow it is found that the resistance (measured in units called ohms) offered by a fixed length of wire of a given material varies inversely as the square of the diameter of the wire. If a wire .01 inches in diameter has a resistance of .4 ohm, what is the resistance of a wire of the same length and material but .03 inches in diameter?
6. It is shown in engineering that the maximum load a cylindrical column of circular cross section can hold varies directly as the fourth power of the diameter and inversely as the square of the height. If a 9 foot column 3 feet in diameter will support a load of 8 tons, how much load will be supported by a column 12 feet high and 2 feet in diameter?

7. The maximum load of a horizontal beam which is supported at both ends varies directly as the width and square of the height, and inversely as the length between supports. If a beam 20 feet long, 5 inches wide, and 6 inches high can support a maximum of 1000 pounds, what is the maximum load of a beam of the same material 40 feet long, 10 inches wide, and 3 inches high?

5

EXPONENTIAL AND LOGARITHMIC FUNCTIONS

Many applications of mathematics, particularly those pertaining to growth and decay of populations, involve exponential functions. Other applications, such as numerical calculations, depend on logarithmic functions. In the third section we shall see how these two types of functions are closely related.

5.1 Inverse Relations

As defined in Chapter 3, a relation is a set of ordered pairs, while a function is defined as a special type of relation in which each value from the domain corresponds to exactly one value from the range. We define the **inverse** of relation R, written R^{-1} (read "R inverse"), as

$$R^{-1} = \{(y, x) \mid (x, y) \in R\}.$$

For example, given $f = \{(x, y) \mid y = 3x + 1\}$, the ordered pair (2, 7) belongs to the function (if $x = 2$, then $y = 7$), and so by the definition of inverse, we must have (7, 2) as an element of f^{-1}. Also, since $(-3, -8) \in f$, $(-8, -3) \in f^{-1}$.

The definition of inverse relation provides a relatively simple method for graphing an inverse, given the original relation. If (a, b) belongs to the original relation, then (b, a) belongs to the inverse. As shown in Figure 5.1, a line from

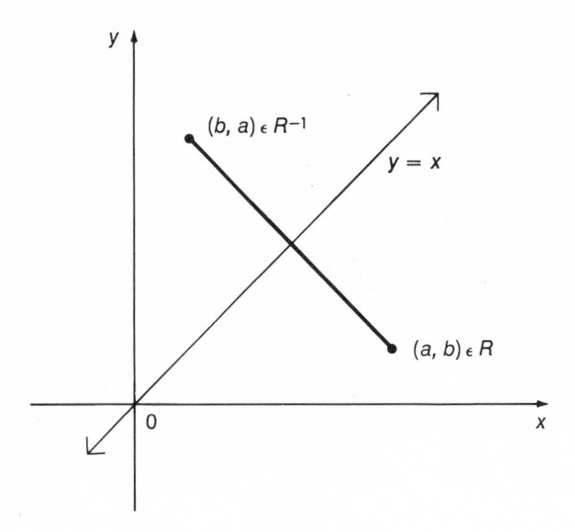

Figure 5.1

(a, b) to (b, a) is bisected by and is perpendicular to the line $y = x$. Hence, to graph R^{-1}, we reflect the graph of R about the line $y = x$. Figure 5.2 shows the graph of $f = \{(x, y) \mid y = 3x + 1\}$ together with the graph of its inverse f^{-1}. Note that in this case both f and f^{-1} are functions.

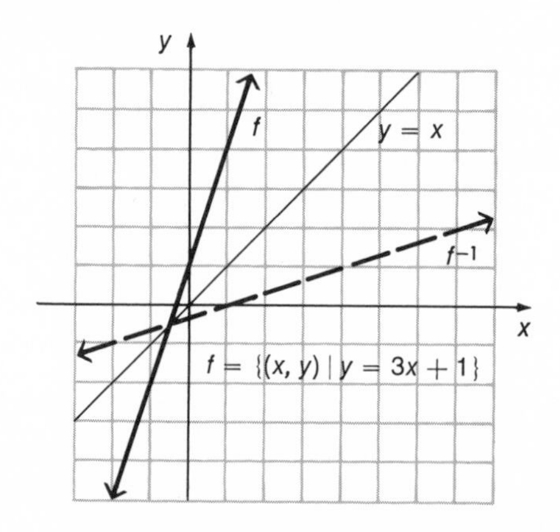

Figure 5.2

Example 1 Figure 5.3 shows the graph of three relations and their inverses. Note that in (a) both R and R^{-1} are functions. In (b), R is a function, but R^{-1} is not, while in (c) R is not a function but R^{-1} is.

Recall that we can identify a function by passing a vertical line through the graph. If any vertical line cuts the graph in more than one point, the graph

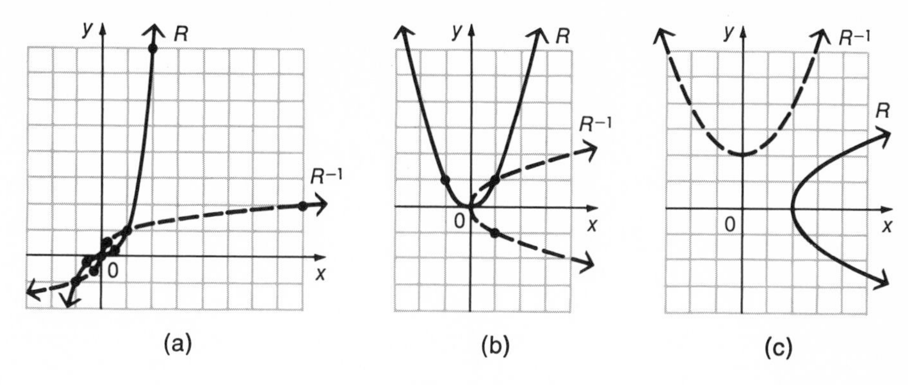

Figure 5.3

is not the graph of a function. There is a similar test to determine if a given relation has an inverse which is a function. If a horizontal line cuts the original graph in more than one point, then the inverse cannot satisfy the definition of a function. (Why?)

We have written the equation defining a function in the form $y = f(x)$, where x is a domain element. We want to express inverse functions in the same way, as $y = f^{-1}(x)$, where x is a domain element of the inverse function. To do this, we note that if $(x, y) \in f$, then $(y, x) \in f^{-1}$. Hence, to get the equation of f^{-1}, given the equation of f, we exchange x and y in the defining equation of f. For example, if $f = \{(x, y) \mid y = 3x + 1\}$, then we get f^{-1} by exchanging x and y in the defining equation for f. This gives

$$x = 3y + 1.$$

If we now solve for y, we get

$$y = \frac{x - 1}{3}.$$

Hence, $f^{-1}(x) = \dfrac{x - 1}{3}$, and we can now write f^{-1} as

$$f^{-1} = \left\{(x, y) \mid y = \frac{x - 1}{3}\right\}.$$

Example 2 Let $f = \{(x, y) \mid y = x^3 - 1\}$. Find f^{-1}. Graph f and f^{-1}.

We obtain the defining equation for f^{-1} by exchanging x and y in the equation for f, and solving for y, as follows.

$$x = y^3 - 1$$
$$y = \sqrt[3]{x + 1}$$

The graphs of f and f^{-1} are shown in Figure 5.4.

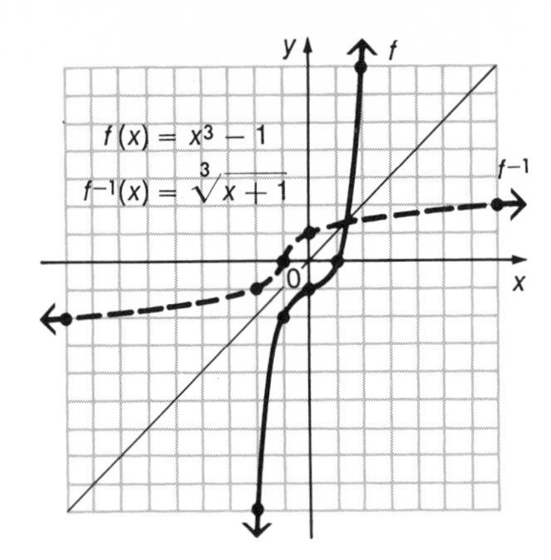

Figure 5.4

We have seen that if $f(x) = 3x + 1$, then $f^{-1}(x) = \dfrac{x-1}{3}$. Note that

$$f[f^{-1}(x)] = f\left[\frac{x-1}{3}\right] = 3\left(\frac{x-1}{3}\right) + 1 = x,$$

while
$$f^{-1}[f(x)] = f^{-1}[3x+1] = \frac{(3x+1)-1}{3} = x.$$

This example illustrates the next result, which can be proven from the definition of inverse relation.

THEOREM 5.1 If both f and f^{-1} are functions, then

$$f[f^{-1}(x)] = x \qquad \text{and} \qquad f^{-1}[f(x)] = x.$$

5.1 Exercises

For each of the following relations, (a) tell whether or not the relation is a function, (b) identify the domain and range of the relation, (c) sketch the graph of the relation, (d) sketch the graph of the inverse, (e) give the domain and range of the inverse, and (f) write an equation for the inverse.

1. $y = 3x - 4$
2. $y = 4x - 5$
3. $x + y = 0$
4. $y = 6 - x$
5. $y = x^2$
6. $y = (x-2)^2$
7. $y = \sqrt[3]{x}$
8. $y = x^3$

Sketch the graph of the inverse of each of the following relations.

9. $x^2 + y^2 = 9$
10. $4x^2 + 9y^2 = 36$
11. $x^2 + 4y^2 = 16$
12. $x^2 - y^2 = 1$

13. $y = |x|$ 16. $y = -\sqrt{4 + x^2}$
14. $x = |y|$ 17. $x = \sqrt{4 - y}$
15. $y = \sqrt{25 - x^2}$ 18. $y = \sqrt{2 + x}$

5.2 Exponential Functions

We know that if $a > 0$, we can define the symbol a^m for any rational value of m. In this section we want to extend the definition of a^m to include all real (and not just rational) values of the exponent m. To do this we shall need the following results.

THEOREM 5.2 If $a > 1$ and m is a positive rational number, then $a^m > 1$.

To prove this, let $m = p/q$, where p and q are positive integers. Since $a > 1$, we can use Exercises 22 and 25 of Section 1.4, to write $a^2 > 1$, and $a^3 > 1$, and in general, $a^p > 1$ for any positive integer p.

Now we must prove $a^{p/q} > 1$. Suppose it is not. Since $a^{p/q}$ is positive, we have

$$0 < a^{p/q} \leq 1.$$

If we now multiply both sides of $a^{p/q} \leq 1$ by the positive quantity $a^{p/q}$, we get

$$(a^{p/q})^2 \leq a^{p/q}.$$

Hence, $(a^{p/q})^2 \leq a^{p/q} \leq 1.$
In the same way, $(a^{p/q})^3 \leq (a^{p/q})^2 \leq a^{p/q} \leq 1,$
and in general, $(a^{p/q})^r \leq 1,$

for any positive integer r. If $r = q$, we have

$$(a^{p/q})^q \leq 1,$$

or

$$a^p \leq 1,$$

which contradicts the result of the previous paragraph. Hence, the assumption is false, and we must have $a^{p/q} > 1$, which completes the proof.

Suppose now $a > 1$ and m and n are rational numbers with $m < n$. If $m < n$, then $n - m > 0$, and by the result above, we can write

$$a^{n-m} > 1.$$

Multiplying both sides by a^m (which is positive) gives

$$a^n > a^m$$

or

$$a^m < a^n,$$

which proves the following result.

THEOREM 5.3 If $a > 1$ and m and n are rational numbers with $m < n$, then $a^m < a^n$.

Now we are ready to define a^x for any real value of x, $(a > 0)$. If x is a real number, then there exists some integer k such that $x \leq k$ (see Exercise 17). We know that a^k is defined and exists. If m is any rational number, with $m \leq x \leq k$, then by Theorem 5.3, $a^m \leq a^k$. Hence, a^k is an upper bound for the set

$$\{a^m \mid m \text{ is rational, } m \leq x\}.$$

By the completeness axiom of Section 1.4, this set has a *least* upper bound. This least upper bound we define as the number a^x, for any real number x. That is, if x is any real number, then we make the following definition.

- ▶ If $a > 1$, then a^x is the least upper bound of the set $\{a^m \mid m \text{ is rational, } m \leq x\}$.
- ▶ If $a = 1$, then $a^x = 1$.
- ▶ If $0 < a < 1$, then $\frac{1}{a} > 1$, and $a^x = \frac{1}{(1/a)^x}$.

Note that we make no attempt to define a^x for $a < 0$. Since, for example, $(-6)^x$ is not a real number if $x = \frac{1}{2}$, we do not define a^x for $a < 0$.

Using the work above, we can define a^x for any real number x and any number $a > 0$. The exponential a^x defined in this way still satisfies all properties of exponents from Chapter 2. Hence, the domain of a function such as $y = 2^x$ now includes all real numbers. We call any such function $y = a^x$ $(a > 0, a \neq 1)$ an **exponential function.** Figure 5.5 shows the graph of $f(x) = 2^x$, which is typical of the graphs of functions of the form $f(x) = a^x$, for $a > 1$,

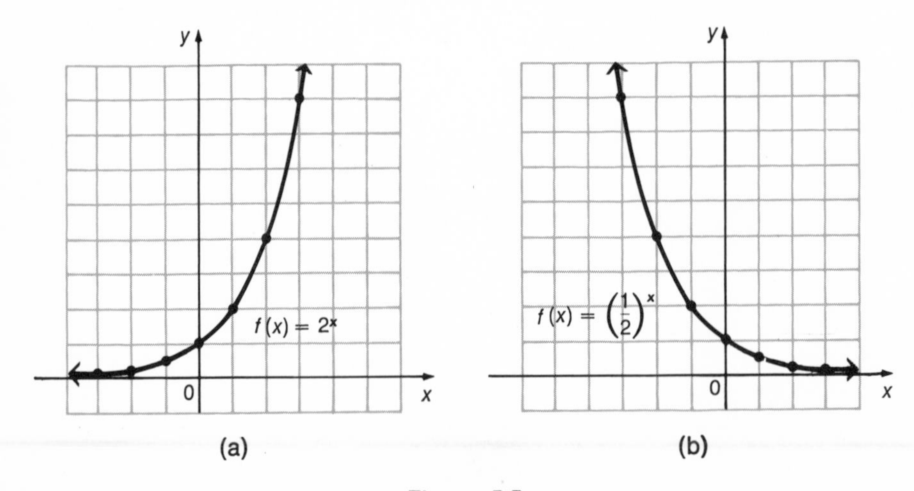

Figure 5.5

and the graph of $f(x) = (\frac{1}{2})^x$ which is typical of the graphs of functions of the form $f(x) = a^x$, for $0 < a < 1$.

Example 3 Graph $y = 2^{-x^2}$.

Since $-x^2 \leq 0$ for all real x, it follows that $0 < y \leq 1$ for real values of x. By plotting several points, we get the result shown in Figure 5.6. Functions of this type are important in probability theory.

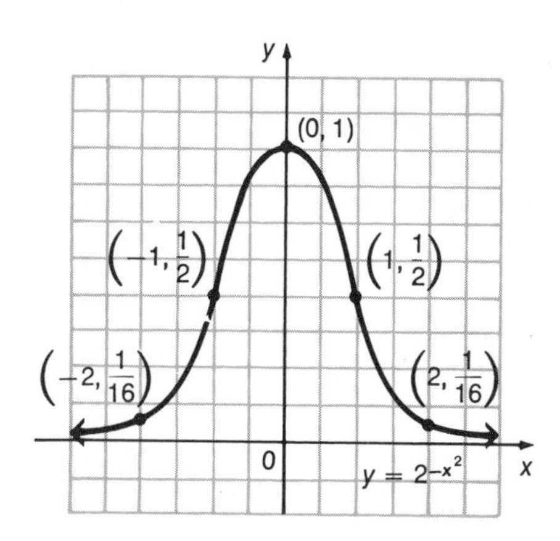

Figure 5.6

5.2 Exercises

Graph each of the following functions.

1. $y = 3^x$
2. $y = 4^x$
3. $y = (\frac{3}{2})^x$
4. $y = 3^{-x}$
5. $y = 10^{-x}$
6. $y = 10^x$
7. $y = 2^{x+1}$
8. $y = 2^{-x+1}$
9. $y = 2^{|x|}$
10. $y = 2^{-|x|}$
11. $y = 2^x + 2^{-x}$
12. $y = (\frac{1}{2})^x + (\frac{1}{2})^{-x}$

13. Using the information given in the table of values below, graph $y = 3^x$ on a large scale for $0 \leq x \leq 1$. Then estimate from the graph (a) $3^{1/3}$, (b) $3^{2/3}$.

x	0	1/4	1/2	3/4	1
3^x	1	1.3	1.7	2.3	3

14. Graph $y = a^x$ for $a = 1$.
15. Suppose the domain of $y = 2^x$ includes only rational values of x (which are the only values discussed prior to this section). Describe in words the resulting graph.

16. Prove: if $0 < a < 1$ and m and n are rational numbers such that $m < n$, then $a^m > a^n$. $\left(\text{Hint: if } 0 < a < 1, \text{ then } \frac{1}{a} > 1.\right)$
17. Let x be any real number. Prove that there exists an integer k such that $x \leq k$. (Hint: use an indirect proof.)

5.3 Logarithmic Functions

In the previous section we discussed exponential functions of the form $y = a^x$, for all positive a, $a \neq 1$. In this section we shall discuss the inverses of such functions. Recall that to obtain the inverse of a function such as $y = a^x$, we exchange x and y. Doing this, we obtain

$$x = a^y,$$

as the inverse of the exponential function $y = a^x$. In order to solve the equation $x = a^y$ for y, we use the following definition. For all real numbers y, all positive numbers a, $a \neq 1$,

▶ $y = \log_a x$ means the same as $x = a^y$.

Log is an abbreviation for "logarithm." Read $\log_a x$ as "the logarithm of x to the base a." A **logarithmic function** is a function of the form $y = \log_a x$ ($a > 0$, $a \neq 1$).

Example 4 In the chart of this example we show several pairs of equivalent statements. The same statement is written in both exponential and logarithmic forms.

Exponential Form	*Logarithmic Form*
$2^3 = 8$	$\log_2 8 = 3$
$(1/2)^{-4} = 16$	$\log_{1/2} 16 = -4$
$10^5 = 100{,}000$	$\log_{10} 100{,}000 = 5$
$3^{-4} = \frac{1}{81}$	$\log_3 \frac{1}{81} = -4$

Since $b^1 = b$ and $b^0 = 1$, for all positive b, we have

▶ $\log_b b = 1$ and $\log_b 1 = 0$.

Exponential and logarithmic functions are inverses of each other. Since the domain of an exponential function is the set of all real numbers, the range of a log function will also be the set of all real numbers. In the same way, the range of an exponential function and the domain of a log function are both the set of positive real numbers. Using the technique of graphing inverses discussed in Section 5.1, we can graph logarithmic functions by reflecting the graphs of exponential functions about the line $y = x$ as shown in Figure 5.7.

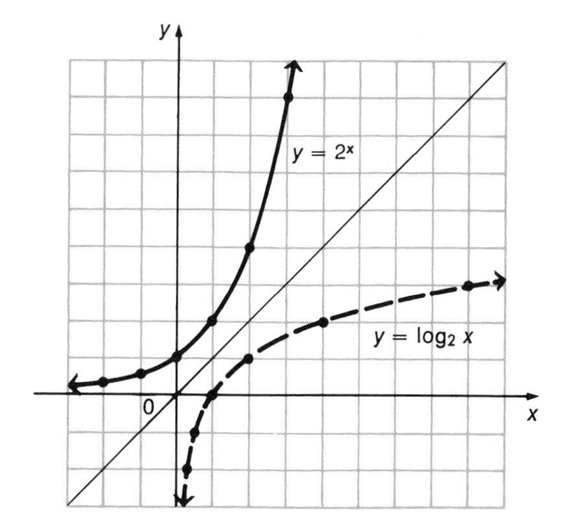

Figure 5.7

This graph suggests the following **axiom of logarithms.**

▶ If $a > 1$ and $m < n$, then $\log_a m < \log_a n$.

Example 5 Graph $y = \log_3 |x|$.

Using the equivalent expression $|x| = 3^y$, we can plot points to obtain the graph (see Figure 5.8).

x	-3	-1	$-\frac{1}{3}$	1	3
y	1	0	-1	0	1

Alternatively, we could graph $|y| = 3^x$, and then find the graph of its inverse.

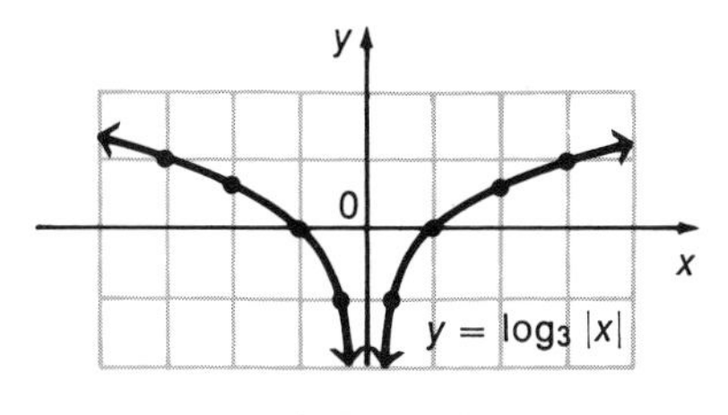

Figure 5.8

Logarithms were originally important as an aid for numerical calculations. The use of logarithms for this purpose is based on the properties discussed in the next theorem.

THEOREM 5.4 If x and y are any positive real numbers, r is any real number, and a is any positive real number, $a \neq 1$, then

(a) $\log_a xy = \log_a x + \log_a y$

(b) $\log_a \frac{x}{y} = \log_a x - \log_a y$

(c) $\log_a x^r = r \log_a x$.

To prove (a), let $m = \log_a x$, and $n = \log_a y$. Then, by the definition of logarithm,

$$a^m = x \qquad \text{and} \qquad a^n = y.$$

Hence, we can write

$$a^m \cdot a^n = xy.$$

By a property of exponents, we have

$$a^{m+n} = xy.$$

Now we can use the definition of logarithm to write

$$\log_a xy = m + n.$$

Since $m = \log_a x$ and $n = \log_a y$, we have

$$\log_a xy = \log_a x + \log_a y.$$

Parts (b) and (c) can be proved similarly.

Example 6 Using the results of Theorem 5.4, we have

(a) $\log_a \frac{mnp}{q^2} = \log_a m + \log_a n + \log_a p - 2 \log_a q$.

(b) $\log_a \sqrt[3]{m^2} = \frac{2}{3} \log_a m$.

Example 7 Assume $\log_{10} 2 = .3010$, and $\log_{10} 3 = .4771$. Find the base 10 logs of 4, 5, 6, and 12.

Using the results of Theorem 5.4, we have

(a) $\log_{10} 4 = \log_{10} 2^2 = 2 \log_{10} 2 = 2(.3010) = .6020$.

(b) $\log_{10} 5 = \log_{10} \frac{10}{2} = \log_{10} 10 - \log_{10} 2 = 1 - .3010 = .6990$.

(c) $\log_{10} 6 = \log_{10} (2 \cdot 3) = \log_{10} 2 + \log_{10} 3 = .3010 + .4771 = .7781$.

(d) $\log_{10} 12 = \log_{10} (4 \cdot 3) = \log_{10} 4 + \log_{10} 3 = .6020 + .4771 = 1.0791$.

5.3 Exercises

Change each of the following to logarithmic form.

1. $3^4 = 81$
2. $2^5 = 32$
3. $10^4 = 10{,}000$
4. $8^2 = 64$
5. $(\frac{1}{2})^{-4} = 16$
6. $(\frac{2}{3})^{-3} = \frac{27}{8}$

7. $10^{-4} = .0001$
8. $(\frac{1}{100})^{-2} = 10{,}000$

Change each of the following to exponential form.

9. $\log_6 36 = 2$
10. $\log_5 5 = 1$
11. $\log_{\sqrt{3}} 81 = 8$
12. $\log_4 \frac{1}{64} = -3$
13. $\log_m k = n$
14. $\log_2 r = y$

Write each of the following expressions as a single logarithm.

15. $\log_a x + \log_a y - \log_a m$
16. $(\log_b k - \log_b m) - \log_b a$
17. $2 \log_m a - 3 \log_m b^2$
18. $\frac{1}{2} \log_y p^3q^4 - \frac{2}{3} \log_y p^4q^3$
19. $-\frac{3}{4} \log_x a^6b^8 + \frac{2}{3} \log_x a^9b^3$
20. $\log_a (pq^2) + 2 \log_a \left(\frac{p}{q}\right)$

5.4 Common Logarithms

Since the number system we use is base 10, logarithms to base 10 are most convenient for numerical calculations. Base 10 logarithms are called **common logarithms.** For simplicity, we shall abbreviate $\log_{10} x$ as $\log x$. Using this notation, we have

$$\begin{aligned}
\log 1000 &= \log 10^3 = 3 \\
\log 10 &= \log 10^1 = 1 \\
\log 1 &= \log 10^0 = 0 \\
\log .1 &= \log 10^{-1} = -1 \\
\log .001 &= \log 10^{-3} = -3,
\end{aligned}$$

and so on.

While it can be shown that there is no rational number x such that $10^x = 6$ (and hence no rational number x such that $x = \log 6$) we can use a table of common logarithms to find a rational approximation for log 6. From the excerpt of the log table included on the next page, we see that a decimal approximation for log 6 is given by .7782. Hence,

$$\log 6 \approx .7782$$

(or, equivalently, $10^{.7782} \approx 6$, where $\approx$ means "approximately equal to"). Since most logs are approximations, we shall replace $\approx$ with $=$ and write $\log 6 = .7782$.

To find log 6240, we first note that

$$1000 < 6240 < 10{,}000,$$

so that $\log 1000 < \log 6240 < \log 10{,}000$, by the axiom of logarithms. Since $\log 1000 = 3$ and $\log 10{,}000 = 4$, we have

$$3 < \log 6240 < 4. \tag{1}$$

To find the decimal part of log 6240 in the table, locate the first two significant figures, 62 in the left column of the table. Then find the third digit, 4, across the top.

	0	1	2	3	4	5	6	7	8	9
5.7	.7559	.7566	.7574	.7582	.7589	.7597	.7604	.7612	.7619	.7627
5.8	.7634	.7642	.7649	.7657	.7664	.7672	.7679	.7686	.7694	.7701
5.9	.7709	.7716	.7723	.7731	.7738	.7745	.7752	.7760	.7767	.7774
6.0	.7782	.7789	.7796	.7803	.7810	.7818	.7825	.7832	.7839	.7846
6.1	.7853	.7860	.7868	.7875	.7882	.7889	.7896	.7903	.7910	.7917
6.2	.7924	.7931	.7938	.7945	.7952	.7959	.7966	.7973	.7980	.7987
6.3	.7993	.8000	.8007	.8014	.8021	.8028	.8035	.8041	.8048	.8055
6.4	.8062	.8069	.8075	.8082	.8089	.8096	.8102	.8109	.8116	.8122
6.5	.8129	.8136	.8142	.8149	.8156	.8162	.8169	.8176	.8182	.8189
6.6	.8195	.8202	.8209	.8215	.8222	.8228	.8235	.8241	.8248	.8254

The intersection of this row and column gives .7952, and hence from (1) we have

$$\log 6240 = 3.7952.$$

The decimal part is called the **mantissa,** and the integer part the **characteristic.**

To find log .00587, we note

$$.001 < .00587 < .01$$

and

$$\log .001 < \log .00587 < \log .01$$

or

$$-3 < \log .00587 < -2.$$

From the log table, the mantissa of log .00587 is .7686. Thus,

$$\log .00587 = -3 + .7686.$$

It is often desirable to write the characteristic, -3, as $7 - 10$, so that the log becomes

$$\log .00587 = 7.7686 - 10.$$

Let us now use the table of common logarithms in the Appendix and the properties of logarithms in Theorem 5.4 to aid in a numerical calculation. For example, to evaluate

$$(\sqrt[3]{42})(76.9)(.00283),$$

we can use the properties of logarithms and the table to write

$$\begin{aligned}\log (\sqrt[3]{42})(76.9)(.00283) &= \tfrac{1}{3}\log 42 + \log 76.9 + \log .00283\\ &= \tfrac{1}{3}(1.6232) + 1.8859 + (7.4518 - 10)\\ &= .5411 + 1.8859 + (7.4518 - 10)\\ &= 9.8788 - 10.\end{aligned}$$

From the log table we find that .756 is the number (to three places) whose logarithm is $9.8788 - 10$, and so

$$(\sqrt[3]{42})(76.9)(.00283) \approx .756.$$

Example 8 Calculate $\sqrt[3]{.00762}$.

We find that

$$\log \sqrt[3]{.00762} = \tfrac{1}{3}(7.8820 - 10).$$

This calculation can be simplified if we write the characteristic, -3, as $27 - 30$. This gives

$$\log \sqrt[3]{.00762} = \tfrac{1}{3}(27.8820 - 30) = 9.2940 - 10.$$

Hence, from the log table,

$$\sqrt[3]{.00762} \approx .197.$$

The table of logs included in this text contains decimal approximations of common logs to four places of accuracy. More accurate tables are available; however, more accuracy can be obtained from the table included in this text by the process of **linear interpolation.** As an example, let us use linear interpolation to approximate log 75.37. Note that

$$\log 75.3 < \log 75.37 < \log 75.4.$$

Figure 5.9 shows the portion of the curve $y = \log x$ between $x = 75.3$ and $x = 75.4$. We shall use the line segment PR to approximate the log curve (this approximation is adequate for values of x relatively close to one another).

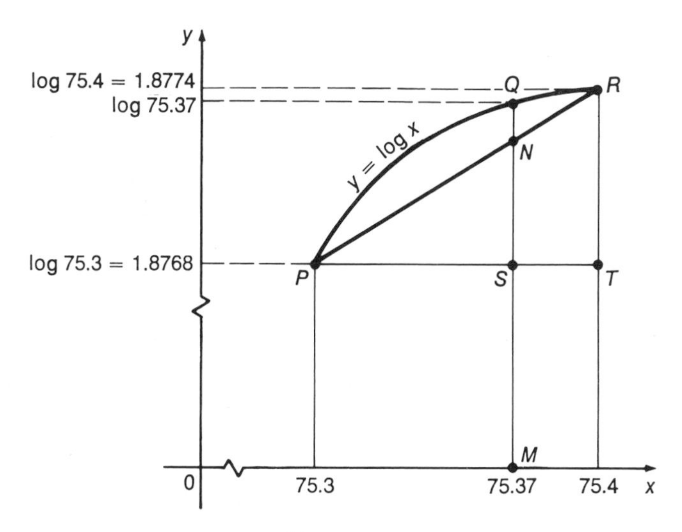

Figure 5.9

From the figure, note that log 75.37 is given by the length of segment MQ, which we cannot find. We can, however, find MN, which we shall use as our approximation to log 75.37. By properties of similar triangles, we have

$$\frac{PS}{PT} = \frac{SN}{TR}.$$

In this case, $PS = 75.37 - 75.3 = .07$, $PT = 75.4 - 75.3 = .1$, and $RT = \log 75.4 - \log 75.3 = 1.8774 - 1.8768 = .0006$. Hence,

$$\frac{.07}{.1} = \frac{SN}{.0006}$$

or

$$SN = .7(.0006) \approx .0004.$$

Since $\log 75.37 = MN = MS + SN$, and since $MS = \log 75.3 = 1.8768$, we have

$$\log 75.37 = 1.8768 + .0004 = 1.8772.$$

Example 9 Find log .008726.

Here $\log .00872 = 7.9405 - 10$ and $\log .00873 = 7.9410 - 10$. Since .008726 is .6 of the distance between .00872 and .00873, we take .6 of the difference between log .00873 and log .00872. We have

$$.6[(7.9410 - 10) - (7.9405 - 10)] = .6(.0005) \approx .0003.$$

Hence,

$$\begin{aligned} \log .008726 &= (7.9405 - 10) + .0003 \\ &= 7.9408 - 10. \end{aligned}$$

Example 10 Find x such that $\log x = .3275$.

From the log table, we see

$$\log 2.12 = .3263 < .3275 < .3284 = \log 2.13.$$

The distance between .3263 and .3275 is .0012, while the distance between .3263 and .3284 is .0021. Hence, .3275 is

$$\frac{.0012}{.0021} \approx .6$$

of the way from .3263 to .3284. From this last result, we see that x should be .6 of the way from 2.12 to 2.13, so that

$$\log 2.126 = .3275.$$

5.4 Exercises

Find the characteristic of the logarithms of each of the following.

1. 875
2. 9462
3. 2,400,000
4. 875000
5. .00023

6. .098
7. .000042
8. .000000257

Find the logarithms of the following numbers.

9. 875
10. 3750
11. 12800
12. 653000
13. 7.63
14. 9.37
15. .000893
16. .00376
17. 3.876
18. 2.975
19. 68250
20. 103200
21. .0003824
22. .00008632

Find the numbers having each of the following logarithms.

23. 1.5366
24. 2.9253
25. 8.8733 − 10
26. 9.2504 − 10
27. 3.4947
28. 4.6863
29. 2.8039
30. 8.8078 − 10

Use logarithms (and interpolation) to find approximations to three-place accuracy for each of the following.

31. $\dfrac{(79.6)(4.83)(9.3)}{43.8}$
32. $\dfrac{(123)(7.892)(.436)}{2.81}$
33. $\dfrac{(21.3)^2}{(.86)^3}$
34. $\dfrac{(.893)^2}{(.624)^3}$
35. $(8.16)^{\frac{1}{3}}$
36. $\sqrt{276}$
37. $(.463)^{.4}$
38. $8^{.614}$
39. π^{π} (Use $\pi = 3.14$)
40. $\dfrac{(7.06)^3}{(31.7)(\sqrt{1.09})}$
41. $\dfrac{\log 21}{\log 3}$
42. $(21.9)^2 + (4.13)^3$

43. Suppose the pressure p (pounds per cubic foot) and volume v (cubic feet) of a certain gas are related by the formula

$$pv^{1.6} = 800.$$

Find p if $v = 7.6$ cubic feet.

44. The approximate period T (in seconds) of a simple pendulum of length L (in feet) is

$$T \approx 2\pi\sqrt{\frac{L}{32}}.$$

Find the period of a pendulum of length $L = 26$ feet.

45. The number of years, n, since two independently evolving languages split off from a common ancestral language is approximated by

$$n \approx \frac{1000 \log r}{2 \log .86},$$

where r is the proportion of words from the ancestral language common to both languages.

(a) Find n if $r = .9$.

(b) Find n if $r = .3$.

(c) How many years have elapsed since the split if half of the words of the ancestral language are common to both languages?

5.5 Exponential and Logarithmic Equations

To solve equations involving logarithms and exponents, it is often helpful to state the following property of logarithms, which follows from the definition of function.

THEOREM 5.5 If $x > 0$, $y > 0$, $b > 0$, $b \neq 1$, then $x = y$ if and only if $\log_b x = \log_b y$.

We can use this result to solve the equation

$$7^x = 12.$$

If we take base 10 logs of both sides, we have

$$\begin{aligned} \log 7^x &= \log 12 \\ x \log 7 &= \log 12 \\ x &= \frac{\log 12}{\log 7}. \end{aligned}$$

To get a decimal approximation for x, we can write

$$x = \frac{\log 12}{\log 7} = \frac{1.0792}{.8451} \approx 1.3.$$

Example 11 Solve $3^{2x-1} = 4^{x+2}$.

If we take logs of both sides, we have

$$\log 3^{2x-1} = \log 4^{x+2}$$

or

$$(2x - 1) \log 3 = (x + 2) \log 4,$$

which leads to

$$2x \log 3 - \log 3 = x \log 4 + 2 \log 4$$

or

$$2x \log 3 - x \log 4 = 2 \log 4 + \log 3.$$

Since the common factor of the left member is x, we have

$$x(2 \log 3 - \log 4) = 2 \log 4 + \log 3$$

or

$$x = \frac{2 \log 4 + \log 3}{2 \log 3 - \log 4}.$$

Using the properties of logarithms, this can be expressed as

$$x = \frac{\log 16 + \log 3}{\log 9 - \log 4}$$

or finally,

$$x = \frac{\log 48}{\log \frac{9}{4}}.$$

This quotient could be approximated by a decimal, if desired.

Example 12 Solve $\log (x + 4) - \log (x + 2) = \log x$.

Here we have

$$\log \frac{x + 4}{x + 2} = \log x$$

$$\frac{x + 4}{x + 2} = x$$

$$x + 4 = x(x + 2)$$

$$x + 4 = x^2 + 2x$$

$$x^2 + x - 4 = 0.$$

By the quadratic formula, we have

$$x = \frac{-1 \pm \sqrt{1 + 16}}{2}$$

$$x = \frac{-1 + \sqrt{17}}{2} \quad \text{or} \quad x = \frac{-1 - \sqrt{17}}{2}.$$

We cannot evaluate $\log x$ for

$$x = \frac{-1 - \sqrt{17}}{2},$$

since this number is negative and hence not in the domain of $\log x$. Thus the only valid solution is the positive number

$$x = \frac{-1 + \sqrt{17}}{2}.$$

Example 13 Solve $\log (3x + 2) + \log (x - 1) = 1$.

Since $1 = \log 10$, we have

$$\log (3x + 2)(x - 1) = \log 10$$

$$(3x + 2)(x - 1) = 10$$

$$3x^2 - x - 2 = 10$$

$$3x^2 - x - 12 = 0.$$

If we now use the quadratic formula, we have

$$x = \frac{1 \pm \sqrt{1 + 144}}{6}$$

$$x = \frac{1 + \sqrt{145}}{6} \quad \text{or} \quad x = \frac{1 - \sqrt{145}}{6}.$$

If $x = \dfrac{1 - \sqrt{145}}{6}$, then $x - 1 < 0$, and so $\log(x - 1)$ does not exist. Hence, this proposed solution must be discarded and the only solution is

$$x = \frac{1 + \sqrt{145}}{6}.$$

5.5 Exercises

Solve each of the following equations.

1. $3^x = 6$
2. $4^x = 12$
3. $6^x = 8$
4. $1^x = 3$
5. $2^x = -3$
6. $(\frac{1}{4})^x = 100$
7. $3^{x-1} = 4$
8. $5^{2-x} = 12$
9. $6^{2x-1} = 8$
10. $2^{x-7} = 5$
11. $1.8^{x+4} = 9.31$
12. $3^{2x-5} = 6$
13. $\log(x - 1) = 1$
14. $\log x^2 = 1$
15. $\log(x - 3) = 1 - \log x$
16. $\log(x - 6) = 2 - \log(x + 15)$
17. $\log(x + 2) = \log 6$
18. $\log x - \log(x + 1) = \log 5$
19. $\log_2 x = 3$
20. $\log_x 10 = 3$
21. $2 + \log x = 0$
22. $\log_3 x + \log_3 (2x + 5) = 1$
23. $\log(x - 3) = 1 + \log(x + 1)$
24. $\log x + \log(3x - 13) = 1$
25. $\log x = \log(x - 4)$
26. $\left(1 + \frac{r}{2}\right)^5 = 9$
27. $100(1 + .02)^{3+n} = 150$

5.6 Natural Logarithms

Natural logarithms, written $\ln x$, are logs to the base e. The number e is an irrational number, approximated by $e \approx 2.71828$, which occurs in many practical applications. Several examples of such applications are presented in this section. To convert from base 10 logs, with which we are familiar, to base e logs, we use the next theorem.

THEOREM 5.6 If x is any positive number, and if a and b are positive real numbers, $a \neq 1$, $b \neq 1$, then

$$\log_a x = \frac{\log_b x}{\log_b a}.$$

To prove this result, we can use the fact that for positive x and positive a, $a \neq 1$, we can write $y = \log_a x$, from which $x = a^y$, or $x = a^{\log_a x}$. Verify this statement from the definition of logarithm, or Theorem 5.1. If we now take base b logs of both sides of this last equation, we have

$$\log_b x = \log_b a^{\log_a x}$$

or

$$\log_b x = (\log_a x)(\log_b a),$$

from which we obtain

$$\log_a x = \frac{\log_b x}{\log_b a}.$$

We can use this result to find the natural logarithm of a number. For example, to find ln 47.2, we can let $x = 47.2$, $a = e$, and $b = 10$. Using these values, we have

$$\ln 47.2 = \log_e 47.2 = \frac{\log_{10} 47.2}{\log_{10} e}$$

$$\ln 47.2 = \frac{\log 47.2}{\log e}.$$

Since $e \approx 2.71828$, we can calculate $\log e$: $\log e = .4343$. Thus,

$$\ln 47.2 = \frac{\log 47.2}{.4343}.$$

Since $1/(.4343) \approx 2.3026$, this last statement becomes

$$\begin{aligned} \ln 47.2 &= (2.3026)(\log 47.2) \\ &= (2.3026)(1.6739) \\ &= 3.8543. \end{aligned}$$

Using this same principle, we can prove the next result.

THEOREM 5.7 For all positive x, $\ln x = (2.3026) \log x$.

Example 14

$$\begin{aligned} \ln .427 &= (2.3026)(\log .427) \\ &= (2.3026)(-1 + .6304) \\ &= (2.3026)(-.3696) \\ &= -.8510 \end{aligned}$$

Natural logarithms are often used in discussions of growth and decay, as illustrated by the next examples.

Example 15 Suppose the amount, y, of a certain radioactive substance present at a time t is given by

$$y = y_0 e^{-.1t},$$

where y_0 is the amount present initially and t is measured in days. Find the half-life of the substance. (The half-life of a radioactive substance is the time it takes for exactly half the sample to decay.)

We want to know the time t that must elapse for y to be reduced to a value equal to $\frac{1}{2}y_0$. That is, we want to solve the equation

$$\tfrac{1}{2}y_0 = y_0e^{-.1t}.$$

Here we have $\frac{1}{2} = e^{-.1t}$, and if we now take natural logs of both sides, we get

$$\ln \tfrac{1}{2} = \ln e^{-.1t}$$
$$\ln \tfrac{1}{2} = -.1t(\ln e).$$

Since $\ln e = 1$, we have

$$\ln \tfrac{1}{2} = -.1t$$
$$t = \frac{-\ln \frac{1}{2}}{.1}$$
$$t = -10 \ln \tfrac{1}{2}.$$

By the earlier results of this section, we have

$$\ln \tfrac{1}{2} = (2.3026) \log \tfrac{1}{2} = 2.3026(-.3010) = -.6931.$$

Hence,

$$t \approx (-10)(-.6932)$$
$$t \approx 6.9 \text{ days}.$$

Example 16 Psychologists tell us that under certain conditions the total number of a certain kind of fact remembered is approximated by

$$y = y_0\left(\frac{1+e}{1+e^{t+1}}\right),$$

where y is the number of facts remembered at a time t measured in days, and y_0 is the initial number remembered. Graph the function.

Here we can plot some points as an aid in graphing the function. If $t = 0$, we have

$$y = y_0\left(\frac{1+e}{1+e^{0+1}}\right) = y_0(1) = y_0.$$

If $t = 1$, we have

$$y = y_0\left(\frac{1+e}{1+e^2}\right) \approx y_0\left(\frac{3.718}{8.389}\right) \approx .44y_0.$$

By plotting several such points, we get the graph of Figure 5.10. This graph, typical of the remembering of certain things under certain conditions, is often called a **forgetting curve.**

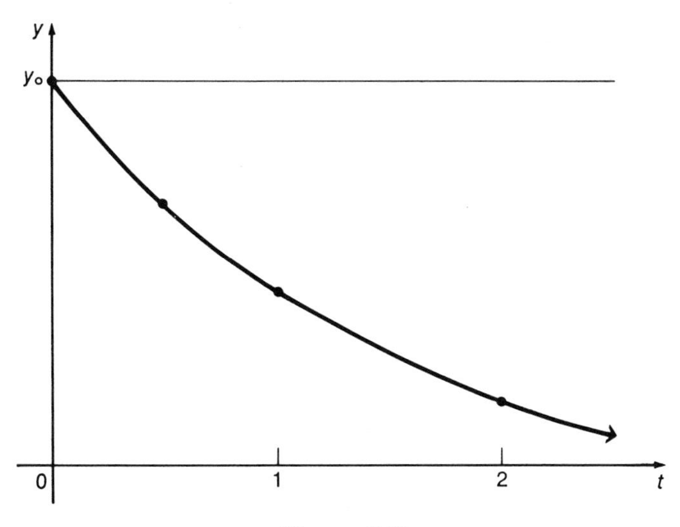

Figure 5.10

Example 17 How long will it take in Example 16 for exactly half the items to be forgotten?

We want to find the value of t that makes $y = \frac{1}{2}y_0$. That is, we want

$$\tfrac{1}{2}y_0 = y_0\left(\frac{1+e}{1+e^{t+1}}\right)$$

or

$$\tfrac{1}{2} = \frac{1+e}{1+e^{t+1}}.$$

This gives $1 + e^{t+1} = 2(1 + e) = 2 + 2e$

or

$$e^{t+1} = 1 + 2e.$$

Taking natural logs of both sides gives

$$\ln e^{t+1} = \ln(1 + 2e)$$

or

$$(t + 1) \ln e = \ln(1 + 2e).$$

Since $\ln e = 1$, we have

$$t = [\ln(1 + 2e)] - 1.$$

We know $e \approx 2.718$, and so

$$t \approx [\ln(1 + 5.436)] - 1$$
$$t \approx 1.86 - 1$$
$$t \approx .86.$$

5.6 Exercises

Find each of the following to the nearest hundredth.

1. $\ln 125$
2. $\ln 63.1$
3. $\ln .9$
4. $\ln 1.5$
5. $\ln 98.6$
6. $\ln 355$
7. $\log_5 10$
8. $\log_9 12$
9. $\log_{15} 5$
10. $\log_6 8$
11. $\log_6 5$
12. $\log_3 15$
13. $\log_{1/2} 3$
14. $\log_{12} 62$
15. $\log_{100} 83$
16. $\log_{200} 375$

Suppose the number of rabbits in a colony of rabbits increases according to the relationship

$$y = y_0 e^{.4t},$$

where t represents time in months and y_0 is the initial population of rabbits.

17. Find the number of rabbits present at time $t = 4$ if $y_0 = 100$.
18. How long will it take for the number of rabbits to triple?

A certain city finds that its population is declining according to the relationship

$$P = P_0 e^{-.04t},$$

where t is time measured in years and P_0 is the population at time $t = 0$.

19. If $P_0 = 1{,}000{,}000$, find the population at time $t = 1$.
20. If $P_0 = 1{,}000{,}000$, estimate the time it will take for the population to be reduced to 750,000.
21. How long will it take for the population to be cut in half?

Carbon 14 is a radioactive isotope of carbon which has a half-life of about 5600 years. (That is, in about 5600 years half of any given sample of carbon 14 will have decayed.) The atmosphere contains much carbon, mostly in the form of carbon dioxide, with small traces of carbon 14. Most of this atmospheric carbon is in the form of the nonradioactive isotope carbon 12. The ratio of carbon 14 to carbon 12 is virtually constant in the atmosphere. However, as a plant absorbs carbon dioxide from the air in the process of photosynthesis, the carbon 12 stays in the plant while the carbon 14 is converted to nitrogen. Thus, the ratio of carbon 14 to carbon 12 is smaller in the plant than it is in the atmosphere. Even when the plant is eaten by an animal, this ratio will continue to decrease. Based on these facts, a method of dating objects called *carbon 14 dating* has been developed. It is explained in the following problems.

22. Suppose an Egyptian mummy is discovered in which the ratio of carbon 14 to carbon 12 is only about half the ratio found in the atmosphere. About how long ago did the Egyptian die?
23. Let R be the (nearly constant) ratio of carbon 14 to carbon 12 found in the atmosphere, and let r be the ratio found in an observed specimen.

It can be shown that the relationship between R and r is given by

$$\frac{R}{r} = e^{(t \ln 2)/5600},$$

where t is the age of the specimen in years. Verify the formula for $t = 0$.

24. Verify the formula in Exercise 23 for $t = 5600$ and then for $t = 11{,}200$.
25. Solve the formula of Exercise 23 for t.
26. Suppose a specimen is found in which $r = \frac{2}{3}R$. Estimate the age of the specimen.
27. If an object is fired vertically upward and is subject only to the force of gravity, g, and to air resistance, then the maximum height, H, attained by the object is

$$H = \frac{1}{K}\left(V_0 - \frac{g}{K} \ln \frac{g + V_0 K}{g}\right),$$

where V_0 is the initial velocity of the object and K is a constant. Find H if $K = 2.5$, $V_0 = 1000$ feet per second, and $g = 32$ feet per second per second.

28. The following formula* can be used to estimate the population of the United States, where t is time in years measured from 1914 (times before 1914 are negative):

$$N = \frac{197{,}273{,}000}{1 + e^{-.03134t}}.$$

Complete the following chart.

Year	*Observed Population*	*Predicted Population*
1790	3,929,000	3,929,000
1810	7,240,000	
1860	31,443,000	30,412,000
1900	75,995,000	
1970	204,000,000	

29. Turnage† has shown that the pull, P, of a tracked vehicle on dry sand under certain conditions is approximated by

$$P = W\left[.2 + .16 \ln \frac{G(bl)^{3/2}}{W}\right]$$

where G is an index of sand strength, W is the load on the vehicle, b is the width of the track, and l is the length of the track. Find P if $W = 10$, $G = 5$, $b = 30.5$ cm, and $l = 61.0$ cm.

* The formula is given in Alfred J. Lotka, *Elements of Mathematical Biology*, reprinted by Dover Press, New York, 1957, page 67.
† Gerald W. Turnage, *Prediction of Track Pull Performance in a Desert Sand*, unpublished MS thesis, The Florida State University, 1971.

5.7 Geometric Progressions

Recall that a sequence is a function whose domain is a subset of the set of positive integers. In Chapter 3 of this text we discussed arithmetic sequences, which are sequences where each term is obtained by adding the same constant number to the preceeding term.

A **geometric sequence** or **geometric progression** is a sequence in which each term after the first is obtained by multiplying the preceeding term by a constant nonzero real number, called the **common ratio.** Hence,

$$2, 8, 32, 128, \ldots$$

is a geometric progression whose first term is 2 and whose common ratio is 4. If the common ratio of a geometric progression is r, then by the definition of geometric progression we have

$$a_{n+1} = r \cdot a_n,$$

for every positive integer n. (Recall: a_n represents the nth term of a sequence.)

Note that in the geometric progression

$$2, 8, 32, 128, \ldots$$

we have

$$\begin{aligned} 8 &= 2 \cdot 4 \\ 32 &= 8 \cdot 4 = (2 \cdot 4) \cdot 4 = 2 \cdot 4^2 \\ 128 &= 32 \cdot 4 = (2 \cdot 4^2) \cdot 4 = 2 \cdot 4^3. \end{aligned}$$

To generalize this, consider a geometric progression whose first term is a_1 and whose common ratio is r. The second term can be written as $a_2 = a_1 r$, the third as $a_3 = a_2 r = (a_1 r)r = a_1 r^2$, and so on. Thus, in general, the nth term of a geometric progression is given by

▶ $$\boldsymbol{a_n = a_1 r^{n-1}}.$$

For example, the sixth term of the progression whose first term is -2 and whose common ratio is $-\frac{1}{2}$ is given by

$$\begin{aligned} a_6 &= -2(-\tfrac{1}{2})^{6-1} \\ a_6 &= -2(-\tfrac{1}{2})^5 \\ a_6 &= \tfrac{1}{16}. \end{aligned}$$

Example 18 Suppose the third term of a geometric progression is 20, while the sixth term of the same progression is 160. Find the first term a_1 and the common ratio, r.

We can use the formula for the nth term of a geometric progression. Here we have

$$\begin{aligned} a_3 &= a_1 r^2 = 20 \\ a_6 &= a_1 r^5 = 160. \end{aligned}$$

Since $a_1 r^2 = 20$, then $a_1 = \frac{20}{r^2}$. Substituting this in the second equation we have

$$\begin{aligned} a_1 r^5 &= 160 \\ \left(\frac{20}{r^2}\right) r^5 &= 160 \\ 20r^3 &= 160 \\ r^3 &= 8 \\ r &= 2. \end{aligned}$$

Since $a_1 r^2 = 20$, and $r = 2$, we have $a_1 = 5$.

Applications of geometric progressions frequently require the sum of a certain number of terms of the progression. To find a formula for the sum, S_n, of the first n terms of a geometric progression, we can first write

$$S_n = a_1 + a_2 + a_3 + \ldots + a_n.$$

This can also be written as

$$S_n = a_1 + a_1 r + a_1 r^2 + \ldots + a_1 r^{n-1}. \tag{2}$$

If $r = 1$, we have $S_n = na_1$, which is the correct result for this case. If $r \neq 1$, we can multiply both sides of (2) by r, obtaining

$$rS_n = a_1 r + a_1 r^2 + a_1 r^3 + \ldots + a_1 r^n. \tag{3}$$

If we subtract (3) from (2), we have

$$S_n - rS_n = a_1 - a_1 r^n$$

or

$$S_n(1 - r) = a_1(1 - r^n)$$

which yields

$$S_n = \frac{a_1(1 - r^n)}{1 - r}.$$

This proves the following result.

THEOREM 5.8 The sum of the first n terms of a geometric progression whose first term is a_1 and whose common ratio is r ($r \neq 1$) is

$$S_n = \sum_{i=1}^{n} a_i = \frac{a_1(1 - r^n)}{1 - r}.$$

Example 19 Find the sum of the first four terms of the sequence

$$10, 2, 2/5, \ldots.$$

Here $a_1 = 10$, $r = 1/5$, and $n = 4$. Using the formula of Theorem 5.8, we have

$$S_4 = \frac{10\left[1 - \left(\frac{1}{5}\right)^4\right]}{1 - \frac{1}{5}}$$

$$S_4 = \frac{10\left[1 - \frac{1}{625}\right]}{\frac{4}{5}}$$

$$S_4 = 10\left(\frac{624}{625}\right) \cdot \frac{5}{4}$$

$$S_4 = \frac{312}{25}.$$

Example 20 Evaluate $\sum_{i=1}^{6} 4 \cdot 3^{i-1}$.

We can write the sum of this example as

$$\begin{aligned}\sum_{i=1}^{6} 4 \cdot 3^{i-1} &= 4 \cdot 3^{1-1} + 4 \cdot 3^{2-1} + 4 \cdot 3^{3-1} + \cdots + 4 \cdot 3^{6-1} \\ &= 4 \cdot 3^0 + 4 \cdot 3^1 + 4 \cdot 3^2 + \cdots + 4 \cdot 3^5.\end{aligned}$$

These numbers represent the terms of a geometric progression with $a_1 = 4 \cdot 3^0 = 4$, $r = 3$, and $n = 6$. Using the formula from Theorem 5.8, we have

$$\begin{aligned}\sum_{i=1}^{6} 4 \cdot 3^{i-1} &= \frac{4(1 - 3^6)}{1 - 3} \\ &= \frac{4(1 - 729)}{-2} \\ &= 1456.\end{aligned}$$

5.7 Exercises

Find the fifth term and the sum of the first five terms for each of the following geometric progressions.

1. 3, 6, 12, 24, ...
2. 5, 20, 80, 320, ...
3. 12, -6, 3, $-\frac{3}{2}$, ...
4. 18, -3, $\frac{1}{2}$, $-\frac{1}{12}$, ...
5. 9, -6, 4, ...
6. 50, -10, 2, ...
7. $a_1 = 4$, $r = 2$
8. $a_1 = 3$, $r = 3$
9. $a_2 = \frac{1}{3}$, $r = 3$
10. $a_2 = -1$, $r = 2$
11. The second term of a geometric progression is 3 and the fifth term is $\frac{8}{9}$. Find the eighth term.
12. Find r and a_6 in a geometric progression where $a_2 = 12$ and $a_5 = \frac{3}{2}$.

The k numbers $b_1, b_2, \ldots, b_k$ are called ***k*th geometric means** of the numbers a and c if

$$a, b_1, b_2, \ldots b_k, c$$

form a geometric progression.

13. Find a geometric mean between 4 and 16.
14. Find a geometric mean between -2 and -8.
15. Insert three geometric means between 2 and 32.
16. Insert four geometric means between 8 and $-\frac{1}{4}$.

Evaluate each of the following.

17. $\sum_{i=1}^{4} 2^i$
18. $\sum_{i=1}^{6} (-3)^i$
19. $\sum_{i=1}^{8} (\frac{1}{2})^i$
20. $\sum_{i=1}^{4} (-\frac{3}{4})^i$
21. $\sum_{i=3}^{7} 3 \cdot 2^{i-1}$
22. $\sum_{i=3}^{8} 2 \cdot (\frac{3}{2})^{i-2}$
23. $\sum_{i=4}^{7} 5 \cdot (\frac{2}{3})^{i-3}$
24. $\sum_{i=3}^{8} \frac{2}{3}(\frac{3}{4})^{i-3}$

5.8 Sums of Infinite Geometric Sequences

In the previous section we found that the sum of the first n terms of a geometric progression is given by

$$S_n = \sum_{i=1}^{n} a_i = \frac{a_1(1 - r^n)}{1 - r}$$

where a_1 is the first term and r $(r \neq 1)$ is the common ratio. Consider now the infinite sequence

$$2, 1, \frac{1}{2}, \frac{1}{4}, \frac{1}{8}, \frac{1}{16}, \ldots$$

whose first term is 2 and whose common ratio is $\frac{1}{2}$. Using the formula above, we can show

$$S_1 = 2$$
$$S_2 = 3$$
$$S_3 = \frac{7}{2}$$
$$S_4 = \frac{15}{4}$$
$$S_5 = \frac{31}{8}$$
$$S_6 = \frac{63}{16},$$

and so on. Note that this sequence of sums,

$$2,\ 3,\ \frac{7}{2},\ \frac{15}{4},\ \frac{31}{8},\ \frac{63}{16},\ \cdots$$

gets closer and closer to the number 4. In fact, by selecting a value of n large enough, we can make S_n as close as we wish to 4. We express this by saying

$$\lim_{n\to\infty} S_n = 4.$$

(Read: "the limit of S_n as n increases without bound is 4.") For no value of n is $S_n = 4$. However, if n is large enough, then S_n is as close to 4 as we might wish.*

Since $$\lim_{n\to\infty} S_n = 4,$$

we say that 4 is the sum of the infinite geometric progression

$$2,\ 1,\ \frac{1}{2},\ \frac{1}{4},\ \cdots$$

and we write $$2 + 1 + \frac{1}{2} + \frac{1}{4} + \frac{1}{8} + \cdots = 4.$$

Example 21 Find $1 + \frac{1}{3} + \frac{1}{9} + \frac{1}{27} + \cdots.$

Here we can use the formula for the first n terms of a geometric progression to write

$$S_1 = 1$$

$$S_2 = \frac{4}{3}$$

$$S_3 = \frac{13}{9}$$

$$S_4 = \frac{40}{27},$$

and in general $$S_n = \frac{1\left[1 - \left(\frac{1}{3}\right)^n\right]}{1 - \frac{1}{3}}.$$

Note that as n gets larger and larger, $\left(\frac{1}{3}\right)^n$ gets closer and closer to 0. That is,

$$\lim_{n\to\infty} \left(\frac{1}{3}\right)^n = 0.$$

* These phrases "large enough" and "as close as we wish" are not nearly precise enough for mathematicians; much of a standard calculus course is devoted to making them more precise.

Thus, it seems reasonable to write

$$\lim_{n\to\infty} S_n = \frac{1(1-0)}{1-\frac{1}{3}} = \frac{3}{2}.$$

Hence, $$1 + \frac{1}{3} + \frac{1}{9} + \frac{1}{27} + \cdots = \frac{3}{2}.$$

In general, if a geometric progression has a first term a_1 and a common ratio r, we have

$$S_n = \frac{a_1(1-r^n)}{1-r}$$

for every positive integer n. If $|r| < 1$, we have

$$\lim_{n\to\infty} r^n = 0.$$

In this case, we can write

$$\lim_{n\to\infty} S_n = \frac{a_1(1-0)}{1-r} \qquad \text{for } |r| < 1,$$

▶ $$\boldsymbol{\lim_{n\to\infty} S_n = \frac{a_1}{1-r} \qquad \text{for } |r| < 1.}$$

This quantity we define as the **sum of the infinite geometric progression** with first term a_1 and common ratio r, $|r| < 1$. We often express

$$\lim_{n\to\infty} S_n$$

by writing S_∞, or $\sum_{i=1}^{\infty} a_i$, so that

$$S_\infty = \sum_{i=1}^{\infty} a_i = \lim_{n\to\infty} S_n = \frac{a_1}{1-r}, \qquad |r| < 1.$$

For example, the sum of the geometric progression with first term $a_1 = -\frac{3}{4}$ and common ratio $r = -\frac{1}{2}$,

$$-\frac{3}{4} + \frac{3}{8} - \frac{3}{16} + \frac{3}{32} - \frac{3}{64} + \cdots,$$

is given by

$$S_\infty = \frac{-\frac{3}{4}}{1-\left(-\frac{1}{2}\right)}$$

$$S_\infty = -\frac{1}{2}.$$

We can use the formula of this section to convert repeating decimals (which represent rational numbers) to the form $\frac{p}{q}$, p and q integers, as shown in the following examples.

Example 22 Write .090909... in the form $\frac{p}{q}$, where p and q are integers.

We can write this decimal as

$$.09 + .0009 + .000009 + \ldots,$$

which is a geometric progression with $a_1 = .09$ and $r = .01$. The sum of this progression is given by

$$\begin{aligned} S_\infty &= \frac{a_1}{1 - r} \\ &= \frac{.09}{1 - .01} \\ &= \frac{.09}{.99} \\ S_\infty &= \frac{1}{11}. \end{aligned}$$

Example 23 Write 2.5121212... in the form $\frac{p}{q}$.

We have

$$2.5121212\ldots = 2.5 + .012 + .00012 + .0000012 + \ldots.$$

Beginning with the second term, we have a geometric progression with $a_1 = .012$ and $r = .01$. Thus, the decimal can be written as

$$2.5 + \frac{.012}{1 - .01} = 2.5 + \frac{.012}{.990} = 2.5 + \frac{2}{165} = \frac{829}{330}.$$

Example 24 Write the repeating decimal 0.99999... in the form $\frac{p}{q}$.

We can write the decimal as an infinite geometric progression,

$$0.99999\ldots = .9 + .09 + .0009 + .00009 + \ldots,$$

where $a_1 = \frac{9}{10}$ and $r = \frac{1}{10}$. This gives

$$S_\infty = \frac{\frac{9}{10}}{1-\frac{1}{10}}$$

$$= \frac{\frac{9}{10}}{\frac{9}{10}}$$

$$S_\infty = 1.$$

Hence, $0.99999\ldots = 1$.

5.8 Exercises

Find each of the following sums which exist by using the formula of this section, where applicable.

1. $\frac{3}{4} + \frac{3}{8} + \frac{3}{16} + \cdots$
2. $\frac{4}{5} + \frac{2}{5} + \frac{1}{5} + \cdots$
3. $3 - \frac{3}{2} + \frac{3}{4} - \cdots$
4. $9 - 3 + 1 - \cdots$
5. $\frac{1}{3} - \frac{2}{9} + \frac{4}{27} - \frac{8}{81} + \cdots$
6. $1 + \frac{1}{1.01} + \frac{1}{(1.01)^2} + \cdots$
7. $\frac{1}{36} + \frac{1}{30} + \frac{1}{25} + \cdots$
8. $1 + \frac{1}{2^2} + \frac{1}{2^4} + \cdots$
9. $\sum_{i=1}^{\infty} \left(\frac{1}{4}\right)^i$
10. $\sum_{i=1}^{\infty} \left(\frac{9}{10}\right)^i$
11. $\sum_{i=1}^{\infty} (1.2)^i$
12. $\sum_{i=1}^{\infty} (1.001)^{i-1}$
13. $\sum_{i=2}^{\infty} (-\frac{1}{4})^i$
14. $\sum_{i=4}^{\infty} (.3)^i$
15. $\sum_{i=1}^{\infty} \frac{1}{5^i}$
16. $\sum_{i=1}^{\infty} \frac{-1}{(-2)^{i+1}}$

Express each of the following in the form $\frac{p}{q}$, with p and q integers.

17. $0.5555\ldots$
18. $0.3333\ldots$
19. $0.313131\ldots$
20. $0.909090\ldots$
21. $0.345555\ldots$
22. $0.568888\ldots$
23. Joann drops a ball from a height of 10 feet and notices that on each bounce the ball returns to about $\frac{3}{4}$ of its previous height. About how far will the ball travel before it comes to rest? (Hint: consider the sum of two progressions.)

MATRIX THEORY

A **matrix** (plural: matrices) is a rectangular ordered array of numbers which are called the **elements** of the matrix. The study of matrices as a mathematical system has been of interest to mathematicians for some time. Recently however, the use of matrices, along with other mathematical methods, has assumed greater importance in the fields of business administration and the social sciences. This, together with the advent of the computer, has led to a new emphasis on the use of matrices as a mathematical tool, particularly for working with systems of equations.

6.1 Basic Properties of Matrices

We shall write a matrix by enclosing the array of numbers in parentheses. Matrices are classified by their **dimension** or **order,** that is, the number of rows and columns they contain. Thus, an $m \times n$ (read "m by n") matrix has m rows and n columns, and is said to be of order $m \times n$. For example,

$$\begin{pmatrix} 2 & 3 & 5 \\ 7 & 1 & 2 \end{pmatrix}, \quad \begin{pmatrix} -1 \\ 1 \\ 2 \\ 0 \end{pmatrix}, \quad \text{and} \quad \begin{pmatrix} 8 & -1 & 0 \\ 2 & 1 & 6 \\ 0 & 5 & -3 \end{pmatrix}$$

are, respectively, a 2×3 matrix, a 4×1 matrix, and a 3×3 matrix. (Note that the number of rows is named first, and the number of columns second.) A **square matrix** has the same number of rows as columns.

A matrix with only one row or column, such as

$$\begin{pmatrix} 2 \\ -1 \\ 4 \end{pmatrix}, \qquad (4 \quad 5), \qquad \text{or} \qquad (2 \quad 8 \quad -7 \quad 6 \quad -3)$$

is called a **row** or **column matrix.**

It is customary to use capital letters to denote matrices and subscript notation to represent the elements of a matrix as follows.

$$A = \begin{pmatrix} a_{11} & a_{12} & a_{13} & \cdots & a_{1n} \\ a_{21} & a_{22} & a_{23} & \cdots & a_{2n} \\ a_{31} & a_{32} & a_{33} & \cdots & a_{3n} \\ \cdot & \cdot & \cdot & & \cdot \\ \cdot & \cdot & \cdot & & \cdot \\ \cdot & \cdot & \cdot & & \cdot \\ a_{m1} & a_{m2} & a_{m3} & & a_{mn} \end{pmatrix}$$

Using this notation, the first row, first column element is denoted a_{11}, the second row, third column element is denoted a_{23}, and in general, the ith row, jth column element is denoted a_{ij}.

Two matrices are **equal** if they are of the same order and if their corresponding elements are equal. By this definition, the matrices

$$\begin{pmatrix} 2 & 1 \\ 3 & -5 \end{pmatrix} \quad \text{and} \quad \begin{pmatrix} 1 & 2 \\ -5 & 3 \end{pmatrix}$$

are not equal (even though they contain the same elements and are of the same order), since the corresponding elements differ. From the definition of equality, we see that if

$$\begin{pmatrix} 2 & 1 \\ m & n \end{pmatrix} = \begin{pmatrix} x & y \\ -1 & 0 \end{pmatrix},$$

then $x = 2$, $y = 1$, $m = -1$, and $n = 0$.

To add two matrices, we use the following definition. The **sum** of two $m \times n$ matrices X and Y is the $m \times n$ matrix $X + Y$, in which each element of $X + Y$ is the sum of the corresponding elements of X and Y. For example, if

$$A = \begin{pmatrix} 1 & 2 & 3 \\ 0 & -1 & 5 \end{pmatrix} \quad \text{and} \quad B = \begin{pmatrix} -2 & 3 & 0 \\ 1 & -7 & 2 \end{pmatrix},$$

then

$$A + B = \begin{pmatrix} -1 & 5 & 3 \\ 1 & -8 & 7 \end{pmatrix}.$$

Note that by this definition only matrices of the same order can be added.

Each real number has an additive inverse. The additive inverse of a matrix can be defined using the definition of the additive inverse of a real number. The **additive inverse,** or **negative,** of a matrix X is the matrix $-X$, in which each element of $-X$ is the additive inverse of the corresponding element of X. By this definition, the additive inverses of matrices A and B above are

$$-A = \begin{pmatrix} -1 & -2 & -3 \\ 0 & 1 & -5 \end{pmatrix} \quad \text{and} \quad -B = \begin{pmatrix} 2 & -3 & 0 \\ -1 & 7 & -2 \end{pmatrix}.$$

Note that

$$A + (-A) = \begin{pmatrix} 0 & 0 & 0 \\ 0 & 0 & 0 \end{pmatrix}.$$

We can now define subtraction for matrices in a manner analogous to the way we defined subtraction for real numbers. That is, for matrices X and Y, we shall define

▶ $$X - Y = X + (-Y).$$

Using A and B as defined above we have

$$A - B = A + (-B) = \begin{pmatrix} 1 & 2 & 3 \\ 0 & -1 & 5 \end{pmatrix} + \begin{pmatrix} 2 & -3 & 0 \\ -1 & 7 & -2 \end{pmatrix}$$

$$= \begin{pmatrix} 3 & -1 & 3 \\ -1 & 6 & 3 \end{pmatrix}.$$

6.1 Exercises

Mark each of the following true or false. If false, tell why.

1. $\begin{pmatrix} 1 & 3 \\ 5 & 7 \end{pmatrix} = \begin{pmatrix} 1 & 5 \\ 3 & 7 \end{pmatrix}$

2. $\begin{pmatrix} 1 \\ 2 \\ 3 \end{pmatrix} = (1 \quad 2 \quad 3)$

3. $\begin{pmatrix} x \\ y \end{pmatrix} = \begin{pmatrix} 3 \\ 5 \end{pmatrix}$ if $x = 3$ and $y = 5$.

4. $\begin{pmatrix} 3 & 5 & 2 & 8 \\ 1 & -1 & 4 & 0 \end{pmatrix}$ is a 4×2 matrix.

5. $\begin{pmatrix} 1 & 9 & -4 \\ 3 & 7 & 2 \\ -1 & 1 & 0 \end{pmatrix}$ is a square matrix.

6. $\begin{pmatrix} 2 & 4 & -1 \\ 3 & 7 & 2 \\ 0 & 0 & 0 \end{pmatrix} = \begin{pmatrix} 2 & 4 & -1 \\ 3 & 7 & 2 \end{pmatrix}$

Perform the indicated operations.

7. $\begin{pmatrix} 1 & 2 & 5 & -1 \\ 3 & 0 & 2 & -4 \end{pmatrix} + \begin{pmatrix} 8 & 10 & -5 & 3 \\ -2 & -1 & 0 & 0 \end{pmatrix}$

8. $\begin{pmatrix} 1 & 5 \\ 2 & -3 \\ 3 & 7 \end{pmatrix} + \begin{pmatrix} 2 & 3 \\ 8 & 5 \\ -1 & 9 \end{pmatrix}$

9. $\begin{pmatrix} 1 & 5 & 7 \\ 2 & 2 & 3 \end{pmatrix} + \begin{pmatrix} 4 & 8 \\ 1 & -1 \end{pmatrix}$

10. $\begin{pmatrix} 2 & 4 \\ -8 & 1 \end{pmatrix} + \begin{pmatrix} 9 & -3 \\ 8 & 5 \end{pmatrix}$

11. $\begin{pmatrix} 1 & 3 & -2 \\ 4 & 7 & 1 \end{pmatrix} - \begin{pmatrix} 3 & 6 & -5 \\ 0 & 4 & 2 \end{pmatrix}$

12. $\begin{pmatrix} 2 & 8 & 12 & 0 \\ 7 & 4 & -1 & 5 \\ 1 & 2 & 0 & 10 \end{pmatrix} - \begin{pmatrix} 1 & 3 & 6 & 9 \\ 2 & -3 & -3 & 4 \\ 8 & 0 & -2 & 17 \end{pmatrix}$

Show that each of the following statements is true, where

$$P = \begin{pmatrix} m & n \\ p & q \end{pmatrix},\ T = \begin{pmatrix} r & s \\ t & u \end{pmatrix},\ X = \begin{pmatrix} x & y \\ z & w \end{pmatrix},\ \textit{and}\ 0 = \begin{pmatrix} 0 & 0 \\ 0 & 0 \end{pmatrix}.$$

13. $X + T = T + X$
14. $X + (T + P) = (X + T) + P$
15. $X + (-X) = 0$
16. $P + 0 = P$
17. Would the statements above still be true if P, T, X and 0 were not square matrices? Why?

6.2 Multiplication of Matrices

Finding the product of two matrices is more complicated than finding the sum. However, this complicated kind of multiplication is justified by applications which are helpful in solving practical problems. The **product** XY of an $m \times n$ matrix X and an $n \times k$ matrix Y is defined as follows. Multiply each element of the first row of X by the corresponding element in the first column of Y. The sum of these n products is the first row, first column element of XY. Similarly, the sum of the products found by multiplying the elements of the first row of X times the corresponding elements of the second column of Y

gives the first row, second column element of XY, and so on. In general, to find the ith row, jth column element of XY, multiply each element in the ith row of X by the corresponding element in the jth column of Y. The sum of these products will give the desired ijth element of XY.

To illustrate this definition, let us find the product AB of the matrices

$$A = \begin{pmatrix} 1 & -3 \\ 7 & 2 \end{pmatrix} \quad \text{and} \quad B = \begin{pmatrix} 1 & 0 & -1 & 2 \\ 3 & 1 & 4 & -1 \end{pmatrix}.$$

We first note that A is a 2×2 matrix and B is a 2×4 matrix so that the number of elements in each row of A equals the number of elements in each column of B as required. Also note that the product matrix AB will be a 2×4 matrix, as illustrated in Figure 6.1.

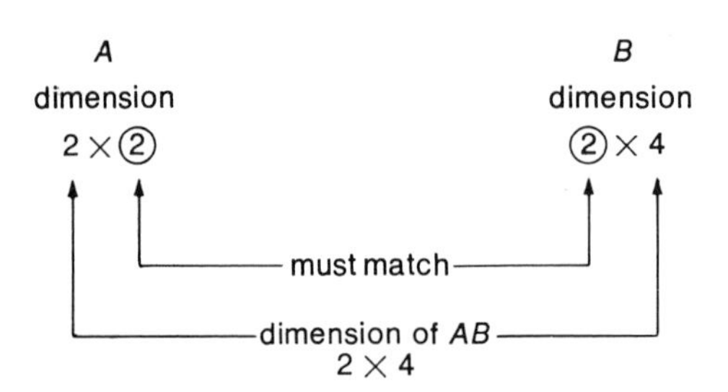

Figure 6.1

Multiplying the first row elements of A times the corresponding first column elements of B and summing up the products, we get the first row, first column element of AB.

$$\begin{pmatrix} [1(1) + (-3)3] & & & \\ & & & \end{pmatrix}$$

Now we multiply the first row elements of A times the corresponding second column elements of B, again summing up the products, to get the first row, second column element of AB.

$$\begin{pmatrix} [1(1) + (-3)3] & [1(0) + (-3)1] & & \\ & & & \end{pmatrix}$$

Continuing in this manner, we complete the product matrix as shown below.

$$\begin{pmatrix} [1(1) + (-3)3] & [1(0) + (-3)1] & [1(-1) + (-3)4] & [1(2) + (-3)(-1)] \\ [7(1) + 2(3)] & [7(0) + 2(1)] & [7(-1) + 2(4)] & [7(2) + 2(-1)] \end{pmatrix}$$

$$= \begin{pmatrix} -8 & -3 & -13 & 5 \\ 13 & 2 & 1 & 12 \end{pmatrix}$$

Can we find the product BA? Since B is 2×4 while A is 2×2, the answer is no.

Example 1 If

$$M = \begin{pmatrix} 1 & 2 & 3 \\ -1 & 0 & -2 \end{pmatrix} \quad \text{and} \quad N = \begin{pmatrix} -3 & -1 \\ 2 & 1 \\ 1 & 2 \end{pmatrix},$$

find MN.

Using the definition from above, we have

$$MN = \begin{pmatrix} [1(-3) + 2(2) + 3(1)] & [1(-1) + 2(1) + 3(2)] \\ [(-1)(-3) + 0(2) + (-2)1] & [(-1)(-1) + 0(1) + (-2)2] \end{pmatrix}$$

$$= \begin{pmatrix} 4 & 7 \\ 1 & -3 \end{pmatrix}.$$

Example 2 A contractor builds three kinds of houses, models A, B, and C, with a choice of two exteriors, colonial or contemporary. Matrix P below shows the number of each kind of house the contractor is planning to build for a new 100 home subdivision. The amounts for each of the main materials he uses depend primarily on the exterior of the house. These amounts are shown in matrix Q below, while matrix R gives the cost for each kind of material. Concrete is measured here in cubic yards, lumber in 1000 board feet, brick in 1000's, and shingles in 100 square feet.

	colonial	contemporary
model A	0	30
model B	10	20
model C	20	20

$$\begin{pmatrix} 0 & 30 \\ 10 & 20 \\ 20 & 20 \end{pmatrix} = P$$

	concrete	lumber	brick	shingles
colonial	10	2	0	2
contemporary	50	1	20	2

$$\begin{pmatrix} 10 & 2 & 0 & 2 \\ 50 & 1 & 20 & 2 \end{pmatrix} = Q$$

	cost per unit
concrete	20
lumber	180
brick	60
shingles	25

$$\begin{pmatrix} 20 \\ 180 \\ 60 \\ 25 \end{pmatrix} = R$$

(a) What is the total cost for each model house?
(b) How much of each of the four kinds of material must be ordered?
(c) What is the total cost of the material?

(a) To find the cost for each model, we must first find matrix PQ, which will show the total amount of each material needed for each model house.

$$PQ = \begin{pmatrix} 0 & 30 \\ 10 & 20 \\ 20 & 20 \end{pmatrix} \begin{pmatrix} 10 & 2 & 0 & 2 \\ 50 & 1 & 20 & 2 \end{pmatrix} = \begin{matrix} & \text{concrete} & \text{lumber} & \text{brick} & \text{shingles} & \\ & \begin{pmatrix} 1500 & 30 & 600 & 60 \\ 1100 & 40 & 400 & 60 \\ 1200 & 60 & 400 & 80 \end{pmatrix} & & & & \begin{matrix} \text{model } A \\ \text{model } B \\ \text{model } C \end{matrix} \end{matrix}$$

If we now multiply PQ times the cost matrix, R, we will get the total cost for each model.

$$\begin{pmatrix} 1500 & 30 & 600 & 60 \\ 1100 & 40 & 400 & 60 \\ 1200 & 60 & 400 & 80 \end{pmatrix} \begin{pmatrix} 20 \\ 180 \\ 60 \\ 25 \end{pmatrix} = \overset{\text{cost}}{\begin{pmatrix} 72{,}900 \\ 54{,}700 \\ 60{,}800 \end{pmatrix}} \begin{matrix} \text{model } A \\ \text{model } B \\ \text{model } C \end{matrix}$$

(b) The totals of the columns of matrix PQ will give a matrix whose elements represent the total amounts of each material needed for the subdivision. Let us call this matrix T and write it as a row matrix.

$$T = (3800 \quad 130 \quad 1400 \quad 200).$$

(c) To find the total cost of all the materials, we need the product of the cost matrix, R, and matrix T, the total amounts matrix. To multiply these and get a 1×1 matrix, representing the total cost, we must multiply a 1×4 matrix times a 4×1 matrix. This is why in (b) we chose to write a row matrix. Since the total materials cost is given by TR, we can write

$$TR = (3800 \quad 130 \quad 1400 \quad 200) \begin{pmatrix} 20 \\ 180 \\ 60 \\ 25 \end{pmatrix} = (188{,}400).$$

In work with matrices, a real number is called a **scalar** to distinguish it from a matrix. The product of a scalar k and a matrix X is the matrix kX each of whose elements is k times the corresponding element of X. Thus,

$$-3\begin{pmatrix} 2 & -5 \\ 1 & 7 \end{pmatrix} = \begin{pmatrix} -6 & 15 \\ -3 & -21 \end{pmatrix}.$$

6.2 Exercises

Answer true or false for each of the following. If false, tell why.

1. The product of a 1×2 matrix and a 2×5 matrix is a 1×5 matrix.
2. The product of a 3×2 matrix and a 4×3 matrix is a 2×4 matrix.
3. If both AB and BA exist, then A and B must both be square matrices.
4. For any matrix A with ith row all zeros and any matrix B for which AB exists, the ith row of AB will contain all zeros.

5. If A and B are $n \times n$ matrices and 0 is the $n \times n$ matrix whose entries are all zeros, and if $AB = 0$, then $A = 0$ or $B = 0$.

Compute the following products.

6. $\begin{pmatrix} 1 & 2 \\ 3 & 4 \end{pmatrix}\begin{pmatrix} -1 & 5 \\ 7 & 0 \end{pmatrix}$

7. $\begin{pmatrix} -1 & 5 \\ 7 & 0 \end{pmatrix}\begin{pmatrix} 1 & 2 \\ 3 & 4 \end{pmatrix}$

8. $\begin{pmatrix} -2 & -3 & 7 \\ 1 & 5 & 6 \end{pmatrix}\begin{pmatrix} 1 \\ 2 \\ 3 \end{pmatrix}$

9. $\begin{pmatrix} 6 \\ 5 \\ 4 \end{pmatrix}\begin{pmatrix} -1 & 1 & 1 \end{pmatrix}$

10. $\begin{pmatrix} 4 & 3 \\ 1 & 2 \\ 0 & -5 \end{pmatrix}\begin{pmatrix} 2 & -2 \\ 1 & -1 \end{pmatrix}$

11. $\begin{pmatrix} -1 & 1 & 1 \end{pmatrix}\begin{pmatrix} 6 \\ 5 \\ 4 \end{pmatrix}$

12. $\begin{pmatrix} 2 & -2 \\ 1 & -1 \end{pmatrix}\begin{pmatrix} 4 & 3 \\ 1 & 2 \\ 0 & -5 \end{pmatrix}$

13. $\begin{pmatrix} 1 & 0 & 1 & 5 & 2 \\ 0 & 1 & -3 & 4 & -1 \end{pmatrix}\begin{pmatrix} 1 & 0 & 0 \\ 0 & 1 & 0 \\ 0 & 0 & 1 \end{pmatrix}$

Using matrices $P = \begin{pmatrix} m & n \\ p & q \end{pmatrix}$, $X = \begin{pmatrix} x & y \\ z & w \end{pmatrix}$, *and* $T = \begin{pmatrix} r & s \\ t & u \end{pmatrix}$, *show that the following statements are true.*

14. $(PX)T = P(XT)$
15. $P(X + T) = PX + PT$
16. $P(X - T) = PX - PT$
17. $k(X + T) = kX + kT$ for any real number k
18. $(k + h)P = kP + hP$ for any real numbers k and h

19. The Bread Box, a small neighborhood bakery, sells four main items: sweet rolls, bread, cakes, and pies. The amount of each ingredient required for these items is given by matrix A.

$$A = \begin{array}{l} \text{rolls (doz)} \\ \text{bread (loaves)} \\ \text{cakes} \\ \text{pies (crust)} \end{array} \overset{\begin{array}{ccccc} \text{eggs} & \text{flour*} & \text{sugar*} & \text{shortening*} & \text{milk*} \end{array}}{\begin{pmatrix} 1 & 4 & \frac{1}{4} & \frac{1}{4} & 1 \\ 0 & 3 & 0 & \frac{1}{4} & 0 \\ 4 & 3 & 2 & 1 & 1 \\ 0 & 1 & 0 & \frac{1}{3} & 0 \end{pmatrix}}$$

The cost (in cents) for each ingredient when purchased in large lots or small lots is given by matrix B.

$$\begin{array}{l} \text{eggs} \\ \text{flour} \\ \text{sugar} \\ \text{shortening} \\ \text{milk} \end{array} \overset{\begin{array}{cc} & \text{cost} \\ \text{large lot} & \text{small lot} \end{array}}{\begin{pmatrix} 1 & 1 \\ 2 & 3 \\ 3 & 4 \\ 12 & 15 \\ 5 & 6 \end{pmatrix}} = B$$

(a) Use matrix multiplication to find a matrix representing the comparative cost per item for the two purchase options.

Suppose a day's orders consist of 20 dozen sweet rolls, 200 loaves of bread, 50 cakes and 60 pies.
(b) Writing the orders as a 1×4 matrix and using matrix multiplication write as a matrix the amount of each ingredient required to fill the day's orders.
(c) Use matrix multiplication to find a matrix representing the costs under the two purchase options to fill the day's orders.

6.3 The Algebra of Matrices

Now that we have defined matrix addition and multiplication, let us see whether or not a mathematical system whose elements are matrices of the same order satisfies the definition of a field, as given in Section 1.4. To begin, let us consider the operation of addition of matrices. The sum of any two matrices of the same order is a matrix of that order by the definition of addition, so that the closure property for addition is satisfied. It is easy to confirm that the commutative and associative properties for addition are also satisfied, since addition of matrices depends only on addition of real numbers (see Exercises 13 and 14 of Section 6.1). We have defined an additive inverse, $-A$, for

* in cups

each matrix A, and to complete the field axioms for addition, we need only to define an identity matrix for addition. For any matrix A, the sum $A + (-A)$ gives a matrix whose elements are all zeros. We shall define this matrix to be a **zero matrix** and denote it 0. Note that there is a zero matrix for each order matrix and that it serves as the identity element for that order, since if 0 and A are of the same order, $0 + A = A + 0 = A$ for any matrix A.

Now let us consider the operation of multiplication for matrices. From the definition of multiplication, it follows that only square matrices can be considered, since the closure axiom does not hold for other kinds of matrices for which multiplication is defined. (Recall that an $m \times n$ matrix must multiply an $n \times k$ matrix with the resulting product an $m \times k$ matrix.) For any square matrices of the same order, the closure axiom will hold, since an $n \times n$ matrix multiplied by an $n \times n$ matrix results in an $n \times n$ matrix.

Does multiplication of matrices satisfy the commutative axiom? Let us consider an example. If

$$A = \begin{pmatrix} 1 & 2 \\ 3 & 4 \end{pmatrix} \quad \text{and} \quad B = \begin{pmatrix} 0 & -1 \\ -2 & 1 \end{pmatrix},$$

we can verify that

$$AB = \begin{pmatrix} -4 & 1 \\ -8 & 1 \end{pmatrix} \quad \text{while} \quad BA = \begin{pmatrix} -3 & -4 \\ 1 & 0 \end{pmatrix}.$$

This one counterexample is enough to show that multiplication of $n \times n$ matrices is noncommutative. The associative property however, does hold true for multiplication of $n \times n$ matrices (see Exercise 14 of Section 6.2).

Is there an identity element for multiplication of $n \times n$ matrices? If so, the product of such an $n \times n$ matrix with any $n \times n$ matrix A should be A itself. That is, if

$$\begin{pmatrix} x_{11} & x_{12} \\ x_{21} & x_{22} \end{pmatrix}$$

is to be the identity matrix, then we must have

$$\begin{pmatrix} x_{11} & x_{12} \\ x_{21} & x_{22} \end{pmatrix}\begin{pmatrix} a_{11} & a_{12} \\ a_{21} & a_{22} \end{pmatrix} = \begin{pmatrix} a_{11} & a_{12} \\ a_{21} & a_{22} \end{pmatrix} \tag{1}$$

for any matrix

$$A = \begin{pmatrix} a_{11} & a_{12} \\ a_{21} & a_{22} \end{pmatrix}.$$

Multiplying the two matrices on the left side of equation (1) and setting the elements of this product matrix equal to the corresponding elements of A, we have the following four restrictions on x_{11}, x_{12}, x_{21}, and x_{22}.

$$\begin{aligned} x_{11}a_{11} + x_{12}a_{21} &= a_{11} \\ x_{21}a_{11} + x_{22}a_{21} &= a_{21} \\ x_{11}a_{12} + x_{12}a_{22} &= a_{12} \\ x_{21}a_{12} + x_{22}a_{22} &= a_{22} \end{aligned}$$

These equations are solved when $x_{11} = 1$, $x_{12} = x_{21} = 0$, and $x_{22} = 1$. (It can be shown that these are the only solutions.) Hence, the 2×2 identity matrix is

$$\begin{pmatrix} 1 & 0 \\ 0 & 1 \end{pmatrix}.$$

Generalizing for a system of $n \times n$ matrices, the **identity matrix,** which we shall denote I, is

$$I = \begin{pmatrix} 1 & 0 & \cdots & 0 \\ 0 & 1 & \cdots & 0 \\ \cdot & \cdot & & \cdot \\ \cdot & \cdot & a_{ij} & \cdot \\ \cdot & \cdot & & \cdot \\ 0 & 0 & \cdots & 1 \end{pmatrix}$$

where $a_{ij} = 1$ when $i = j$, and $a_{ij} = 0$ otherwise.

Now that we have defined an identity matrix, we can investigate to see whether or not every $n \times n$ matrix A has a multiplicative inverse, which we shall denote A^{-1}. For 2×2 matrices, if the inverse matrix for

$$A = \begin{pmatrix} a_{11} & a_{12} \\ a_{21} & a_{22} \end{pmatrix}$$

exists, it should satisfy the equation $AA^{-1} = I$, or

$$\begin{pmatrix} a_{11} & a_{12} \\ a_{21} & a_{22} \end{pmatrix} \begin{pmatrix} x_{11} & x_{12} \\ x_{21} & x_{22} \end{pmatrix} = \begin{pmatrix} 1 & 0 \\ 0 & 1 \end{pmatrix},$$

where

$$\begin{pmatrix} x_{11} & x_{12} \\ x_{21} & x_{22} \end{pmatrix} = A^{-1}.$$

Multiplying the two matrices on the left side and setting the elements of the product matrix equal to the corresponding elements of I, we have

$$\begin{aligned} a_{11}x_{11} + a_{12}x_{21} &= 1 \\ a_{11}x_{12} + a_{12}x_{22} &= 0 \\ a_{21}x_{11} + a_{22}x_{21} &= 0 \\ a_{21}x_{12} + a_{22}x_{22} &= 1. \end{aligned}$$

Solving these four equations for x_{11}, x_{12}, x_{21}, and x_{22}, we can identify the **multiplicative inverse** A^{-1} as the matrix

$$A^{-1} = \begin{pmatrix} \dfrac{a_{22}}{a_{11}a_{22} - a_{21}a_{12}} & \dfrac{-a_{12}}{a_{11}a_{22} - a_{21}a_{12}} \\ \dfrac{-a_{21}}{a_{11}a_{22} - a_{21}a_{12}} & \dfrac{a_{11}}{a_{11}a_{22} - a_{21}a_{12}} \end{pmatrix}$$

which can be written as follows,

$$\frac{1}{a_{11}a_{22} - a_{21}a_{12}} \begin{pmatrix} a_{22} & -a_{12} \\ -a_{21} & a_{11} \end{pmatrix}.$$

We shall denote $a_{11}a_{22} - a_{21}a_{12}$ as δ (the Greek letter "delta") and write

$$A^{-1} = \frac{1}{\delta} \begin{pmatrix} a_{22} & -a_{12} \\ -a_{21} & a_{11} \end{pmatrix}.$$

Example 3 Let

$$M = \begin{pmatrix} 2 & -1 \\ 4 & -1 \end{pmatrix}.$$

Find M^{-1} and show $MM^{-1} = I$.

Here $a_{11} = 2$, $a_{12} = -1$, $a_{21} = 4$, and $a_{22} = -1$. Thus, by definition, $\delta = (2)(-1) - 4(-1) = -2 + 4 = 2$, so that

$$M^{-1} = \tfrac{1}{2} \begin{pmatrix} -1 & 1 \\ -4 & 2 \end{pmatrix} = \begin{pmatrix} -\frac{1}{2} & \frac{1}{2} \\ -2 & 1 \end{pmatrix}.$$

Using the definition of matrix multiplication, we have

$$MM^{-1} = \begin{pmatrix} 2 & -1 \\ 4 & -1 \end{pmatrix} \begin{pmatrix} -\frac{1}{2} & \frac{1}{2} \\ -2 & 1 \end{pmatrix} = \begin{pmatrix} 1 & 0 \\ 0 & 1 \end{pmatrix} = I.$$

Note that whenever $\delta = 0$, A^{-1} does not exist, so that not every 2×2 matrix has an inverse. Although the method shown here for finding the inverse can be extended to any $n \times n$ matrix, it rapidly becomes complex. For example, to find the inverse of a 3×3 matrix we would need to solve a system of nine equations.

We know that for all real numbers a, b, and c, $a(b + c) = ab + ac$. Is there a similar distributive property for matrices? That is, does

$$A(B + C) = AB + AC$$

for all square matrices A, B, and C of the same order? Let us consider an example. If

$$A = \begin{pmatrix} 1 & -1 \\ 2 & 1 \end{pmatrix}, \quad B = \begin{pmatrix} 2 & -1 \\ 0 & 3 \end{pmatrix}, \quad \text{and} \quad C = \begin{pmatrix} -1 & 3 \\ 2 & 1 \end{pmatrix},$$

then we have

$$A(B+C) = \begin{pmatrix} 1 & -1 \\ 2 & 1 \end{pmatrix} \left[\begin{pmatrix} 2 & -1 \\ 0 & 3 \end{pmatrix} + \begin{pmatrix} -1 & 3 \\ 2 & 1 \end{pmatrix} \right]$$
$$= \begin{pmatrix} 1 & -1 \\ 2 & 1 \end{pmatrix} \begin{pmatrix} 1 & 2 \\ 2 & 4 \end{pmatrix}$$
$$= \begin{pmatrix} -1 & -2 \\ 4 & 8 \end{pmatrix}.$$

On the other hand,

$$AB + AC = \begin{pmatrix} 1 & -1 \\ 2 & 1 \end{pmatrix} \begin{pmatrix} 2 & -1 \\ 0 & 3 \end{pmatrix} + \begin{pmatrix} 1 & -1 \\ 2 & 1 \end{pmatrix} \begin{pmatrix} -1 & 3 \\ 2 & 1 \end{pmatrix}$$
$$= \begin{pmatrix} 2 & -4 \\ 4 & 1 \end{pmatrix} + \begin{pmatrix} -3 & 2 \\ 0 & 7 \end{pmatrix}$$
$$= \begin{pmatrix} -1 & -2 \\ 4 & 8 \end{pmatrix}.$$

While such an example does not, of course, prove anything, it can be shown that

$$A(B+C) = AB + AC$$

for all square matrices A, B, and C of the same order. This property of matrices is called the left distributive property of matrix multiplication over matrix addition, while

$$(B+C)A = BA + CA$$

is called the right distributive property.

In summary, for any system of square matrices of a given order, the following axioms hold.

Addition	*Multiplication*
closure	closure
commutative	
associative	associative
identity	identity
inverse	inverse ($\delta \neq 0$)*

distributive

Since multiplication of $n \times n$ matrices is noncommutative, we see that the system is not a field.

* We have defined δ only for 2×2 matrices; an analogous definition can be given for larger square matrices. (See Section 6.4.)

The properties which do hold true for $n \times n$ matrices, however, enable us to solve matrix equations in much the same way that we solve algebraic equations. To solve matrix equations, in addition to the properties above, we need to assume a multiplication axiom of equality for matrices.

Example 4 Given $A = \begin{pmatrix} 2 & 2 \\ -1 & -2 \end{pmatrix}$ and $B = \begin{pmatrix} 2 \\ 3 \end{pmatrix}$, find a matrix X so that $AX = B$.

To solve the matrix equation $AX = B$, we can use the fact that $A^{-1}A = I$ and $IX = X$ as follows.

$AX = B$	given
$A^{-1}(AX) = A^{-1}B$	multiplication axiom of equality for matrices
$(A^{-1}A)X = A^{-1}B$	associative axiom of multiplication for matrices
$IX = A^{-1}B$	multiplication inverse axiom for matrices
$X = A^{-1}B$	multiplication identity axiom for matrices

In using the multiplication axiom of equality for matrices we must be careful to multiply on the left on both sides of the equation, since, unlike multiplication of real numbers, multiplication of matrices is not commutative. Since $X = A^{-1}B$, we must find A^{-1}.

$$A^{-1} = \frac{1}{-4+2}\begin{pmatrix} -2 & -2 \\ 1 & 2 \end{pmatrix} = \begin{pmatrix} 1 & 1 \\ -\frac{1}{2} & -1 \end{pmatrix}$$

Now we can find X.

$$X = A^{-1}B = \begin{pmatrix} 1 & 1 \\ -\frac{1}{2} & -1 \end{pmatrix}\begin{pmatrix} 2 \\ 3 \end{pmatrix} = \begin{pmatrix} 5 \\ -4 \end{pmatrix}$$

6.3 Exercises

Find the inverse, if it exists, of each of the following matrices.

1. $\begin{pmatrix} 4 & 3 \\ 2 & 1 \end{pmatrix}$
2. $\begin{pmatrix} 1 & 1 \\ -1 & -1 \end{pmatrix}$
3. $\begin{pmatrix} 1 & 0 \\ 0 & 2 \end{pmatrix}$
4. $\begin{pmatrix} -1 & -5 \\ 2 & 3 \end{pmatrix}$
5. $\begin{pmatrix} 3 & 4 \\ 9 & 12 \end{pmatrix}$
6. $\begin{pmatrix} 2 & 7 \\ -2 & 7 \end{pmatrix}$
7. $\begin{pmatrix} 1 & 7 \\ 4 & 2 \end{pmatrix}$

Solve the matrix equation $AX = B$ for X, given matrices A and B as follows.

8. $A = \begin{pmatrix} 1 & 3 \\ -2 & 4 \end{pmatrix}$, $B = \begin{pmatrix} 2 \\ 6 \end{pmatrix}$

9. $A = \begin{pmatrix} 2 & -2 \\ 1 & 0 \end{pmatrix}$, $B = \begin{pmatrix} -4 \\ 3 \end{pmatrix}$

10. $A = \begin{pmatrix} 2 & -4 \\ 1 & -2 \end{pmatrix}$, $B = \begin{pmatrix} 8 \\ 5 \end{pmatrix}$

11. $A = \begin{pmatrix} 10 & -3 \\ 2 & -1 \end{pmatrix}$, $B = \begin{pmatrix} 4 \\ 8 \end{pmatrix}$

12. $A = \begin{pmatrix} 5 & 7 \\ 8 & 9 \end{pmatrix}$, $B = \begin{pmatrix} -4 \\ 33 \end{pmatrix}$

Using $X = \begin{pmatrix} x & y \\ z & w \end{pmatrix}$, *and* $0 = \begin{pmatrix} 0 & 0 \\ 0 & 0 \end{pmatrix}$, *show that the following are true.*

13. $IX = XI = X$
14. $X \cdot X^{-1} = X^{-1} \cdot X = I$ if X^{-1} exists.
15. $X \cdot 0 = 0 \cdot X = 0$

16. Using the definitions and properties of matrix algebra, prove that for square matrices A and B, if $AB = 0$, and if A^{-1} exists, then $B = 0$. (Assume 0 is a matrix with the appropriate dimension and with all entries zero.)

6.4 Determinants

For the 2×2 matrix

$$X = \begin{pmatrix} x_{11} & x_{12} \\ x_{21} & x_{22} \end{pmatrix},$$

we have defined the number $\delta = x_{11}x_{22} - x_{21}x_{12}$. We call δ the **determinant** of matrix X, and write

$$|X| = \begin{vmatrix} x_{11} & x_{12} \\ x_{21} & x_{22} \end{vmatrix} = x_{11}x_{22} - x_{21}x_{12},$$

or

$$\delta(X) = x_{11}x_{22} - x_{21}x_{12}.$$

In this section, we shall investigate methods that can be used to find determinants for any square matrix.

As we shall see, for each square matrix X, with real number elements, there is a unique real number determinant $\delta(X)$. The set of ordered pairs of the form $(X, \delta(X))$ defines a function with domain the set of square matrices X, and range the set of real numbers $\delta(X)$.

The determinant of a 3×3 matrix, which we shall refer to as a 3×3 determinant, can be defined in a manner similar to that given above. Thus, for the matrix

$$A = \begin{pmatrix} a_{11} & a_{12} & a_{13} \\ a_{21} & a_{22} & a_{23} \\ a_{31} & a_{32} & a_{33} \end{pmatrix},$$

we define the determinant as

$$\delta(A) = \begin{vmatrix} a_{11} & a_{12} & a_{13} \\ a_{21} & a_{22} & a_{23} \\ a_{31} & a_{32} & a_{33} \end{vmatrix} = \begin{aligned}&(a_{11}a_{22}a_{33} + a_{12}a_{23}a_{31} + a_{13}a_{21}a_{32}) \\ &- (a_{31}a_{22}a_{13} + a_{32}a_{23}a_{11} + a_{33}a_{21}a_{12}).\end{aligned}$$

From this definition of the determinant of a 3×3 matrix, we can rearrange terms and factor to get

$$\begin{vmatrix} a_{11} & a_{12} & a_{13} \\ a_{21} & a_{22} & a_{23} \\ a_{31} & a_{32} & a_{33} \end{vmatrix} = \begin{aligned}&a_{11}(a_{22}a_{33} - a_{32}a_{23}) - a_{21}(a_{12}a_{33} - a_{32}a_{13}) \\ &+ a_{31}(a_{12}a_{23} - a_{22}a_{13}).\end{aligned} \tag{2}$$

Note that each of the quantities in parentheses above represents a 2×2 determinant which is the determinant of that part of the 3×3 array remaining when the row and column of the multiplier are eliminated.

$$a_{11}(a_{22}a_{33} - a_{32}a_{23}) \begin{pmatrix} a_{11} & a_{12} & a_{13} \\ a_{21} & a_{22} & a_{23} \\ a_{31} & a_{32} & a_{33} \end{pmatrix}$$

$$a_{21}(a_{12}a_{33} - a_{32}a_{13}) \begin{pmatrix} a_{11} & a_{12} & a_{13} \\ a_{21} & a_{22} & a_{23} \\ a_{31} & a_{32} & a_{33} \end{pmatrix}$$

$$a_{31}(a_{12}a_{23} - a_{22}a_{13}) \begin{pmatrix} a_{11} & a_{12} & a_{13} \\ a_{21} & a_{22} & a_{23} \\ a_{31} & a_{32} & a_{33} \end{pmatrix}$$

Such a 2×2 determinant is called a **minor** of an element in the 3×3 determinant. Thus, the minors of a_{11}, a_{21}, and a_{31} are as shown in the chart below.

element	*minor*
a_{11}	$\begin{vmatrix} a_{22} & a_{23} \\ a_{32} & a_{33} \end{vmatrix}$
a_{21}	$\begin{vmatrix} a_{12} & a_{13} \\ a_{32} & a_{33} \end{vmatrix}$
a_{31}	$\begin{vmatrix} a_{12} & a_{13} \\ a_{22} & a_{23} \end{vmatrix}$

We see that the 3×3 determinant can be evaluated by multiplying each element in the first column by its minor and combining the products as indicated in (2) above. This is called the **expansion** of the determinant by minors about the first column.

Example 5 Evaluate the determinant

$$\begin{vmatrix} 1 & 3 & -2 \\ -1 & -2 & -3 \\ 1 & 1 & 2 \end{vmatrix}$$

by expanding by minors about the first column.

Using the procedure above, we have

$$\begin{vmatrix} 1 & 3 & -2 \\ -1 & -2 & -3 \\ 1 & 1 & 2 \end{vmatrix} = 1\begin{vmatrix} -2 & -3 \\ 1 & 2 \end{vmatrix} - (-1)\begin{vmatrix} 3 & -2 \\ 1 & 2 \end{vmatrix} + 1\begin{vmatrix} 3 & -2 \\ -2 & -3 \end{vmatrix}$$

$$\begin{aligned} &= 1[(-2)(2) - (1)(-3)] + 1[(3)(2) - (1)(-2)] \\ &\qquad + 1[(3)(-3) - (-2)(-2)] \\ &= 1(-1) + 1(8) + 1(-13) \\ &= -1 + 8 - 13 \\ &= -6. \end{aligned}$$

To get equation (2) we could have rearranged terms and factored differently by factoring out the three elements of the second or third columns or of any of the three rows of the array. Therefore, expanding by minors about any row or any column results in the same value for the determinant. To determine the correct signs for the terms of other expansions, the following array of signs is helpful.

$$\begin{matrix} + & - & + \\ - & + & - \\ + & - & + \end{matrix}$$

The signs alternate for each row and column, beginning with + in the first row, first column position. Thus, the array of signs is easy to reproduce at will. If the expansion is to be about the second column, for example, the first term would have a minus sign associated with it, the second a plus sign, and the third a minus sign.

Example 6 Evaluate the determinant of Example 5 by expansion by minors about the second column.

Using the methods described above, we have

$$\begin{vmatrix} 1 & 3 & -2 \\ -1 & -2 & -3 \\ 1 & 1 & 2 \end{vmatrix} = -3\begin{vmatrix} -1 & -3 \\ 1 & 2 \end{vmatrix} + (-2)\begin{vmatrix} 1 & -2 \\ 1 & 2 \end{vmatrix} - 1\begin{vmatrix} 1 & -2 \\ -1 & -3 \end{vmatrix}$$

$$= -3(1) - 2(4) - 1(-5)$$
$$= -3 - 8 + 5$$
$$= -6.$$

This method of expansion by minors can be extended to find determinants of any $n \times n$ matrix, as shown in the next example.

Example 7 Evaluate

$$\begin{vmatrix} -1 & -2 & 3 & 2 \\ 0 & 1 & 4 & -2 \\ 3 & -1 & 4 & 0 \\ 2 & 1 & 0 & 3 \end{vmatrix}.$$

Expanding by minors about the fourth row gives

$$-2\begin{vmatrix} -2 & 3 & 2 \\ 1 & 4 & -2 \\ -1 & 4 & 0 \end{vmatrix} + 1\begin{vmatrix} -1 & 3 & 2 \\ 0 & 4 & -2 \\ 3 & 4 & 0 \end{vmatrix} - 0\begin{vmatrix} -1 & -2 & 2 \\ 0 & 1 & -2 \\ 3 & -1 & 0 \end{vmatrix} + 3\begin{vmatrix} -1 & -2 & 3 \\ 0 & 1 & 4 \\ 3 & -1 & 4 \end{vmatrix}$$

$$= -2(6) + 1(-50) - 0 + 3(-41)$$
$$= -185.$$

6.4 Exercises

Evaluate the following determinants.

1. $\begin{vmatrix} 5 & 8 \\ 2 & -4 \end{vmatrix}$

2. $\begin{vmatrix} -3 & 0 \\ 0 & 9 \end{vmatrix}$

3. $\begin{vmatrix} -1 & -2 \\ 5 & 3 \end{vmatrix}$

4. $\begin{vmatrix} 6 & -4 \\ 0 & -1 \end{vmatrix}$

5. $\begin{vmatrix} 9 & 3 \\ -3 & -1 \end{vmatrix}$

6. $\begin{vmatrix} 0 & 2 \\ 1 & 5 \end{vmatrix}$

7. $\begin{vmatrix} 1 & 2 & 0 \\ -1 & 2 & -1 \\ 0 & 1 & 4 \end{vmatrix}$

8. $\begin{vmatrix} 2 & 1 & -1 \\ 4 & 7 & -2 \\ 2 & 4 & 0 \end{vmatrix}$

9. $\begin{vmatrix} 10 & 2 & 1 \\ -1 & 4 & 3 \\ -3 & 8 & 10 \end{vmatrix}$

10. $\begin{vmatrix} 7 & -1 & 1 \\ 1 & -7 & 2 \\ -2 & 1 & 1 \end{vmatrix}$

11. $\begin{vmatrix} 1 & -2 & 3 \\ 0 & 0 & 0 \\ 1 & 10 & -12 \end{vmatrix}$

12. $\begin{vmatrix} 2 & 3 & 0 \\ 1 & 9 & 0 \\ -1 & -2 & 0 \end{vmatrix}$

13. $\begin{vmatrix} 3 & 3 & -1 \\ 2 & 6 & 0 \\ -6 & -6 & 2 \end{vmatrix}$

14. $\begin{vmatrix} 5 & -3 & 2 \\ -5 & 3 & -2 \\ 1 & 0 & 1 \end{vmatrix}$

15. $\begin{vmatrix} 1 & 1 & 0 & 1 \\ 2 & 1 & 0 & 2 \\ 0 & 1 & -1 & 1 \\ 1 & -1 & 1 & 1 \end{vmatrix}$

16. $\begin{vmatrix} 2 & 7 & 0 & -1 \\ 1 & 0 & 1 & 3 \\ 2 & 4 & -1 & -1 \\ -1 & 1 & 0 & 8 \end{vmatrix}$

6.5 Properties of Determinants

To evaluate the determinant of a matrix by expansion by minors often requires a great deal of computation. For a 3×3 matrix, it is necessary to evaluate the determinant of three 2×2 matrices; for a 4×4, twelve determinants must be evaluated, and, in general, for an $n \times n$ matrix, it can be shown that one must evaluate $n!/2$ determinants.* This work is considerably simplified if a determinant has many zero elements. In this section we present a theorem which is useful in transforming matrices so that their determinants are easier to evaluate.

THEOREM 6.1 For any square matrix X and any real number k:

(a) If every element in a row (or column) of X is zero, then $\delta(X) = 0$.

(b) If corresponding rows and columns of X are interchanged to form the matrix Y, then $\delta(Y) = \delta(X)$.

(c) If any two rows (or columns) of X are interchanged to form the matrix Y, then $\delta(Y) = -\delta(X)$.

(d) If every element of a row (or column) of X is multiplied by k to form the matrix Y, then $\delta(Y) = k \cdot \delta(X)$.

(e) If k times any row (or column) of X is added to another row (or column) of X to form the matrix Y, then $\delta(Y) = \delta(X)$.

(f) If two rows (or columns) of X are identical, then $\delta(X) = 0$.

* $n! = n(n-1)(n-2) \cdots (2)(1)$.

We shall not prove Theorem 6.1 in general. Proofs are given below for some parts of the theorem for 3×3 matrices and others are included in the exercises. These proofs indicate the general line of reasoning for the general case.

It is easy to see that Theorem 6.1(a) holds in general, since each term of the expansion by minors about the zero row (or column) would have a zero factor, so that $\delta(X) = 0$. For example,

$$\begin{vmatrix} a & b & c \\ 0 & 0 & 0 \\ d & e & f \end{vmatrix} = -0 \begin{vmatrix} b & c \\ e & f \end{vmatrix} + 0 \begin{vmatrix} a & c \\ d & f \end{vmatrix} - 0 \begin{vmatrix} a & b \\ d & e \end{vmatrix} = 0.$$

To demonstrate Theorem 6.1(b) for a 3×3 matrix, note the following. Let

$$X = \begin{vmatrix} a & b & c \\ d & e & f \\ g & h & i \end{vmatrix} \quad \text{and} \quad Y = \begin{vmatrix} a & d & g \\ b & e & h \\ c & f & i \end{vmatrix},$$

where Y was obtained from X by interchanging the corresponding rows and columns of X. Expanding about row 1 of X, we have

$$\begin{aligned} \delta(X) &= a \begin{vmatrix} e & f \\ h & i \end{vmatrix} - b \begin{vmatrix} d & f \\ g & i \end{vmatrix} + c \begin{vmatrix} d & e \\ g & h \end{vmatrix} \\ &= a(ei - hf) - b(di - gf) + c(dh - ge). \end{aligned}$$

Expanding about column 1 of Y gives

$$\begin{aligned} \delta(Y) &= a \begin{vmatrix} e & h \\ f & i \end{vmatrix} - b \begin{vmatrix} d & g \\ f & i \end{vmatrix} + c \begin{vmatrix} d & g \\ e & h \end{vmatrix} \\ &= a(ei - fh) - b(di - fg) + c(dh - eg). \end{aligned}$$

Thus, $\delta(Y) = \delta(X)$.

The proof of Theorem 6.1(e) for a 3×3 matrix in which the first column of Y is formed by adding k times the second column of X to the first column of X is as follows. If

$$\delta(X) = \begin{vmatrix} a & b & c \\ d & e & f \\ g & h & i \end{vmatrix} \quad \text{and} \quad \delta(Y) = \begin{vmatrix} a + kb & b & c \\ d + ke & e & f \\ g + kh & h & i \end{vmatrix},$$

then $\delta(X) = a(ei - hf) - d(bi - hc) + g(bf - ec)$, where we expanded about column 1 of X. Expanding Y about column 1 gives

$$\delta(Y) = (a + kb)\begin{vmatrix} e & f \\ h & i \end{vmatrix} - (d + ke)\begin{vmatrix} b & c \\ h & i \end{vmatrix} + (g + kh)\begin{vmatrix} b & c \\ e & f \end{vmatrix}$$

$$\begin{aligned} &= (a + kb)(ei - hf) - (d + ke)(bi - hc) + (g + kh)(bf - ec) \\ &= a(ei - hf) + kb(ei - hf) - d(bi - hc) - ke(bi - hc) \\ &\quad + g(bf - ec) + kh(bf - ec). \end{aligned}$$

By combining terms two, four, and six, and factoring out the k, we have

$$\begin{aligned} &= a(ei - hf) - d(bi - hc) + g(bf - ec) \\ &\quad + k[bei - bhf - ebi + ehc - hbf - hec] \\ &= a(ei - hf) - d(bi - hc) + g(bf - ec) + k(0) \\ &= \delta(X). \end{aligned}$$

Theorem 6.1(f) is a consequence of (e) and (a), since, by (e), one of the identical rows (or columns) can be multiplied by -1 and added to the other to produce a zero row (or column) without changing the value of the determinant. Then, by (a) the value of the determinant is zero.

The following examples illustrate the use of the properties of determinants to simplify the evaluation of determinants.

Example 8 Without expanding, show that the value of the following determinant is zero.

$$\begin{vmatrix} 2 & 5 & -1 \\ 1 & -15 & 3 \\ -2 & 10 & -2 \end{vmatrix}$$

Examining the columns of the array, we note that each element in column two is -5 times the corresponding element in column three. Thus, if we use Theorem 6.1(e) and multiply column three by 5, and then add the results to the corresponding elements in column two, we have the equivalent determinant

$$\begin{vmatrix} 2 & 0 & -1 \\ 1 & 0 & 3 \\ -2 & 0 & -2 \end{vmatrix}.$$

By Theorem 6.1(a), the value of this determinant is zero.

Example 9 Find $\delta(A)$ if

$$\delta(A) = \begin{vmatrix} 4 & 2 & 1 & 0 \\ -2 & 4 & -1 & 7 \\ -5 & 2 & 3 & 1 \\ 6 & 4 & -3 & 2 \end{vmatrix}.$$

Our goal is to change row one (we could have selected any row or column) to a row in which every element but one is zero, using Theorem 6.1(e). To begin, we multiply column two by -2 and add the results to column one, replacing column one with this new column.

$$\begin{vmatrix} 0 & 2 & 1 & 0 \\ -10 & 4 & -1 & 7 \\ -9 & 2 & 3 & 1 \\ -2 & 4 & -3 & 2 \end{vmatrix} \quad \begin{array}{l} \text{column one replaced by} \\ -2 \text{ times column two,} \\ \text{plus column one} \end{array}$$

We then replace column two by the results we get when we multiply column three by -2 and add to column two.

$$\begin{vmatrix} 0 & 0 & 1 & 0 \\ -10 & 6 & -1 & 7 \\ -9 & -4 & 3 & 1 \\ -2 & 10 & -3 & 2 \end{vmatrix} \quad \begin{array}{l} \text{column two replaced by} \\ -2 \text{ times column three,} \\ \text{plus column two} \end{array}$$

The first row has only one nonzero number and so we expand about row one.

$$\delta(A) = 1 \begin{vmatrix} -10 & 6 & 7 \\ -9 & -4 & 1 \\ -2 & 10 & 2 \end{vmatrix}$$

Now let us try to change column three to a column with two zeros.

$$\begin{vmatrix} 53 & 34 & 0 \\ -9 & -4 & 1 \\ -2 & 10 & 2 \end{vmatrix} \quad \begin{array}{l} \text{row one replaced by} \\ -7 \text{ times row two,} \\ \text{added to row one} \end{array}$$

$$\begin{vmatrix} 53 & 34 & 0 \\ -9 & -4 & 1 \\ 16 & 18 & 0 \end{vmatrix} \quad \begin{array}{l} \text{row three replaced by} \\ -2 \text{ times row two,} \\ \text{added to row three} \end{array}$$

Now we can easily expand about column three to find the value of $\delta(A)$.

$$\delta(A) = -1 \begin{vmatrix} 53 & 34 \\ 16 & 18 \end{vmatrix} = -1(954 - 544) = -410$$

Note that in applying Theorem 6.1(e) we worked with sums of *rows* to get a *column* with only one nonzero number and with sums of *columns* to get a *row* with one nonzero number.

6.5 Exercises

Without evaluating, use Theorem 6.1 to verify that each of the following determinants is zero.

1. $\begin{vmatrix} 1 & 3 & -1 \\ 0 & 0 & 0 \\ 2 & 0 & 1 \end{vmatrix}$

2. $\begin{vmatrix} -1 & 2 & 4 \\ 4 & -8 & -16 \\ 3 & 0 & 5 \end{vmatrix}$

3. $\begin{vmatrix} 3 & 6 & 6 \\ 2 & 0 & 4 \\ 1 & 4 & 2 \end{vmatrix}$

4. $\begin{vmatrix} 1 & 0 & 0 \\ 1 & 0 & 1 \\ 3 & 0 & 0 \end{vmatrix}$

Use the appropriate parts of Theorem 6.1 to verify each of the following.

5. $\begin{vmatrix} 2 & 3 & 4 \\ 3 & 4 & 5 \\ 4 & 5 & 6 \end{vmatrix} = 0$

6. $\begin{vmatrix} 3 & 4 & -1 \\ 1 & 0 & 2 \\ -2 & 3 & 5 \end{vmatrix} = \begin{vmatrix} 5 & 4 & 3 \\ 1 & 0 & 2 \\ -2 & 3 & 5 \end{vmatrix}$

7. $\begin{vmatrix} 1 & 0 & 5 \\ 3 & 2 & -1 \\ 6 & 0 & 2 \end{vmatrix} = \begin{vmatrix} 1 & 3 & 6 \\ 0 & 2 & 0 \\ 5 & -1 & 2 \end{vmatrix}$

8. $\begin{vmatrix} 3 & 4 & 5 \\ -1 & 1 & 2 \\ 0 & 4 & -3 \end{vmatrix} = -\begin{vmatrix} 4 & 3 & 5 \\ 1 & -1 & 2 \\ 4 & 0 & -3 \end{vmatrix}$

9. $\begin{vmatrix} 2 & 0 & 4 \\ 1 & 3 & 2 \\ -1 & 1 & 0 \end{vmatrix} = \begin{vmatrix} 2 & 0 & 4 \\ 1 & 3 & 2 \\ 1 & 1 & 4 \end{vmatrix}$

10. $-2\begin{vmatrix} 1 & -1 & 2 \\ 3 & 0 & 4 \\ 2 & 5 & 1 \end{vmatrix} = \begin{vmatrix} -2 & 2 & -4 \\ 3 & 0 & 4 \\ 2 & 5 & 1 \end{vmatrix}$

11. (a) Prove Theorem 6.1(c) for

$$X = \begin{vmatrix} a & b & c \\ d & e & f \\ g & h & i \end{vmatrix} \quad \text{and} \quad Y = \begin{vmatrix} d & e & f \\ a & b & c \\ g & h & i \end{vmatrix}.$$

(b) Prove Theorem 6.1(d) for

$$X = \begin{vmatrix} a & b & c \\ d & e & f \\ g & h & i \end{vmatrix} \quad \text{and} \quad Y = \begin{vmatrix} ka & b & c \\ kd & e & f \\ kg & h & i \end{vmatrix}.$$

12. Evaluate the following determinants after using Theorem 6.1 to introduce zeros as illustrated in Example 9.

(a) $\begin{vmatrix} 1 & 0 & 2 & 2 \\ 2 & 4 & 1 & -1 \\ 1 & -3 & 1 & 0 \\ 1 & 1 & 0 & 1 \end{vmatrix}$ (b) $\begin{vmatrix} 2 & -1 & 1 & 0 \\ 1 & 1 & 0 & 1 \\ 0 & -1 & 1 & 1 \\ 1 & 2 & 1 & 2 \end{vmatrix}$

Determinants can be used to find the area of a triangle given the coordinates of its vertices. Given a triangle PQR with vertices (x_1, y_1), (x_3, y_3) and (x_2, y_2), as shown in the figure, we can introduce segments PM, RN, and QS perpendicular to the x-axis, forming trapezoids PMNR, NSQR, and PMSQ. Recall that the area of a trapezoid is given by $\frac{1}{2}$ the sum of the parallel bases times the altitude. For example, the area of trapezoid PMSQ equals $\frac{1}{2}(y_1 + y_3)(x_3 - x_1)$. The area of triangle PQR can be found by subtracting the area of PMSQ from the sum of the areas of PMNR and RNSQ. Thus, for the area A of triangle PQR, we have

$$A = \tfrac{1}{2}(x_3y_1 - x_1y_3 + x_2y_3 - x_3y_2 + x_1y_2 - x_2y_1).$$

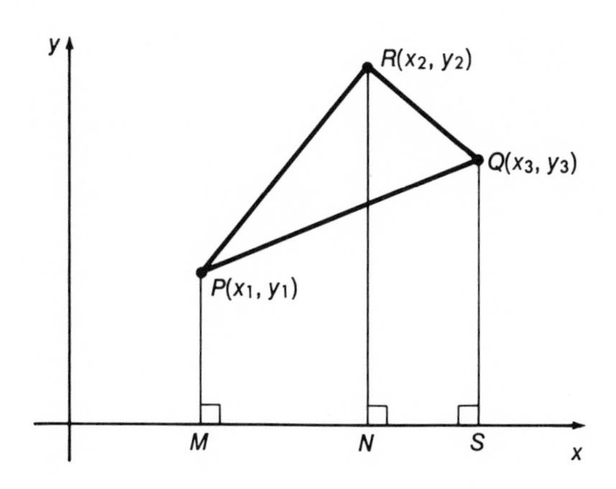

By evaluating the determinant

$$\begin{vmatrix} x_1 & y_1 & 1 \\ x_2 & y_2 & 1 \\ x_3 & y_3 & 1 \end{vmatrix}$$

it can be shown that

$$A = \tfrac{1}{2}\begin{vmatrix} x_1 & y_1 & 1 \\ x_2 & y_2 & 1 \\ x_3 & y_3 & 1 \end{vmatrix}.$$

Use the formula given to find the area of the following triangles.

13. $P(0, 1)$, $Q(2, 0)$, $R(1, 3)$
14. $P(2, 5)$, $Q(-1, 3)$, $R(4, 0)$
15. $P(2, -2)$, $Q(0, 0)$, $R(-3, -4)$
16. $P(4, 7)$, $Q(5, -2)$, $R(1, 1)$
17. $P(3, 8)$, $Q(-1, 4)$, $R(0, 1)$
18. $P(-3, -1)$, $Q(4, 2)$, $R(3, -3)$

7

SYSTEMS OF EQUATIONS AND INEQUALITIES

Systems of equations or inequalities have become important in many applications. With the advent of the computer, it has become practical to solve large systems with many variables so that new uses for this branch of mathematics have been developed.

A **system of equations** is a set of equations whose solution set is the intersection of the solution sets of the member equations. If some of the members of the system are inequalities, the system is called a **system of inequalities.** It is customary to write a system with only one set brace on the left, or when the context is clear, to omit set braces entirely. For example, the compound sentence $2x + y = 4$ and $x - y = 6$ gives a system which may be written as

$$\begin{cases} 2x + y = 4 \\ \ \ x - y = 6 \end{cases} \qquad \text{or simply} \qquad \begin{aligned} 2x + y &= 4 \\ x - y &= 6. \end{aligned}$$

The solution set of this system is $\{(x, y) \mid 2x + y = 4\} \cap \{(x, y) \mid x - y = 6\}$. In this chapter we shall discuss methods for solving several kinds of systems.

7.1 Linear Systems

A **first degree equation in *n* unknowns** is any equation of the form

$$a_1x_1 + a_2x_2 + \cdots + a_nx_n = k,$$

where $a_1, a_2, \cdots, a_n$ and k are constants, and $x_1, x_2, \cdots, x_n$ are variables. Such equations are also called **linear equations.** Generally, we shall confine our discussion of linear systems to those with two or three variables, although the methods we use can be extended to systems with more variables.

Recall from Chapter 3 that the solution set of a linear equation in two variables is an infinite set of ordered pairs. Since the graph of such an equation is a straight line, there are three possibilities for the solution set of a system of two linear equations in two variables, such as

$$a_1x + b_1y = c_1$$
$$a_2x + b_2y = c_2.$$

1. The two graphs intersect in a single point. The coordinates of this point give the solution of the system. This is the most common case (see Figure 7.1(a)).

2. The graphs are distinct parallel lines. When this is the case, the equations are said to be **inconsistent,** that is, there is no solution common to both equations. The solution set of the linear system is $\varnothing$ (see Figure 7.1(b)).

3. The graphs are the same line. In this case, the equations are said to be **dependent,** since any solution of one equation is also a solution of the other. Here the solution set of the system can be written as either of the two identical sets $\{(x, y) \mid a_1x + b_1y = c_1\}$ or $\{(x, y) \mid a_2x + b_2y = c_2\}$ (see Figure 7.1(c)).

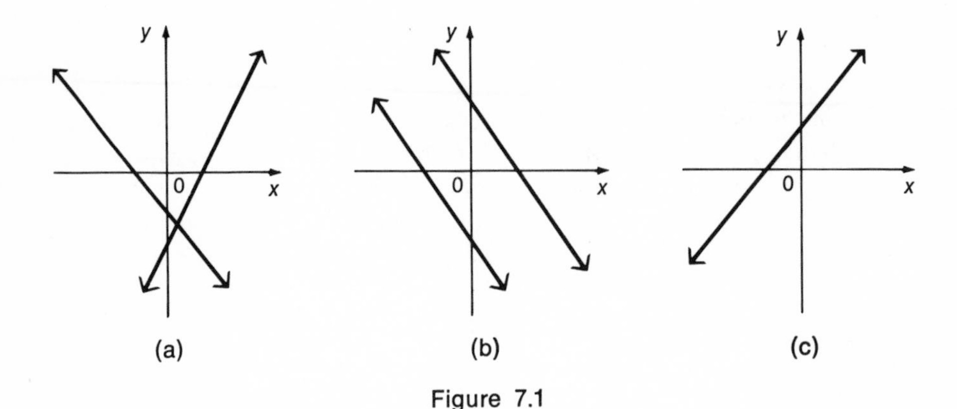

Figure 7.1

A solution of a linear equation $ax + by + cz = k$ with three variables is an **ordered triple** (x, y, z). Thus, the solution set of such an equation is an infinite set of ordered triples. It is shown in geometry that the graph of a linear equation in three variables is a plane. By considering the possible intersections of the planes representing three equations in three unknowns, we see that the solution set of such a system may again be either a single ordered triple (x, y, z), an infinite set of ordered triples, or $\varnothing$.

The most general method of solving a linear system depends on the following theorem, which is stated for equations in three variables. The theorem can be generalized to linear equations with any number of variables.

THEOREM 7.1 The ordered triple (x, y, z) satisfies each of the equations

$$a_1x + b_1y + c_1z = k_1$$
$$a_2x + b_2y + c_2z = k_2,$$

if and only if (x, y, z) satisfies the equation

$$m_1(a_1x + b_1y + c_1z) + m_2(a_2x + b_2y + c_2z) = m_1k_1 + m_2k_2, \quad (1)$$

for any nonzero real numbers m_1 and m_2.

To prove Theorem 7.1, we note that k_1 and k_2 can be substituted respectively for $a_1x + b_1y + c_1z$ and $a_2x + b_2y + c_2z$ in equation (1) to give

$$m_1k_1 + m_2k_2 = m_1k_1 + m_2k_2,$$

while from equation (1), setting corresponding terms equal gives

$$k_1 = a_1x + b_1y + c_1z$$

and

$$k_2 = a_2x + b_2y + c_2z.$$

To demonstrate how Theorem 7.1 is used, let us solve the system

$$3x - 4y = 1 \quad (2)$$

$$2x + 3y = 12. \quad (3)$$

We choose the multipliers m_1 and m_2 of Theorem 7.1 so that one of the variables will be eliminated in the expression

$$m_1(3x - 4y) + m_2(2x + 3y) = m_1(1) + m_2(12).$$

Since $a_1 = 3$ and $a_2 = 2$, we can eliminate x in the sum by choosing $m_1 = -2$ and $m_2 = 3$ as follows.

$$\begin{aligned} -2(3x - 4y) + 3(2x + 3y) &= -2(1) + 3(12) \quad (4) \\ -6x + 8y + 6x + 9y &= -2 + 36 \\ 17y &= 34 \\ y &= 2 \end{aligned}$$

To find the value of x, we can now either choose new multipliers m_1 and m_2 so as to eliminate y by again applying Theorem 7.1, or we can substitute 2 for y in one of the given equations. By substituting 2 for y in equation (2), we have

$$\begin{aligned} 3x - 4(2) &= 1 \\ 3x &= 9 \\ x &= 3. \end{aligned}$$

Thus, the solution set of the system is $\{(3, 2)\}$, which can be verified by checking in equation (3). For obvious reasons, this method of solution is often called the **elimination method.**

The solution of systems with three or more variables requires repeated use of Theorem 7.1, as illustrated in the following example.

Example 1 Solve the system

$$2x + y - z = 2 \quad (5)$$

$$x + 3y + 2z = 1 \quad (6)$$

$$x + y + z = 2. \quad (7)$$

We need to use several steps to obtain an equation in one variable. First, we must eliminate the same variable from each of two pairs of equations and then eliminate another variable from the two resulting equations. Suppose we wish to eliminate x first. Multiplying equations (5) and (6) by -1 and 2 respectively, we have

$$\begin{array}{r} -2x - y + z = -2 \\ 2x + 6y + 4z = 2 \\ \hline 5y + 5z = 0 \end{array}$$

or

$$y + z = 0. \quad (8)$$

(It is often easier to add vertically rather than horizontally when using Theorem 7.1.) We must now eliminate x again from a different pair of equations, say (6) and (7). To do this, we multiply (6) by -1 and (7) by 1.

$$\begin{array}{r} -x - 3y - 2z = -1 \\ x + y + z = 2 \\ \hline -2y - z = 1 \end{array} \quad (9)$$

Equations (8) and (9) can now be added (both multipliers m_1 and m_2 are 1) to find y.

$$\begin{array}{r} -2y - z = 1 \\ y + z = 0 \\ \hline -y \quad = 1 \\ y = -1 \end{array}$$

By substitution in equation (8), we find z. (We could have used equation (9).)

$$\begin{array}{r} y + z = 0 \\ -1 + z = 0 \\ z = 1 \end{array}$$

Now, from equation (7), (we could have used (5) or (6)) using the values for y and z, we find x.

$$\begin{array}{r} x + y + z = 2 \\ x - 1 + 1 = 2 \\ x = 2 \end{array}$$

Thus, the solution set of the system is $\{(2, -1, 1)\}$.

7.1 Exercises

Solve the following systems of equations.

1. $2x - 3y = -7$
   $5x + 4y = 17$

2. $4m + 3n = -1$
   $2m + 5n = 3$

3. $5p + 7q = 6$
   $10p - 3q = 46$

4. $12s - 5t = 9$
   $3s - 8t = -18$

5. $6x + 7y = -2$
   $7x - 6y = 26$

6. $2a + 9b = 3$
   $5a + 7b = -8$

7. $\frac{x}{2} + \frac{y}{3} = 8$
   $\frac{2x}{3} + \frac{3y}{2} = 17$

8. $\frac{x}{5} + 3y = 31$
   $2x - \frac{y}{5} = 8$

9. $\frac{2}{x} + \frac{1}{y} = \frac{3}{2}$ $\left(\text{Hint: let } \frac{1}{x} = t \text{ and } \frac{1}{y} = u.\right)$
   $\frac{3}{x} - \frac{1}{y} = 1$

10. $\frac{1}{x} + \frac{3}{y} = \frac{16}{5}$
    $\frac{5}{x} + \frac{4}{y} = 5$

11. $x + 4y - z = 6$
    $2x - y + z = 3$
    $3x + 2y + 3z = 16$

12. $4x - 3y + z = 9$
    $3x + 2y - 2z = 4$
    $x - y + 3z = 5$

13. $5m + n - 3p = -6$
    $2m + 3n + p = 5$
    $-3m - 2n + 4p = 3$

14. $2r - 5s + 4t = -35$
    $5r + 3s - t = 1$
    $r + s + t = 1$

15. $a - 3b - 2c = -3$
    $3a + 2b - c = 12$
    $-a - b + 4c = 3$

16. $2x + 2y + 2z = 6$
    $3x - 3y - 4z = -1$
    $x + y + 3z = 11$

17. Find a and b so that the line with equation $ax + by = 5$ contains the points $(-2, 1)$ and $(-1, -2)$.
18. Find a, b, and c so that the graph of the equation $y = ax^2 + bx + c$ contains the points $(2, 3)$, $(-1, 0)$, and $(-2, 2)$.
19. Betsy invests \$10,000, received from an endowment fund, in three ways. With one part, she buys mutual funds which offer a return of 8% per year. The second part, which amounts to twice the first, is used to buy government bonds at 9% per year. She puts the remainder in the bank at 5% annual interest. The first year her investments bring a return of \$830. How much did she invest in each way?
20. The Busy Beaver Manufacturing Company makes two products, gadgets and thingamabobs. Both require time on two machines: gadgets, one hour on machine A and two hours on machine B; thingamabobs, three hours on machine A and one hour on machine B. Both machines are operated for 15 hours a day. How many of each product can be produced in a day under these conditions?

7.2 Solution of Linear Systems by Determinants–Cramer's Rule

In the previous section, we discussed the elimination method for solving systems of equations. We can use this method to solve the general system of two equations in two variables,

$$a_1x + b_1y = c_1 \tag{10}$$

$$a_2x + b_2y = c_2. \tag{11}$$

To eliminate y and solve for x, we multiply equation (10) by b_2 and equation (11) by $-b_1$ and then add.

$$\begin{aligned} a_1b_2x + b_1b_2y &= c_1b_2 \\ -a_2b_1x - b_1b_2y &= -c_2b_1 \\ \hline (a_1b_2 - a_2b_1)x \qquad &= c_1b_2 - c_2b_1 \end{aligned}$$

or

$$x = \frac{c_1b_2 - c_2b_1}{a_1b_2 - a_2b_1}.$$

Similarly, to solve for y, we multiply equation (10) by $-a_2$ and equation (11) by a_1 and add.

$$\begin{aligned} -a_1a_2x - a_2b_1y &= -a_2c_1 \\ a_1a_2x + a_1b_2y &= a_1c_2 \\ \hline (a_1b_2 - a_2b_1)y &= a_1c_2 - a_2c_1 \end{aligned}$$

or

$$y = \frac{a_1c_2 - a_2c_1}{a_1b_2 - a_2b_1}.$$

Both numerators and the common denominator of these values for x and y can be written as determinants, since

$$c_1b_2 - c_2b_1 = \begin{vmatrix} c_1 & b_1 \\ c_2 & b_2 \end{vmatrix},$$

$$a_1c_2 - a_2c_1 = \begin{vmatrix} a_1 & c_1 \\ a_2 & c_2 \end{vmatrix},$$

$$a_1b_2 - a_2b_1 = \begin{vmatrix} a_1 & b_1 \\ a_2 & b_2 \end{vmatrix}.$$

Then we can write the solutions for x and y as

$$x = \frac{\begin{vmatrix} c_1 & b_1 \\ c_2 & b_2 \end{vmatrix}}{\begin{vmatrix} a_1 & b_1 \\ a_2 & b_2 \end{vmatrix}} \quad \text{and} \quad y = \frac{\begin{vmatrix} a_1 & c_1 \\ a_2 & c_2 \end{vmatrix}}{\begin{vmatrix} a_1 & b_1 \\ a_2 & b_2 \end{vmatrix}}.$$

We have just derived the following result, which is known as **Cramer's Rule.**

THEOREM 7.2 Given the system

$$a_1x + b_1y = c_1$$
$$a_2x + b_2y = c_2,$$

with $a_1b_2 - a_2b_1 \neq 0$, then

$$x = \frac{\begin{vmatrix} c_1 & b_1 \\ c_2 & b_2 \end{vmatrix}}{\begin{vmatrix} a_1 & b_1 \\ a_2 & b_2 \end{vmatrix}} \quad \text{and} \quad y = \frac{\begin{vmatrix} a_1 & c_1 \\ a_2 & c_2 \end{vmatrix}}{\begin{vmatrix} a_1 & b_1 \\ a_2 & b_2 \end{vmatrix}}.$$

For convenience, we shall denote the three determinants in the solution as

$$\begin{vmatrix} a_1 & b_1 \\ a_2 & b_2 \end{vmatrix} = D, \quad \begin{vmatrix} c_1 & b_1 \\ c_2 & b_2 \end{vmatrix} = D_x, \quad \text{and} \quad \begin{vmatrix} a_1 & c_1 \\ a_2 & c_2 \end{vmatrix} = D_y.$$

Note that D is the determinant of the matrix of coefficients of the given system; D_x is the determinant of a matrix obtained by replacing the coefficients of x in the coefficient matrix by the respective constants; and D_y is defined similarly.

Example 2 Use Cramer's Rule to solve the system

$$5x + 7y = -1$$
$$6x + 8y = 1.$$

By Cramer's Rule, we know

$$x = \frac{D_x}{D} \quad \text{and} \quad y = \frac{D_y}{D}.$$

Thus we need to evaluate D, D_x, and D_y. We have

$$D = \begin{vmatrix} 5 & 7 \\ 6 & 8 \end{vmatrix} = 5(8) - 6(7) = -2,$$

$$D_x = \begin{vmatrix} -1 & 7 \\ 1 & 8 \end{vmatrix} = (-1)(8) - (1)(7) = -15,$$

$$D_y = \begin{vmatrix} 5 & -1 \\ 6 & 1 \end{vmatrix} = 5(1) - (6)(-1) = 11.$$

Thus we have $x = \frac{-15}{-2}$, and $y = \frac{11}{-2}$. The solution set is $\{(\frac{15}{2}, \frac{-11}{2})\}$, as can be verified by substituting within the given system.

In a similar manner, Cramer's Rule can be generalized to systems of three equations in three variables (or n equations in n variables).

THEOREM 7.3 Given the system

$$\begin{aligned} a_1x + b_1y + c_1z &= d_1 \\ a_2x + b_2y + c_2z &= d_2 \\ a_3x + b_3y + c_3z &= d_3, \end{aligned}$$

with

$$D_x = \begin{vmatrix} d_1 & b_1 & c_1 \\ d_2 & b_2 & c_2 \\ d_3 & b_3 & c_3 \end{vmatrix}, \quad D_y = \begin{vmatrix} a_1 & d_1 & c_1 \\ a_2 & d_2 & c_2 \\ a_3 & d_3 & c_3 \end{vmatrix},$$

$$D_z = \begin{vmatrix} a_1 & b_1 & d_1 \\ a_2 & b_2 & d_2 \\ a_3 & b_3 & d_3 \end{vmatrix}, \quad D = \begin{vmatrix} a_1 & b_1 & c_1 \\ a_2 & b_2 & c_2 \\ a_3 & b_3 & c_3 \end{vmatrix} \neq 0,$$

then

$$x = \frac{D_x}{D}, \quad y = \frac{D_y}{D}, \quad \text{and} \quad z = \frac{D_z}{D}.$$

Example 3 Use Cramer's Rule to solve the system

$$\begin{aligned} x + y - z + 2 &= 0 \\ 2x - y + z + 5 &= 0 \\ x - 2y + 3z - 4 &= 0. \end{aligned}$$

To use Cramer's Rule, the system must be rewritten in the form

$$\begin{aligned} x + y - z &= -2 \\ 2x - y + z &= -5 \\ x - 2y + 3z &= 4. \end{aligned}$$

To find D, we can expand the coefficient matrix by minors about row one.

$$D = \begin{vmatrix} 1 & 1 & -1 \\ 2 & -1 & 1 \\ 1 & -2 & 3 \end{vmatrix} = 1\begin{vmatrix} -1 & 1 \\ -2 & 3 \end{vmatrix} - 1\begin{vmatrix} 2 & 1 \\ 1 & 3 \end{vmatrix} + (-1)\begin{vmatrix} 2 & -1 \\ 1 & -2 \end{vmatrix}$$

$$= 1(-1) - 1(5) - 1(-3)$$

$$= -3$$

If we replace the coefficients of x by the respective constants and expand about row one, we find D_x.

$$D_x = \begin{vmatrix} -2 & 1 & -1 \\ -5 & -1 & 1 \\ 4 & -2 & 3 \end{vmatrix} = -2\begin{vmatrix} -1 & 1 \\ -2 & 3 \end{vmatrix} - 1\begin{vmatrix} -5 & 1 \\ 4 & 3 \end{vmatrix} + (-1)\begin{vmatrix} -5 & -1 \\ 4 & -2 \end{vmatrix}$$

$$= -2(-1) - 1(-19) - 1(14)$$

$$= 7$$

Replacing the coefficients of y by the respective constants, and expanding about column one gives D_y.

$$D_y = \begin{vmatrix} 1 & -2 & -1 \\ 2 & -5 & 1 \\ 1 & 4 & 3 \end{vmatrix} = 1\begin{vmatrix} -5 & 1 \\ 4 & 3 \end{vmatrix} - 2\begin{vmatrix} -2 & -1 \\ 4 & 3 \end{vmatrix} + 1\begin{vmatrix} -2 & -1 \\ -5 & 1 \end{vmatrix}$$

$$= 1(-19) - 2(-2) + 1(-7)$$

$$= -22$$

Finally, replacing the coefficients of z by the respective constants, and expanding about column two gives D_z.

$$D_z = \begin{vmatrix} 1 & 1 & -2 \\ 2 & -1 & -5 \\ 1 & -2 & 4 \end{vmatrix} = -1\begin{vmatrix} 2 & -5 \\ 1 & 4 \end{vmatrix} + (-1)\begin{vmatrix} 1 & -2 \\ 1 & 4 \end{vmatrix} - (-2)\begin{vmatrix} 1 & -2 \\ 2 & -5 \end{vmatrix}$$

$$= -1(13) - 1(6) + 2(-1)$$

$$= -21$$

Thus we have

$$x = \frac{D_x}{D} = \frac{7}{-3} = -\frac{7}{3}, \; y = \frac{D_y}{D} = \frac{-22}{-3} = \frac{22}{3}, \text{ and } z = \frac{D_z}{D} = \frac{-21}{-3} = 7,$$

so that the solution set is $\{(-\frac{7}{3}, \frac{22}{3}, 7)\}$.

Example 4 Use Cramer's Rule to solve the system

$$\begin{aligned} 2x - 3y + 4z &= 10 \\ 6x - 9y + 12z &= 24 \\ x + 2y - 3z &= 5. \end{aligned}$$

We need to find D, D_x, D_y, and D_z. Here

$$D = \begin{vmatrix} 2 & -3 & 4 \\ 6 & -9 & 12 \\ 1 & 2 & -3 \end{vmatrix} = 2\begin{vmatrix} -9 & 12 \\ 2 & -3 \end{vmatrix} - 6\begin{vmatrix} -3 & 4 \\ 2 & -3 \end{vmatrix} + 1\begin{vmatrix} -3 & 4 \\ -9 & 12 \end{vmatrix}$$
$$= 2(3) - 6(1) + 1(0)$$
$$= 0.$$

As stated in Theorem 7.3, Cramer's Rule does not apply if $D = 0$ and so to check for this possibility, we calculate D first. In this case, the system contains equations which are either dependent or inconsistent. We could now use the elimination method to tell which. Verify that this system is inconsistent.

7.2 Exercises

Use Cramer's Rule to solve the following systems of equations.

1. $\begin{aligned} 2x - 3y &= 4 \\ x + 5y &= 2 \end{aligned}$

2. $\begin{aligned} 7x + 3y &= 5 \\ 2x + 4y &= 3 \end{aligned}$

3. $\begin{aligned} 5x + 2y &= 7 \\ 6x + y &= 8 \end{aligned}$

4. $\begin{aligned} x - 9y &= 4 \\ 3x + 2y &= 5 \end{aligned}$

5. $\begin{aligned} 12x - 8y &= 3 \\ 15x + 4y &= 9 \end{aligned}$

6. $\begin{aligned} 2x - 10y &= 5 \\ 7x + 6y &= 8 \end{aligned}$

7. $\begin{aligned} x + 2y + 3z &= 4 \\ 4x + 3y + 2z &= 1 \\ -x - 2y - 3z &= 0 \end{aligned}$

8. $\begin{aligned} 2x - y + 3z &= 1 \\ 3x + 2y - z &= 4 \\ 5x - y + z &= 2 \end{aligned}$

9. $\begin{aligned} x - 2y + 3z &= 4 \\ 5x + 7y - z &= 2 \\ 2x + 2y - 5z &= 3 \end{aligned}$

10. $\begin{aligned} -3x - 2y - z &= 4 \\ 4x + y + z &= 5 \\ 3x - 2y + 2z &= 1 \end{aligned}$

11. $\begin{aligned} x + 2y &= 10 \\ 3x + 4z &= 7 \\ - y - z &= 1 \end{aligned}$

12. $\begin{aligned} 5x - 2y &= 3 \\ 4y + z &= 8 \\ x + 2z &= 4 \end{aligned}$

7.3 Solution of Linear Systems by Matrices

Consider the linear system of equations

$$\begin{aligned} a_1x + b_1y + c_1z &= d_1 \\ a_2x + b_2y + c_2z &= d_2 \\ a_3x + b_3y + c_3z &= d_3. \end{aligned}$$

We can write the coefficients as a 3×3 matrix,

$$\begin{pmatrix} a_1 & b_1 & c_1 \\ a_2 & b_2 & c_2 \\ a_3 & b_3 & c_3 \end{pmatrix},$$

which we shall call the **coefficient matrix** of the system. If we adjoin to the coefficient matrix the column of constants, we have the **augmented matrix** of the system,

$$\begin{pmatrix} a_1 & b_1 & c_1 & d_1 \\ a_2 & b_2 & c_2 & d_2 \\ a_3 & b_3 & c_3 & d_3 \end{pmatrix}.$$

We can operate with the rows of this augmented matrix just as we would with the equations of a system of equations, since the augmented matrix is actually an abbreviated form of the system. We can perform any transformation of the matrix which will result in an equivalent system. Operations which produce such transformations are stated in the following theorem.

THEOREM 7.4 For any real number k and any augmented matrix of a system of linear equations, the following operations will result in the matrix of an equivalent system.

(a) Any two rows may be interchanged.

(b) The elements of any row may be multiplied by the nonzero real number k.

(c) Any row may be changed by adding to the elements of the row k times the elements of another row.

We can use Theorem 7.4 to solve a linear system. For example, to solve

$$\begin{aligned} x - y + 5z &= -6 \\ 3x + 3y - z &= 10 \\ x + 3y + 2z &= 5, \end{aligned}$$

we begin by writing the augmented matrix of the linear system,

$$\begin{pmatrix} 1 & -1 & 5 & -6 \\ 3 & 3 & -1 & 10 \\ 1 & 3 & 2 & 5 \end{pmatrix}.$$

We wish to use Theorem 7.4 to transform this matrix into one in which the value of the variables will be easy to see. That is, since each column in the matrix represents the coefficients of one variable, we wish to transform the matrix so that the augmented matrix is of the form

$$\begin{pmatrix} 1 & 0 & 0 & m \\ 0 & 1 & 0 & n \\ 0 & 0 & 1 & p \end{pmatrix}$$

(where m, n, and p are real numbers) from which, by rewriting the matrix as a linear system, we have

$$\begin{aligned} x \qquad\qquad &= m \\ y \qquad &= n \\ z &= p. \end{aligned}$$

In performing this transformation, it is best to work by columns beginning in each column with the element which is to become 1. In our example,

$$\begin{pmatrix} 1 & -1 & 5 & -6 \\ 3 & 3 & -1 & 10 \\ 1 & 3 & 2 & 5 \end{pmatrix},$$

we already have 1 in the desired position of the first column, so we begin by using Theorem 7.4(c) to change the second row so that the first element is zero. We do this by multiplying each element in row one by -3 and adding the result to the corresponding elements in row two to get the matrix

$$\begin{pmatrix} 1 & -1 & 5 & -6 \\ 0 & 6 & -16 & 28 \\ 1 & 3 & 2 & 5 \end{pmatrix}.$$

Now, to change the first element in row three to zero, we use Theorem 7.4(c) and multiply row one by -1 and add the results to row three.

$$\begin{pmatrix} 1 & -1 & 5 & -6 \\ 0 & 6 & -16 & 28 \\ 0 & 4 & -3 & 11 \end{pmatrix}$$

The same procedure is followed to transform columns two and three with one additional step necessary to transform the desired element in each column to 1, which should be done first.

$$\begin{pmatrix} 1 & -1 & 5 & -6 \\ 0 & 1 & \frac{-8}{3} & \frac{14}{3} \\ 0 & 4 & -3 & 11 \end{pmatrix}$$ row two multiplied by $\frac{1}{6}$ (Theorem 7.4(b))

$$\begin{pmatrix} 1 & 0 & \frac{7}{3} & \frac{-4}{3} \\ 0 & 1 & \frac{-8}{3} & \frac{14}{3} \\ 0 & 4 & -3 & 11 \end{pmatrix}$$ row one added to row two (Theorem 7.4(c))

$$\begin{pmatrix} 1 & 0 & \frac{7}{3} & \frac{-4}{3} \\ 0 & 1 & \frac{-8}{3} & \frac{14}{3} \\ 0 & 0 & \frac{23}{3} & \frac{-23}{3} \end{pmatrix}$$ -4 times row two added to row three (Theorem 7.4(b))

$$\begin{pmatrix} 1 & 0 & \frac{7}{3} & \frac{-4}{3} \\ 0 & 1 & \frac{-8}{3} & \frac{14}{3} \\ 0 & 0 & 1 & -1 \end{pmatrix}$$ row three multiplied by $\frac{3}{23}$ (Theorem 7.4(b))

$$\begin{pmatrix} 1 & 0 & 0 & 1 \\ 0 & 1 & \frac{-8}{3} & \frac{14}{3} \\ 0 & 0 & 1 & -1 \end{pmatrix}$$ $-\frac{7}{3}$ times row three added to row one (Theorem 7.4(c))

$$\begin{pmatrix} 1 & 0 & 0 & 1 \\ 0 & 1 & 0 & 2 \\ 0 & 0 & 1 & -1 \end{pmatrix}$$ $\frac{8}{3}$ times row three added to row two (Theorem 7.4(c))

The linear system associated with this final matrix is

$$\begin{aligned} x \qquad &= 1 \\ y \quad &= 2 \\ z &= -1, \end{aligned}$$

and the solution set is $\{(1, 2, -1)\}$.

Example 5 Use Theorem 7.4 to solve the linear system

$$\begin{aligned} 3x - 4y &= 1 \\ 5x + 2y &= 19. \end{aligned}$$

The augmented matrix is

$$\begin{pmatrix} 3 & -4 & 1 \\ 5 & 2 & 19 \end{pmatrix}.$$

We want the final matrix to be of the form

$$\begin{pmatrix} 1 & 0 & k \\ 0 & 1 & j \end{pmatrix}$$ for k and j real numbers.

The transformations are performed as follows.

$$\begin{pmatrix} 1 & \frac{-4}{3} & \frac{1}{3} \\ 5 & 2 & 19 \end{pmatrix} \quad \text{row one multiplied by } \tfrac{1}{3}$$

$$\begin{pmatrix} 1 & \frac{-4}{3} & \frac{1}{3} \\ 0 & \frac{26}{3} & \frac{52}{3} \end{pmatrix} \quad -5 \text{ times row one added to row two}$$

$$\begin{pmatrix} 1 & \frac{-4}{3} & \frac{1}{3} \\ 0 & 1 & 2 \end{pmatrix} \quad \text{row two multiplied by } \tfrac{3}{26}$$

$$\begin{pmatrix} 1 & 0 & 3 \\ 0 & 1 & 2 \end{pmatrix} \quad \tfrac{4}{3} \text{ times row two added to row one}$$

From the last matrix we get the system

$$\begin{aligned} x \quad &= 3 \\ y &= 2, \end{aligned}$$

which gives the solution set $\{(3, 2)\}$. Note that the solution can be read directly from the third column of the final matrix.

The method of solving linear systems described in this section is called the *Gaussian Method*. Since it is easily adapted for use on computers, it is more efficient than Cramer's Rule for solving large systems of equations.

7.3 Exercises

Use matrices to solve the following systems of equations.

1. $\begin{aligned} x + y &= 10 \\ 2x - 5y &= 5 \end{aligned}$

2. $\begin{aligned} 3x - 2y &= 4 \\ 3x + y &= -2 \end{aligned}$

3. $\begin{aligned} 2x - 3y &= 10 \\ 2x + 2y &= 5 \end{aligned}$

4. $\begin{aligned} 6x + y &= 5 \\ 5x + y &= 3 \end{aligned}$

5. $\begin{aligned} 2x - 5y &= 10 \\ 3x + y &= 15 \end{aligned}$

6. $\begin{aligned} 4x - y &= 3 \\ -2x + 3y &= 1 \end{aligned}$

7. $\begin{aligned} x + y - z &= 6 \\ 2x - y + z &= -9 \\ x - 2y + 3z &= 1 \end{aligned}$

8. $\begin{aligned} x + 3y - 6z &= 7 \\ 2x - y + z &= 1 \\ x + 2y + 2z &= -1 \end{aligned}$

9. $\begin{aligned} -x + y \quad &= -1 \\ y - z &= 6 \\ x \quad + z &= -1 \end{aligned}$

10. $\begin{aligned} x + y \quad\quad &= 1 \\ 2x \quad\quad - z &= 0 \\ y + 2z &= -2 \end{aligned}$

11. $\begin{aligned} 2x - y + 3z &= 0 \\ x + 2y - z &= 5 \\ 2y + z &= 1 \end{aligned}$

12. $\begin{aligned} 4x + 2y - 3z &= 6 \\ x - 4y + z &= -4 \\ -x \quad + 2z &= 2 \end{aligned}$

13. $\begin{aligned} 3x + 2y \quad - w &= 0 \\ 2x \quad + z + 2w &= 5 \\ x + 2y - z \quad &= -2 \\ 2x - y + z + w &= 2 \end{aligned}$

14. $\begin{aligned} x + 3y - 2z - w &= 9 \\ 4x + y + z + 2w &= 2 \\ -3x - y + z - w &= -5 \\ x - y - 3z - 2w &= 2 \end{aligned}$

7.4 Solutions of Linear Systems of Equations by Inverses

Another way to use matrices to solve linear systems is to write the system as a matrix equation $AX = B$, where X is a matrix of the variables of the system. As shown in the previous chapter, multiplying both sides of the matrix equation by A^{-1} gives

$$X = A^{-1}B.$$

Thus, to find X, we must find A^{-1} and then the product $A^{-1}B$. Setting equal the corresponding elements of X and $A^{-1}B$ gives the solution of the system.

Example 6 Use the inverse of the coefficient matrix to solve the linear system

$$\begin{aligned} 2x - 3y &= 4 \\ x + 5y &= 2. \end{aligned}$$

To represent the system as a matrix equation, we use one matrix for the coefficients, one for the variables, and one for the constants, as follows.

$$A = \begin{pmatrix} 2 & -3 \\ 1 & 5 \end{pmatrix}, \quad X = \begin{pmatrix} x \\ y \end{pmatrix} \quad \text{and} \quad B = \begin{pmatrix} 4 \\ 2 \end{pmatrix}$$

The system can then be written in matrix form as the equation $AX = B$, since

$$AX = \begin{pmatrix} 2 & -3 \\ 1 & 5 \end{pmatrix}\begin{pmatrix} x \\ y \end{pmatrix} = \begin{pmatrix} 2x - 3y \\ x + 5y \end{pmatrix} = \begin{pmatrix} 4 \\ 2 \end{pmatrix} = B.$$

To solve the system, we first find A^{-1} as follows.

$$A^{-1} = \frac{1}{13}\begin{pmatrix} 5 & 3 \\ -1 & 2 \end{pmatrix} = \begin{pmatrix} \frac{5}{13} & \frac{3}{13} \\ \frac{-1}{13} & \frac{2}{13} \end{pmatrix}$$

Next, we find the product $A^{-1}B$.

$$A^{-1}B = \begin{pmatrix} \frac{5}{13} & \frac{3}{13} \\ \frac{-1}{13} & \frac{2}{13} \end{pmatrix} \begin{pmatrix} 4 \\ 2 \end{pmatrix} = \begin{pmatrix} 2 \\ 0 \end{pmatrix}$$

Since $X = A^{-1}B$,

$$X = \begin{pmatrix} x \\ y \end{pmatrix} = \begin{pmatrix} 2 \\ 0 \end{pmatrix}.$$

Thus, the solution set of the system is $\{(2, 0)\}$.

Example 7 Use the inverse of the coefficient matrix to solve the system

$$\begin{aligned} 2x - 5y &= 20 \\ 3x + 4y &= 7. \end{aligned}$$

The matrices we need are

$$A = \begin{pmatrix} 2 & -5 \\ 3 & 4 \end{pmatrix}, \quad X = \begin{pmatrix} x \\ y \end{pmatrix}, \quad \text{and} \quad B = \begin{pmatrix} 20 \\ 7 \end{pmatrix}.$$

We begin by finding A^{-1}.

$$A^{-1} = \frac{1}{23}\begin{pmatrix} 4 & 5 \\ -3 & 2 \end{pmatrix} = \begin{pmatrix} \frac{4}{23} & \frac{5}{23} \\ -\frac{3}{23} & \frac{2}{23} \end{pmatrix}$$

We now find $A^{-1}B$.

$$A^{-1}B = \begin{pmatrix} \frac{4}{23} & \frac{5}{23} \\ -\frac{3}{23} & \frac{2}{23} \end{pmatrix} \begin{pmatrix} 20 \\ 7 \end{pmatrix} = \begin{pmatrix} 5 \\ -2 \end{pmatrix}$$

Then, since $X = A^{-1}B$, we have

$$X = \begin{pmatrix} x \\ y \end{pmatrix} = \begin{pmatrix} 5 \\ -2 \end{pmatrix},$$

and the solution of the system is $x = 5$, $y = -2$.

7.4 Exercises

Use the inverse of the coefficient matrix to solve the following systems of linear equations.

1. $\begin{aligned} 7x + 3y &= 5 \\ 2x + 4y &= 3 \end{aligned}$

2. $\begin{aligned} 5x + 2y &= 7 \\ 6x + y &= 8 \end{aligned}$

3. $\begin{aligned} x - 9y &= 4 \\ 3x + 2y &= 5 \end{aligned}$

4. $\begin{aligned} 12x - 8y &= 3 \\ 15x + 4y &= 9 \end{aligned}$

5. $2x - 10y = 5$
$7x + 6y = 8$

6. $5x + 3y = 5$
$-4x + y = 13$

7. $x + y = 10$
$2x - 5y = 5$

8. $4x - y = 3$
$-2x + 3y = 1$

9. $2x - 3y = 10$
$2x + 2y = 5$

10. $3x - 2y = 4$
$3x + y = -2$

7.5 Nonlinear Systems of Equations

A **nonlinear system of equations** is one in which at least one of the equations is not a first degree equation. Some nonlinear systems can be solved by the elimination method, using a theorem which is a generalization of Theorem 7.1.

THEOREM 7.5 If R and S are algebraic expressions in x and y, then (x, y) satisfies both $R = k_1$ and $S = k_2$ (k_1 and k_2 real numbers) if and only if (x, y) also satisfies the equation

$$m_1R + m_2S = m_1k_1 + m_2k_2,$$

for nonzero real numbers m_1 and m_2.

The proof parallels the proof of Theorem 7.1. To illustrate the use of Theorem 7.5, consider the system

$$x^2 + y^2 = 4 \qquad (12)$$

$$2x^2 - y^2 = 8. \qquad (13)$$

We can use Theorem 7.5 with $m_1 = m_2 = 1$ to get

$$\begin{aligned} (x^2 + y^2) + (2x^2 - y^2) &= 4 + 8 \\ 3x^2 &= 12 \\ x^2 &= 4 \\ x &= 2 \text{ or } x = -2. \end{aligned}$$

Then, by substitution into equation (12), we can find the corresponding value of y.

$$\begin{aligned} 2^2 + y^2 &= 4 \quad \text{or} \quad & (-2)^2 + y^2 &= 4 \\ y^2 &= 0 & y^2 &= 0 \\ y &= 0 & y &= 0 \end{aligned}$$

From this result, we see that the solution set of the system is $\{(2, 0), (-2, 0)\}$, as shown in the graph of Figure 7.2 on page 176.

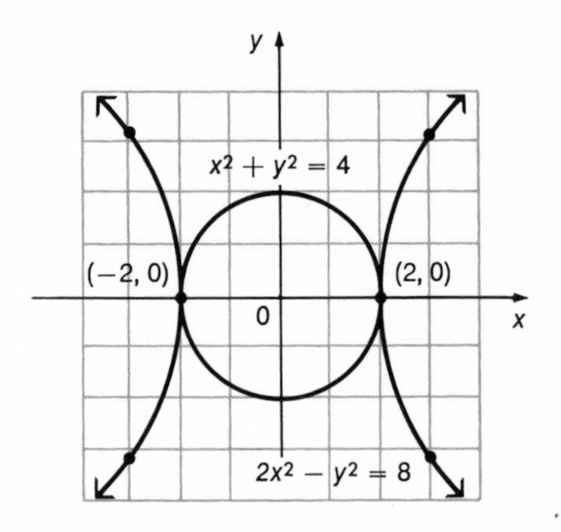

Figure 7.2

Another method for solving systems of equations, the **substitution method,** is particularly useful in solving a nonlinear system which includes a linear equation. For example, in the system

$$3x^2 - 2y = 5 \tag{14}$$

$$x + 3y = -4, \tag{15}$$

equation (15) can be solved for x and the resulting expression substituted into equation (14) as shown below.

$$x + 3y = -4 \tag{15}$$

$$x = -3y - 4$$

Substituting this value of x into equation (14) gives

$$\begin{aligned} 3x^2 - 2y &= 5 \qquad (14)\\ 3(-3y - 4)^2 - 2y &= 5\\ 3(9y^2 + 24y + 16) - 2y &= 5\\ 27y^2 + 70y + 43 &= 0. \end{aligned}$$

Using the quadratic formula, we have

$$y = \frac{-70 \pm \sqrt{4900 - 4644}}{54}$$

$$y = -1 \text{ or } y = \frac{-43}{27}.$$

Substituting these values for y into equation (15), we can find the corresponding values for x.

$$x + 3(-1) = -4 \qquad \text{or} \qquad x + 3\left(\frac{-43}{27}\right) = -4$$

$$x = -1 \qquad\qquad x = \frac{7}{9}$$

The solution set is $\{(-1, -1), (\frac{7}{9}, -\frac{43}{27})\}$, as shown on the graph of Figure 7.3.

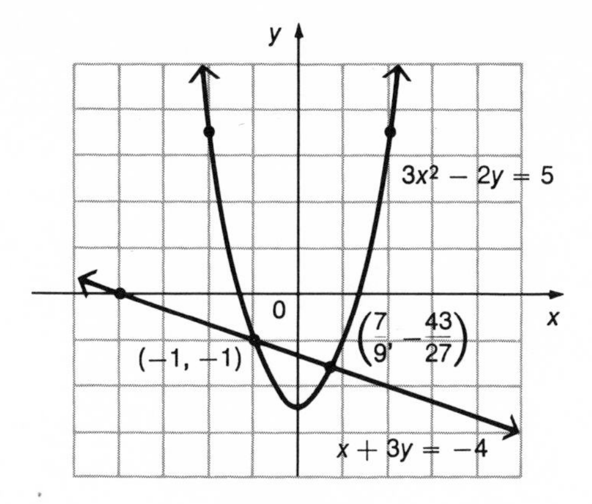

Figure 7.3

Sometimes a combination of the elimination method and the substitution method is effective in solving a system, as illustrated in Example 8.

Example 8 Solve the system

$$x^2 + 3xy + y^2 = 22 \tag{16}$$

$$x^2 - xy + y^2 = 6. \tag{17}$$

By Theorem 7.5, using $m_1 = 1$ and $m_2 = -1$, we have

$$\begin{array}{r} x^2 + 3xy + y^2 = 22 \\ -x^2 + xy - y^2 = -6 \\ \hline 4xy = 16. \end{array} \tag{18}$$

Now we solve equation (18) for either x or y and substitute the resulting expression into one of the given equations. Let us solve for y.

$$y = \frac{4}{x} \qquad (x \neq 0) \tag{19}$$

(Note that if $x = 0$ there is no value of y which satisfies the system.) Substituting for y in equation (17) and simplifying gives

$$x^2 - x\left(\frac{4}{x}\right) + \left(\frac{4}{x}\right)^2 = 6$$

$$x^2 - 4 + \frac{16}{x^2} = 6$$

$$x^4 - 4x^2 + 16 = 6x^2$$

$$x^4 - 10x^2 + 16 = 0$$

$$(x^2 - 2)(x^2 - 8) = 0$$

$$x^2 = 2 \text{ or } x^2 = 8$$

$$x = \sqrt{2} \text{ or } x = -\sqrt{2} \text{ or } x = 2\sqrt{2} \text{ or } x = -2\sqrt{2}.$$

We substitute these values of x into equation (19) to find the corresponding values for y.

$$\text{If } x = \sqrt{2},\ y = \frac{4}{\sqrt{2}} = 2\sqrt{2}.$$

$$\text{If } x = -\sqrt{2},\ y = -\frac{4}{\sqrt{2}} = -2\sqrt{2}.$$

$$\text{If } x = 2\sqrt{2},\ y = \frac{4}{2\sqrt{2}} = \sqrt{2}.$$

$$\text{If } x = -2\sqrt{2},\ y = -\frac{4}{2\sqrt{2}} = -\sqrt{2}.$$

From this we have the solution set of the system,

$$\{(\sqrt{2}, 2\sqrt{2}), (-\sqrt{2}, -2\sqrt{2}), (2\sqrt{2}, \sqrt{2}), (-2\sqrt{2}, -\sqrt{2})\}.$$

Example 9 Solve the system

$$x^2 + y^2 = 16 \qquad (20)$$

$$|x| + y = 4. \qquad (21)$$

The substitution method is required here. From the definition of absolute value, equation (21) can be rewritten as follows.

$$|x| = 4 - y \qquad (21)$$

$$x = 4 - y \text{ or } x = -(4 - y) = y - 4. \qquad (22)$$

(Since $|x| \geq 0$ for all real x, here we must have $4 - y \geq 0$, or $4 \geq y$.) Substituting from each part of the compound sentence into equation (20) we get the same equation.

$$(4 - y)^2 + y^2 = 16 \qquad \text{or} \qquad (y - 4)^2 + y^2 = 16$$
$$(16 - 8y + y^2) + y^2 = 16$$
$$2y^2 - 8y = 0$$
$$2y(y - 4) = 0$$
$$y = 0 \text{ or } y = 4.$$

From equation (22) we have

$$\text{If } y = 0,\ x = 4 - 0 \text{ or } x = 0 - 4$$
$$x = 4 \text{ or } x = -4$$
$$\text{If } y = 4,\ x = 4 - 4 \text{ or } x = 4 - 4$$
$$x = 0.$$

Thus, the solution set is $\{(4, 0), (-4, 0), (0, 4)\}$, as shown in Figure 7.4.

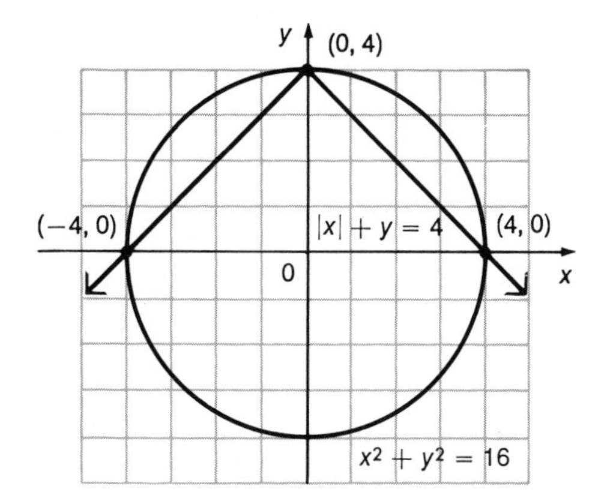

Figure 7.4

7.5 Exercises

Solve the following systems of equations.

1. $x^2 + y^2 = 13$
   $2x + y = 1$
2. $2x^2 - y^2 = 7$
   $y = 3x - 7$
3. $3x^2 + 2y^2 = 5$
   $x - y = -2$
4. $x^2 + y^2 = 5$
   $-3x + 4y = 2$
5. $5x^2 - y^2 = 9$
   $x^2 + y^2 = 45$
6. $x^2 + 3y^2 = 91$
   $2x^2 - y^2 = 119$
7. $2x^2 + 2y^2 = 20$
   $3x^2 + 3y^2 = 30$
8. $x^2 + y^2 = 4$
   $5x^2 + 5y^2 = 28$
9. $9x^2 + 4y^2 = 1$
   $x^2 + y^2 = 1$
10. $2x^2 - 3y^2 = 8$
    $6x^2 + 5y^2 = 24$
11. $x^2 + 2xy - y^2 = 14$
    $x^2 - y^2 = -16$
12. $3x^2 + xy + 3y^2 = 7$
    $x^2 + y^2 = 2$
13. $x^2 + 4y^2 = 25$
    $3xy = 18$
14. $5x^2 - 2y^2 = 6$
    $2xy = 4$
15. $x^2 - xy + y^2 = 5$
    $2x^2 + xy - y^2 = 10$
16. $3x^2 + 2xy - y^2 = 9$
    $x^2 - xy + y^2 = 9$
17. $x^2 + 2xy - y^2 + y = 1$
    $3x + y = 6$
18. $x^2 - 4x + y + xy = -10$
    $2x - y = 10$
19. $y = |x - 1|$
    $y = x^2 - 4$
20. $2x^2 - y^2 = 4$
    $|x| = |y|$

7.6 Systems of Inequalities

In Chapters 3 and 4, we saw that the solution set of an inequality in two variables is usually an infinite set whose graph is one or more regions of the coordinate plane. The solution set of a system of inequalities, such as

$$\begin{aligned} 4y &< -3x + 12 \\ x^2 &< 2y, \end{aligned}$$

is the intersection of the solution sets of its members, or

$$\{(x, y) \mid 4y < -3x + 12\} \cap \{(x, y) \mid x^2 < 2y\}.$$

Since this description of the solution set is not very informative, it is customary to find the graph of the solution set. To do this, we graph both inequalities on the same coordinate axes and identify, by shading, the region common to both graphs, as shown in Figure 7.5. Since the points on the boundary lines are not in the solution sets, these lines are dashed.

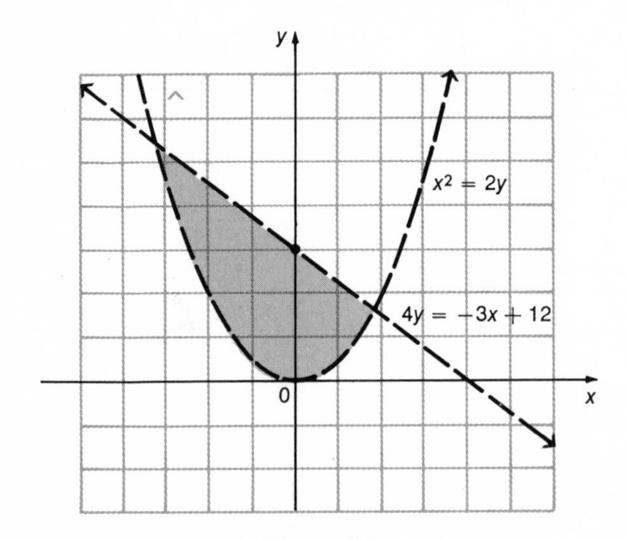

Figure 7.5

Example 10 Graph the solution set of the system

$$\begin{aligned} y &\geq 2^x \\ 9x^2 + 4y^2 &\leq 36 \\ 2x + y &< 1. \end{aligned}$$

The graph is obtained by graphing the three inequalities on the same axes and shading the region common to all three as shown in Figure 7.6. Note that two boundary lines are solid and one is dashed. The points where the dashed line intersects the solid line are shown as open circles since they do not belong to the solution set. The solution set is written as

$$\{(x, y)\,|\,y \geq 2^x\} \cap \{(x, y)\,|\,9x^2 + 4y^2 \leq 36\} \cap \{(x, y)\,|\,2x + y < 1\},$$

although in systems such as this a graph is more readily understood.

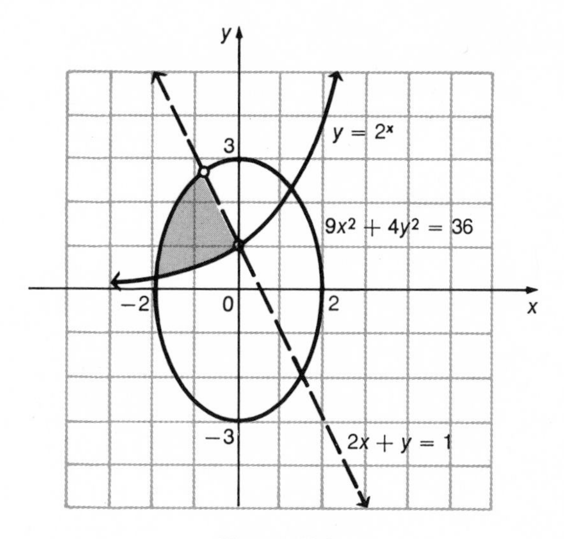

Figure 7.6

Example 11 Graph the solution set of the system

$$|x| \leq 3$$
$$y \leq 0$$
$$y \geq |x| + 1.$$

The graph is shown in Figure 7.7. From the graph, we see that the solution sets of $y \leq 0$ and $y \geq |x| + 1$ have no elements in common. Hence, the solution set of the system is $\varnothing$.

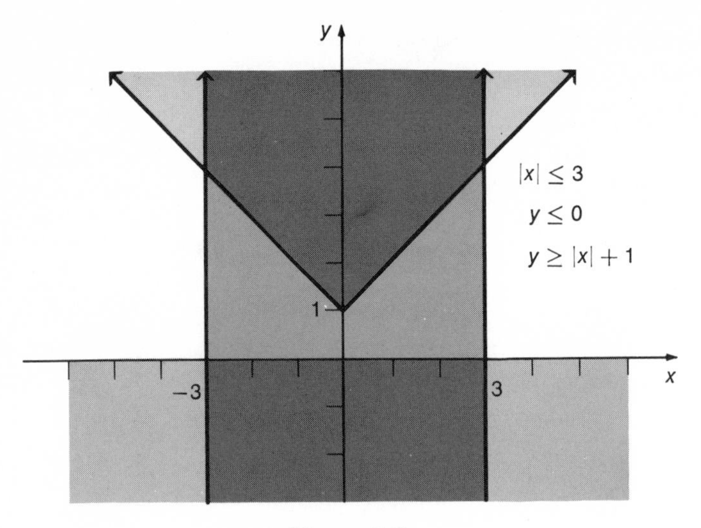

Figure 7.7

7.6 Exercises

Graph the solution set of each of the following systems of inequalities.

1. $x + 2y \leq 4$
   $y \geq x^2 - 1$

2. $x^2 + y^2 \leq 9$
   $-x \geq y^2$

3. $x^2 - y^2 < 1$
   $-1 < y < 1$

4. $2x^2 - y^2 > 4$
   $2y^2 - x^2 > 4$

5. $y \geq x^2 + 4x + 4$
   $y < -x^2$

6. $x^2 + 4y^2 \leq 16$
   $x + y < -1$

7. $x^2 + y^2 \leq 4$
   $x^2 + 4y^2 \geq 16$

8. $y \geq x^2 - 4$
   $y < 2$

9. $4x^2 + 9y^2 > 1$
   $x^2 - y^2 \geq 1$
   $-4 \leq x \leq 4$

10. $2x - 3y < 6$
    $x + y \leq 3$
    $x \geq -2$

11. $y \geq 3^x$
    $y \leq |5x| + 3$
    $|x| \leq 2$

12. $y \leq \log x$
    $|x - 8| \leq 3$
    $y \geq -1$

POLYNOMIAL FUNCTIONS

A **polynomial function** of degree n is a function of the form

$$P(x) = a_n x^n + a_{n-1}x^{n-1} + \cdots + a_1 x + a_0,$$

where $a_n \neq 0$. We call a_n the **leading coefficient** of $P(x)$. We have discussed two special types of polynomial functions, those of degree 1 (linear) and those of degree 2 (quadratic). We shall discuss polynomial functions in general in this chapter.

8.1 Operations on Complex Numbers

Virtually all our work so far in this text has been with real numbers. However, we know that not all polynomial functions have real number zeros ($x^2 + 1$ does not, for example). To find zeros for every polynomial function, we need a new set of numbers, called the set of complex numbers.

Let us define the number i as follows.

▶ $$\boldsymbol{i = \sqrt{-1}}$$

That is, we define i to be a number whose square is $-1(i^2 = -1)$. We define $\sqrt{-a}$, $a > 0$, by saying

▶ $$\boldsymbol{\sqrt{-a} = i\sqrt{a}} \qquad (a > 0)$$

Thus, $\sqrt{-16} = 4i$, $\sqrt{-75} = 5\sqrt{3}i$, and so on.

We shall call $a + bi$, a and b real numbers, a **complex number.** If $b \neq 0$, the complex number is also called an **imaginary number.** If $a = 0$ and $b \neq 0$, then the complex number is a nonzero multiple of i and is called a **pure imaginary number.** If $b = 0$, the complex number is of the form $a + 0i$ or just a, a real number, so that every real number is a complex number. Hence, both the real numbers and the pure imaginary numbers are subsets of the set of complex numbers. When a complex number is written in the form $a + bi$, where a and b are real numbers, the number is said to be written in **standard form.**

Let us define **equality** for complex numbers as

▶ $$a + bi = c + di \text{ if and only if } a = c \text{ and } b = d.$$

Thus, $2 + mi = k + 3i$ if and only if $2 = k$ and $m = 3$.

The complex numbers satisfy the field axioms. Some of these are verified later in this section, with the remainder left for the exercises. Assuming these properties, we can write

▶ $$(a + bi) + (c + di) = (a + c) + (b + d)i,$$

which defines the **sum** of two complex numbers. In a similar way,

$$\begin{aligned}(a + bi)(c + di) &= ac + adi + bic + bidi \\ &= ac + adi + cbi + bdi^2 \\ &= ac + (ad + cb)i + bd(-1)\end{aligned}$$

▶ $$(a + bi)(c + di) = (ac - bd) + (ad + bc)i,$$

which defines the **product** of two complex numbers.

Example 1 Find (a) the sum of $3 - 4i$ and $-2 + 6i$, (b) the product of $2 - 3i$ and $3 + 4i$.

(a) $(3 - 4i) + (-2 + 6i) = [3 + (-2)] + (-4 + 6)i = 1 + 2i$

(b) To use the definition to evaluate $(2 - 3i)(3 + 4i)$, we let $a = 2$, $b = -3$, $c = 3$, and $d = 4$. This gives

$$\begin{aligned}(2 - 3i)(3 + 4i) &= [2 \cdot 3 - (-3)4] + [2 \cdot 4 + (-3)3]i \\ &= (6 + 12) + (8 - 9)i \\ &= 18 - i.\end{aligned}$$

An alternate way to work Example 1(b), is to treat $2 - 3i$ and $3 + 4i$ as binomials, and multiply them as we did binomials, to get

$$\begin{aligned}(2 - 3i)(3 + 4i) &= 2 \cdot 3 + 2(4i) + (-3i)(3) + (-3i)(4i) \\ &= 6 + 8i - 9i - 12i^2 \\ &= 6 - i + 12 \\ &= 18 - i.\end{aligned}$$

Since $(a + bi) + (0 + 0i) = (a + 0) + (b + 0)i = a + bi$, for all complex numbers $a + bi$, we call $0 + 0i$ the **additive identity** for complex numbers. And since

$$\begin{aligned}(a + bi) + (-a - bi) &= [a + (-a)] + [b + (-b)]i \\ &= 0 + 0i,\end{aligned}$$

we call $-a - bi$ the **negative** or **additive inverse** of $a + bi$.

Using the definition of additive inverse, we define **subtraction** of complex numbers as follows.

▶ $$(a + bi) - (c + di) = (a + bi) + (-c - di)$$

Since $$\begin{aligned}(a + bi)(1 + 0i) &= (a \cdot 1 - b \cdot 0) + (b \cdot 1 + a \cdot 0)i \\ &= a + bi,\end{aligned}$$

we call $1 + 0i$ the **multiplicative identity** for complex numbers. What about a multiplicative inverse for $a + bi$? If we rewrite

$$\frac{1}{a + bi}$$

in the standard form for complex numbers (by multiplying numerator and denominator by $a - bi$), we have

$$\begin{aligned}\frac{1(a - bi)}{(a + bi)(a - bi)} &= \frac{a - bi}{a^2 + b^2} \\ &= \frac{a}{a^2 + b^2} - \frac{b}{a^2 + b^2}i.\end{aligned}$$

To check that this is indeed the multiplicative inverse of $a + bi$, we multiply

$$\begin{aligned}(a + bi)\left[\frac{a}{a^2 + b^2} - \frac{b}{a^2 + b^2}i\right] &= \frac{a(a + bi) - b(a + bi)i}{a^2 + b^2} \\ &= \frac{a^2 + abi - abi + b^2}{a^2 + b^2} \\ &= 1.\end{aligned}$$

Hence, the **multiplicative inverse** of $a + bi$ (a and b not both equal to 0) is

▶ $$\frac{a}{a^2 + b^2} - \frac{b}{a^2 + b^2}i.$$

As with real numbers, we write the quotient of two complex numbers $a + bi$ and $c + di$ as

$$\frac{a + bi}{c + di} = (a + bi)\left(\frac{1}{c + di}\right).$$

Since $\frac{1}{c+di}$ is the multiplicative inverse of $c + di$, we can write

$$\frac{a+bi}{c+di} = (a+bi)\left(\frac{c-di}{c^2+d^2}\right)$$

$$= \frac{(a+bi)(c-di)}{c^2+d^2}$$

▶ $$\frac{a+bi}{c+di} = \frac{ac+bd}{c^2+d^2} + \frac{bc-ad}{c^2+d^2}i.$$

This last equality is called the **quotient** of the complex numbers $a + bi$ and $c + di$.

In practice, to divide two complex numbers such as $3 - 2i$ and $-4 + 3i$, it is customary to multiply both numerator and denominator of

$$\frac{3-2i}{-4+3i}$$

by $-4 - 3i$, the **conjugate** of $-4 + 3i$. Doing this, we have

$$\frac{(3-2i)(-4-3i)}{(-4+3i)(-4-3i)} = \frac{-18-i}{25} = \frac{-18}{25} - \frac{1}{25}i.$$

We have seen that we can define addition and multiplication of complex numbers so that the associative, commutative, and distributive properties hold. In the exercises, you are asked to show that the complex numbers are closed under addition and multiplication. The set of complex numbers has identity elements for both addition and multiplication. Every complex number has an additive inverse, and every complex number (except $0 + 0i$) has a multiplicative inverse. Because complex numbers satisfy all these properties, we say that complex numbers form a field, and we speak of the **field of complex numbers.** We saw in Chapter 1 that the set of real numbers, together with the operations of addition and multiplication, forms a field. However, the field of complex numbers, unlike the field of real numbers, cannot be ordered, as shown in Exercise 50 of this section.

8.1 Exercises

Perform the following operations and express all results in standard form.

1. $(3 + 2i) + (4 - 3i)$
2. $(4 - i) + (2 + 5i)$
3. $(6 - 4i) - (3 + 2i)$
4. $(5 - 2i) - (5 + 3i)$
5. $(-2 + 3i) - (-4 + 2i)$
6. $(-3 + 5i) - (-4 + 3i)$
7. $(2 + i)(3 - 2i)$
8. $(-2 + 3i)(4 - 2i)$
9. $(2 + 4i)(-1 + 3i)$
10. $(1 + 3i)(2 - 5i)$
11. $(5 + 2i)(5 - 3i)$
12. $(-3 + 2i)^2$

13. $(2+i)^2$
14. $(\sqrt{6}-i)(\sqrt{6}+i)$
15. $(2-i)(2+i)$
16. $(5+4i)(5-4i)$
17. $i(3-4i)(3+4i)$
18. $i(2+5i)(2-5i)$
19. $i(3-4i)^2$
20. $i(2+6i)^2$
21. $\dfrac{1+i}{1-i}$
22. $\dfrac{2-i}{2+i}$
23. $\dfrac{4-3i}{4+3i}$
24. $\dfrac{5+6i}{5-6i}$
25. $\dfrac{4+i}{6+2i}$
26. $\dfrac{3-2i}{5+3i}$
27. $\dfrac{5-2i}{6-i}$
28. $\dfrac{3-4i}{2-5i}$
29. $\dfrac{1-3i}{1+i}$
30. $\dfrac{-3+4i}{2-i}$

Simplify (Hint: first calculate i^4.)

31. i^6
32. i^7
33. i^8
34. i^{19}
35. i^{85}
36. i^{78}

Simplify.

37. $\sqrt{-9}$
38. $\sqrt{-25}$
39. $\sqrt{-18}$
40. $\sqrt{-45}$
41. $\sqrt{-150}$
42. $\sqrt{-180}$

Prove that complex numbers satisfy the following properties.

43. Commutative property for addition.
44. Commutative property for multiplication.
45. Associative property for addition.
46. Associative property for multiplication.
47. Distributive property of multiplication over addition.
48. Closure property of addition.
49. Closure property of multiplication.
50. We know that the order relation for real numbers, $<$, satisfies three properties:

    (1) Trichotomy: if a and b are real numbers, then either $a < b$, $a = b$, or $a > b$.
    (2) If a, b, and c are real numbers and if $a < b$, then $a + c < b + c$.
    (3) If a and b are real numbers such that $0 < a$ and $0 < b$, then $0 < ab$.

    We shall now show that if a and b are complex numbers it is not possible to give a definition of $a < b$ in such a way that the three properties above still hold. We shall prove this by assuming that a valid meaning can be given to $a < b$ for a and b complex numbers, and then we shall show that such an assumption leads to an absurdity. Fill in the blanks in the following proof with one of the three properties from above.

THEOREM It is not possible to define an order relation $<$ for complex numbers in such a way that the three properties above still hold.

Proof

$i \neq 0$	$i = \sqrt{-1} \neq 0$
Either $0 < i$ or $i < 0$	____________
Let us assume $0 < i$	assumption
$0 < i^2$	Use $a = b = i$ in property ____
$0 < -1$	$i^2 = -1$

(Note: $0 < -1$ seems to be a contradiction. However, we have only assumed that $<$ can be somehow defined for complex numbers; we have not assumed that $<$ would then have the same meaning it does for real numbers.)

$1 < 0$	add 1 to both sides of the previous step by property ____
$0 < 1$	use $a = b = -1$ in property ____
$1 < 0$ and $0 < 1$	contradicts property ____

The assumption that $0 < i$ has led to a contradiction. In the same way, it can be shown that if we assume $i < 0$ we get another contradiction. Since $i \neq 0$, and since $i < 0$ and $0 < i$ lead to contradictions, we see it is not possible to define $<$ for complex numbers in such a way that the three properties still hold.

8.2 Complex Zeros of Quadratic Polynomials

In Chapter 4 we developed the quadratic formula for finding the zeros of a quadratic function. (Recall: a quadratic polynomial is of degree 2, and a zero of a function $f(x)$ is a value of x that makes $f(x) = 0$.) The properties used to prove the quadratic formula in Chapter 4 are equally valid for complex numbers, so we can now state a more general form of the quadratic formula.

THEOREM 8.1 QUADRATIC FORMULA

The zeros of $y = ax^2 + bx + c$ (a, b, and c complex numbers, $a \neq 0$), are given by

$$x = \frac{-b \pm \sqrt{b^2 - 4ac}}{2a}.$$

For example, to find the zeros of $y = x^2 + x + 1$, we note that $a = 1$, $b = 1$, and $c = 1$. Using the quadratic formula, we have

$$x = \frac{-1 \pm \sqrt{1 - 4}}{2}$$

$$= \frac{-1 \pm \sqrt{-3}}{2}.$$

Thus, the zeros of the function above can be written as the complex numbers

$$\frac{-1 + i\sqrt{3}}{2} \quad \text{and} \quad \frac{-1 - i\sqrt{3}}{2}.$$

Example 2 Find the zeros of $y = x^2 + 8x + 25$.

Here we have $a = 1$, $b = 8$, and $c = 25$. Using the quadratic formula, we can write

$$x = \frac{-8 \pm \sqrt{64 - 100}}{2}$$

$$= \frac{-8 \pm 6i}{2}$$

$$x = -4 + 3i \text{ or } x = -4 - 3i.$$

As mentioned in the quadratic formula in Theorem 8.1, the coefficients of the polynomial must be complex numbers, but not necessarily real numbers. For example, to find the zeros of $y = 2ix^2 - ix + 3i$ we first note that $a = 2i$, $b = -i$, and $c = 3i$. This gives

$$x = \frac{i \pm \sqrt{(-i)^2 - 4(2i)(3i)}}{4i}$$

$$= \frac{i \pm \sqrt{-1 - 24i^2}}{4i}$$

$$= \frac{i \pm \sqrt{-1 + 24}}{4i}$$

$$= \frac{i \pm \sqrt{23}}{4i}$$

$$= \frac{(i \pm \sqrt{23})i}{4i \cdot i}$$

$$= \frac{-1 \pm i\sqrt{23}}{-4}$$

$$= \frac{1}{4} \pm \frac{i\sqrt{23}}{4}.$$

8.2 Exercises

Find all zeros of each of the following.

1. $y = x^2 + 9$
2. $y = x^2 + 16$
3. $y = 4m^2 + 25$
4. $y = 9p^2 + 49$
5. $y = p^2 - 4p + 8$
6. $y = r^2 + 8r + 25$
7. $y = 4z^2 - 4z + 5$
8. $y = 9g^2 - 12g + 13$
9. $y = k^2 + 2k + 3$
10. $y = n^2 - 3n + 4$
11. $y = 2m^2 - m + 1$
12. $y = 2x^2 - 3x + 2$
13. $y = 3k^2 - k + 2$
14. $y = 5z^2 - 3z + 1$
15. $y = x^2 + ix + 1$
16. $y = 3m^2 - 2m + i$
17. $y = ip^2 - 2p + i$
18. $y = 2ix^2 - 3x + 3$
19. $y = (1 + i)x^2 - x + (1 - i)$
20. $y = (2 + i)r^2 - 3r + (2 - i)$

Write quadratic equations having the following numbers as zeros. (Hint: $y = (x - r_1)(x - r_2)$ has r_1 and r_2 as solutions.)

21. i and $-i$
22. $1 + i$ and $1 - i$
23. $2 + 3i$ and $2 - 3i$
24. $4 - i$ amd $4 + i$
25. $\sqrt{3}i$ and $-\sqrt{3}i$
26. $\sqrt{6}i$ and $-\sqrt{6}i$
27. $1 + \sqrt{2}i$ and $1 - \sqrt{2}i$
28. $\sqrt{3} + 2i$ and $\sqrt{3} - 2i$
29. $1 + i$ and $2 - 3i$
30. $4 + 2i$ and $1 - 3i$
31. $2 + 4i$ and $3 - 2i$
32. $5 + i$ and $2 + 3i$

8.3 The Factor Theorem

In this section we begin a discussion of the zeros of general polynomial functions having coefficients from the field of complex numbers (polynomials over the complex numbers). Recall from Section 2.2 that the process of synthetic division is useful when dividing a polynomial by a binomial of the form $x - k$, where k is any real number. For example, this process can be used to divide $3x^3 - 4x^2 - 5x - 25$ by $x - 3$, as follows.

$$\begin{array}{r|rrrr} 3 & 3 & -4 & -5 & -25 \\ & & 9 & 15 & 30 \\ \hline & 3 & 5 & 10 & 5 \end{array}$$

Hence: $3x^3 - 4x^2 - 5x - 25 = (3x^2 + 5x + 10)(x - 3) + 5$.

We can now extend this method to cases where k is any complex number, and to polynomials having complex coefficients. For example, to divide $x^3 + ix^2 + 4x + i$ by $x + 2i$, we can write

$$\begin{array}{r|rrrr} -2i & 1 & i & 4 & i \\ & & -2i & -2 & -4i \\ \hline & 1 & -i & 2 & -3i. \end{array}$$

From this, we can write

$$x^3 + ix^2 + 4x + i = (x^2 - ix + 2)(x + 2i) - 3i.$$

These examples suggest the division algorithm for polynomials, given in the following theorem.

THEOREM 8.2 DIVISION ALGORITHM

If $P(x)$ is a polynomial over the complex numbers, and k is any complex number, then there exists a unique polynomial $Q(x)$ over the complex numbers and a unique complex number r such that

$$P(x) = (x - k)Q(x) + r.$$

It is possible to obtain more information about the complex number r (the remainder) of the division algorithm. We know that if we divide $P(x)$ by $x - k$, we can write

$$P(x) = (x - k)Q(x) + r,$$

for some polynomial $Q(x)$ and complex number r. Since this last equality is true for all complex values of x, it is true for $x = k$. Thus, we can write

$$P(k) = (k - k)Q(k) + r$$
$$P(k) = r.$$

Hence we have proved the following theorem.

THEOREM 8.3 REMAINDER THEOREM

If $P(x)$ is a polynomial over the field of complex numbers, and if k is any complex number, then the remainder when $P(x)$ is divided by $x - k$ is equal to $P(k)$.

Thus, instead of evaluating $P(k)$ by substituting k for x in the polynomial $P(x)$, we can find $P(k)$ by dividing $P(x)$ by $x - k$. Then $P(k)$ equals the remainder obtained in the division.

Example 3 Let $P(x) = -x^4 + 3x^2 - 4x - 5$, and find $P(-2)$.

Using the remainder theorem and synthetic division, we have

$$\begin{array}{r|rrrrr} -2 & -1 & 0 & 3 & -4 & -5 \\ & & 2 & -4 & 2 & 4 \\ \hline & -1 & 2 & -1 & -2 & -1. \end{array}$$

Hence, $P(-2) = -1$, the remainder obtained when $P(x)$ is divided by $x - (-2) = x + 2$.

By the remainder theorem, if $P(k) = 0$, then $x - k$ is a factor of $P(x)$, and conversely, if $x - k$ is a factor of $P(x)$, then $P(k)$ must equal 0. Thus, we have proved the following theorem.

THEOREM 8.4 FACTOR THEOREM

If $P(x)$ is a polynomial over the complex numbers, and if k is any complex number, then $x - k$ is a factor of $P(x)$ if and only if $P(k) = 0$.

Example 4 Is $x - i$ a factor of $P(x) = 3x^3 + (-4 - 3i)x^2 + (5 + 4i)x - 5i$?

The only way $x - i$ can be a factor of $P(x)$ is for $P(i)$ to be 0. To see if this is the case, we use synthetic division.

$$\begin{array}{r|rrrr} i & 3 & -4-3i & 5+4i & -5i \\ & & 3i & -4i & 5i \\ \hline & 3 & -4 & 5 & 0 \end{array}$$

Since the remainder is 0, this means $P(i) = 0$, and hence $x - i$ is a factor of $P(x)$.

By the results above, any time we have a zero of $P(x)$ we also have found a factor of $P(x)$. That is, if k is a zero of $P(x)$, then $x - k$ is a factor of $P(x)$, and conversely.

8.3 Exercises

Use synthetic division to decide whether or not the given number is a zero of the given polynomial.

1. $2;\ x^2 + 2x - 8$
2. $-1;\ m^2 + 4m - 5$
3. $1 - i;\ y^2 - 2y + 2$
4. $3 - 2i;\ r^2 - 6r + 13$
5. $2 + i;\ k^2 + 3k + 4$
6. $1 - 2i;\ z^2 - 3z + 5$
7. $2;\ g^3 - 3g^2 + 4g - 4$
8. $-3;\ m^3 + 2m^2 - m + 6$
9. $4;\ 2r^3 - 6r^2 - 9r + 4$
10. $-6;\ 2y^3 + 9y^2 - 16y + 12$
11. $i;\ x^3 + 2ix^2 + 2x + i$
12. $-i;\ p^3 - ip^2 + 3p + 5i$
13. $1 + i;\ p^3 + 3p^2 - p + 1$
14. $2 - i;\ 2r^3 - r^2 + 3r - 5$

Evaluate $P(x)$ for the given polynomial and the given value of x.

15. $x = 3;\ P(x) = x^2 - 4x + 5$
16. $x = -2;\ P(x) = x^2 + 5x + 6$
17. $x = 2 + i;\ P(x) = x^2 - 5x + 1$
18. $x = 3 - 2i;\ P(x) = x^2 - x + 3$
19. $x = 1 - i;\ P(x) = x^3 + x^2 - x + 1$
20. $x = 2 - 3i;\ P(x) = x^3 + 2x^2 + x - 5$

Determine the value of k that makes the first polynomial divisible by the second.

21. $3x^3 - 12x^2 - 11x + k$; $x - 5$
22. $4x^3 + 6x^2 - 5x + k$; $x + 2$
23. $2x^4 + 5x^3 - 2x^2 + kx + 3$; $x + 3$
24. $5x^4 + 16x^3 - 15x^2 + kx + 16$; $x + 4$

8.4 Complex Zeros of General Polynomial Functions

We have seen that if a polynomial $P(x)$ can be divided by $x - k$, then $P(x) = 0$, and conversely. The next theorem shows that every polynomial function of degree greater than or equal to 1 has a zero, and thus that every such polynomial can be factored.

THEOREM 8.5 THE FUNDAMENTAL THEOREM OF ALGEBRA
If $P(x)$ is a polynomial of degree at least 1 over the complex numbers, there exists at least one complex number k such that $P(k) = 0$.

This theorem was first proved by the mathematician Gauss in his doctoral thesis in 1799 when he was 20. Although many proofs of this result have been given, none of them involve only the ideas of this text, and hence no proof is included here. From the fundamental theorem and the factor theorem, we have the following result.

THEOREM 8.6 If $P(x)$ is a polynomial of degree n ($n \geq 1$) over the complex numbers, there exists a complex number k and a polynomial $Q(x)$ over the complex numbers such that

$$P(x) = (x - k)Q(x).$$

We could also use the factor theorem and the fundamental theorem to factor the polynomial $Q(x)$ of Theorem 8.6. If we continue this process a total of n times, we get the result of the next theorem.

THEOREM 8.7 If $P(x)$ is a polynomial of degree n ($n \geq 1$) over the complex numbers, then there exist n complex numbers k_1, $k_2, \cdots, k_n$, which need not be distinct, such that

$$P(x) = a(x - k_1)(x - k_2) \cdots (x - k_n),$$

where a is the leading coefficient in $P(x)$.

Using this result, we can prove that a polynomial of degree $n (n \geq 1)$ has at most n distinct zeros. To prove this, we use an indirect proof. Thus, suppose a polynomial $P(x)$ of degree n $(n \geq 1)$ has $n + 1$ distinct zeros. Let us label the distinct zeros $k_1, k_2, k_3, \cdots, k_n$, and k. By Theorem 8.7, we can use the n zeros $k_1, k_2, \cdots, k_n$ to write

$$P(x) = a(x - k_1)(x - k_2) \cdots (x - k_n),$$

where a is the leading coefficient in $P(x)$. If we now let $x = k$, we have

$$P(k) = a(k - k_1)(k - k_2) \cdots (k - k_n). \tag{1}$$

We have assumed that k is also a zero of $P(x)$, so that $P(k) = 0$. Using this fact, and equation (1), we have

$$0 = a(k - k_1)(k - k_2) \cdots (k - k_n). \tag{2}$$

We assumed $k_1, k_2, \cdots, k_n$, and k were $n + 1$ *distinct* zeros of $P(x)$. Hence, $k - k_1 \neq 0$, $k - k_2 \neq 0$, $\cdots$, $k - k_n \neq 0$. By the definition of $P(x)$, $a \neq 0$. Thus, the product

$$a(k - k_1)(k - k_2) \cdots (k - k_n) \neq 0. \tag{3}$$

(Here we assume without proof that the product of n nonzero complex numbers is nonzero.) Using (2) and (3), we have

$$0 = a(k - k_1)(k - k_2) \cdots (k - k_n) \neq 0.$$

This is a contradiction. Hence, our original assumption is invalid, and we have the following result.

THEOREM 8.8 If $P(x)$ is a polynomial of degree n $(n \geq 1)$ over the complex numbers, then there exist at most n distinct complex zeros of $P(x)$.

Note that there exist *at most n* distinct zeros. For example, the polynomial $P(x) = x^3 + 3x^2 + 3x + 1 = (x + 1)^3$ is of degree 3 but has only one zero, $x = -1$.

Using the results of the previous section, we can show that both $2 + i$ and $2 - i$ are zeros of $P(x) = x^3 - x^2 - 7x + 15$. It is not just coincidence that both $2 + i$ and its conjugate $2 - i$ are zeros of this polynomial. We can prove that if $a + bi$ is a zero of a polynomial function over the reals, then so is $a - bi$.

To see this, we need the following properties of complex conjugates. Let $z = a + bi$, and write $\bar{z}$ for the conjugate of z, $\bar{z} = a - bi$. We leave the proof of the following equalities for the exercises.

▶ For any complex numbers c and d,

(a) $\overline{c + d} = \bar{c} + \bar{d}$

(b) $\overline{cd} = \bar{c} \cdot \bar{d}$

(c) $\overline{c^n} = (\bar{c})^n$.

Now consider the polynomial over the *reals*

$$P(x) = a_n x^n + a_{n-1}x^{n-1} + \cdots + a_1 x + a_0,$$

where $a_n \neq 0$. If $z = a + bi$ is a zero of $P(x)$, we can write

$$P(z) = a_n z^n + a_{n-1}z^{n-1} + \cdots + a_1 z + a_0 = 0.$$

If we take the conjugate of both sides of this last equation, we have

$$\overline{a_n z^n + a_{n-1}z^{n-1} + \cdots + a_1 z + a_0} = \bar{0}.$$

By the properties of complex conjugates mentioned above, this becomes

$$\overline{a_n z^n} + \overline{a_{n-1}z^{n-1}} + \cdots + \overline{a_1 z} + \overline{a_0} = \bar{0}$$
$$\bar{a}_n \cdot \overline{z^n} + \bar{a}_{n-1} \cdot \overline{z^{n-1}} + \cdots + \bar{a}_1 \cdot \bar{z} + \bar{a}_0 = 0$$

(Here we use property (b), and the fact that for a real number a, $\bar{a} = a$.)

$$a_n(\bar{z})^n + a_{n-1}(\bar{z})^{n-1} + \cdots + a_1(\bar{z}) + a_0 = 0.$$

Hence, $\bar{z}$ is also a zero of $P(x)$, and we have proved the following result.

THEOREM 8.9 If $P(x)$ is a polynomial over the reals, and if $a + bi$ is a zero of $P(x)$, then so is $a - bi$.

Note the importance of the requirement that the polynomial have real coefficients. For example, $P(x) = x - (1 + i)$ has $1 + i$ as a zero, but the conjugate $1 - i$ is not a zero.

This last result is important in helping predict the number of real zeros of a polynomial over the reals. A polynomial of odd degree $n (n \geq 1)$ must have at least one real zero (since we have just seen that complex zeros occur in pairs). On the other hand, a polynomial of even degree $n (n \geq 2)$ need have no real zeros, but on the other hand, may have up to n real zeros.

For example, suppose we want to find all zeros of $x^4 - 7x^3 + 18x^2 - 22x + 12$, given that $x = 1 - i$ is a zero. Since $x = 1 - i$ is a zero, we know by Theorem 8.9 that the conjugate $x = 1 + i$ is also a zero. Thus, we should first divide the original polynomial by $1 - i$ and then divide the quotient by $1 + i$, as follows.

$1 - i$	1	-7	18	-22	12
		$1 - i$	$-7 + 5i$	$16 - 6i$	-12
	1	$-6 - i$	$11 + 5i$	$-6 - 6i$	0

$1 + i$	1	$-6 - i$	$11 + 5i$	$-6 - 6i$
		$1 + i$	$-5 - 5i$	$6 + 6i$
	1	-5	6	0

Now we must find the zeros of the quadratic polynomial $x^2 - 5x + 6$. By factoring, we find the solutions to be 2 and 3, so that the four zeros of the original polynomial are $1 - i$, $1 + i$, 2, and 3.

8.4 Exercises

For each of the following, find a polynomial of lowest degree with real coefficients having the given zeros.

1. $5 + i$ and $5 - i$
2. $3 - 2i$ and $3 + 2i$
3. $2, 1 - i$ and $1 + i$
4. $-3, 2 - i$, and $2 + i$
5. $1 + \sqrt{2}, 1 - \sqrt{2}$, and 1
6. $1 - \sqrt{3}, 1 + \sqrt{3}$, and -2
7. $2 + i, 2 - i, 3$, and -1
8. $3 + 2i, 3 - 2i, -1$, and 2
9. 2, and $3 + i$
10. -1 and $4 - 2i$
11. $1 + \sqrt{2}$ and $1 - i$
12. $\sqrt{3} + 2$ and $2 + 3i$

For each of the following polynomials, one zero is given. Find all others.

13. $x^3 - x^2 - 4x - 6;\ x = 3$
14. $x^3 - 5x^2 + 17x - 13;\ x = 1$
15. $x^3 + x^2 - 4x - 24;\ x = -2 + 2i$
16. $x^3 + x^2 - 20x - 50;\ x = -3 + i$
17. $2x^3 - 2x^2 - x - 6;\ x = 2$
18. $2x^3 - 5x^2 + 6x - 2;\ x = 1 + i$
19. $x^4 + 5x^2 + 4;\ x = -i$
20. $x^4 + 10x^3 + 27x^2 + 10x + 26;\ x = i$
21. $x^4 - 3x^3 + 6x^2 + 2x - 60;\ x = 1 + 3i$
22. $x^4 - 6x^3 - x^2 + 86x + 170;\ x = 5 + 3i$

Prove each of the following statements for any complex numbers c and d.

23. $\overline{c + d} = \bar{c} + \bar{d}$
24. $\overline{cd} = \bar{c} \cdot \bar{d}$
25. $\overline{(c^n)} = (\bar{c})^n$

8.5 Rational Zeros of Polynomial Functions

We know by the fundamental theorem of algebra that every polynomial of degree at least 1 has a zero. However, the fundamental theorem merely asserts that such a zero exists, but gives no help at all in identifying zeros. For second degree polynomials we can use the quadratic formula. Similar but more complicated formulas exist for finding zeros of third and fourth degree polynomial functions. It was proved by the Norwegian mathematician Abel, at age 22, that it is not possible to produce a formula that will find zeros of fifth degree or higher polynomials.

In practice, however, it is ordinarily sufficient to find only rational zeros, or decimal approximations of any irrational zeros. In this section we shall develop a necessary condition for a rational number to be a zero of a polynomial function, and then in the next section we shall discuss approximations of irrational zeros.

Now consider a polynomial function $P(x)$ with integer coefficients. The next theorem gives a useful test for determining whether or not a given rational number is a possible zero of $P(x)$.

THEOREM 8.10 Let $P(x) = a_nx^n + a_{n-1}x^{n-1} + \cdots + a_1x + a_0$ $(a_n \neq 0)$ be a polynomial with integer coefficients. If p/q is a rational number written in lowest terms with the property that $P(p/q) = 0$, then p is a factor of a_0 and q is a factor of a_n.

Since p/q is a zero of $P(x)$, we can write

$$a_n\left(\frac{p}{q}\right)^n + a_{n-1}\left(\frac{p}{q}\right)^{n-1} + \cdots + a_1\left(\frac{p}{q}\right) + a_0 = 0.$$

This can also be written as

$$a_n \cdot \frac{p^n}{q^n} + a_{n-1} \cdot \frac{p^{n-1}}{q^{n-1}} + \cdots + a_1 \cdot \frac{p}{q} + a_0 = 0.$$

If we multiply both sides of this last result by q^n, and add $-a_0q^n$ to both sides, we have

$$a_np^n + a_{n-1}p^{n-1}q + \cdots + a_1pq^{n-1} = -a_0q^n.$$

Factoring out p gives

$$p(a_np^{n-1} + a_{n-1}p^{n-2}q + \cdots + a_1q^{n-1}) = -a_0q^n.$$

By this last result, p must be a factor of $-a_0q^n$ (why?). Since we have assumed that p/q is written in lowest terms, p and q have no common factor other than 1. Hence, p is not a factor of q^n. Thus, we must conclude that p is a factor of a_0. In a similar way, we can show that q is a factor of a_n.

Example 5 Find all rational zeros of $P(x) = 2x^4 - 11x^3 + 14x^2 - 11x + 12$.

If p/q is to be a rational zero of $P(x)$, we know by Theorem 8.10 that p must be a factor of $a_0 = 12$, and q must be a factor of $a_4 = 2$. Hence, p must be ± 1, ± 2, ± 3, ± 4, ± 6, or ± 12, while q must be ± 1 or ± 2. From this, we see that any rational zero of $P(x)$ will come from the list

$$\pm 1, \pm \tfrac{1}{2}, \pm 2, \pm 3, \pm \tfrac{3}{2}, \pm 4, \pm 6, \text{ or } \pm 12.$$

Note that we have no assurance that any of these numbers are zeros. However, if $P(x)$ has any rational zeros at all, they will be in the list above. We can check these proposed zeros by synthetic division. By trial and error we can find that $x = 4$ is a zero.

$$\begin{array}{r|rrrrr} 4 & 2 & -11 & 14 & -11 & 12 \\ & & 8 & -12 & 8 & -12 \\ \hline & 2 & -3 & 2 & -3 & 0 \end{array}$$

As a fringe benefit of this calculation, we now only need to look for zeros of the simpler polynomial $Q(x) = 2x^3 - 3x^2 + 2x - 3$. By Theorem 8.10 any rational zero of $Q(x)$ will have a numerator of ± 3 or ± 1, with a denominator of ± 1 or ± 2. Hence, any rational zeros of $Q(x)$ will come from the list

$$\pm 3, \pm \tfrac{3}{2}, \pm 1, \pm \tfrac{1}{2}.$$

Again we use synthetic division and trial and error to find that $x = \frac{3}{2}$ is a zero.

$$\begin{array}{r|rrrr} \frac{3}{2} & 2 & -3 & 2 & -3 \\ & & 3 & 0 & 3 \\ \hline & 2 & 0 & 2 & 0 \end{array}$$

The quotient is $2x^2 + 2$, which, by the quadratic formula, has i and $-i$ as zeros. Hence, the solution set of the polynomial function

$$P(x) = 2x^4 - 11x^3 + 14x^2 - 11x + 12$$

is given by $\{4, \frac{3}{2}, i, -i\}$.

To find any rational zeros of a polynomial with rational coefficients, first multiply the polynomial by a number that will clear it of all fractional coefficients. Then use Theorem 8.10.

8.5 Exercises

Find all rational zeros of the following polynomials.

1. $x^3 - 2x^2 - 13x - 10$
2. $x^3 + 5x^2 + 2x - 8$
3. $x^3 + 6x^2 - x - 30$
4. $x^3 - x^2 - 10x - 8$
5. $x^3 + 9x^2 - 14x - 24$
6. $x^3 + 3x^2 - 4x - 12$

7. $x^4 + 9x^3 + 21x^2 - x - 30$
8. $x^4 + 4x^3 - 7x^2 - 34x - 24$
9. $6x^3 + 17x^2 - 31x - 12$
10. $15x^3 + 61x^2 + 2x - 8$
11. $12x^3 + 20x^2 - x - 6$
12. $12x^3 + 40x^2 + 41x + 12$
13. $2x^3 + 7x^2 + 12x - 8$
14. $2x^3 + 20x^2 + 68x - 40$
15. $x^4 + 4x^3 + 3x^2 - 10x + 50$
16. $x^4 - 2x^3 + x^2 + 18$
17. Show that $x^2 - 2$ has no rational zeros, so that $\sqrt{2}$ must be irrational.

8.6 Approximate Zeros of Polynomial Functions

We know that every polynomial function of degree at least 1 has a zero. However, we do not know whether or not the function has real zeros; and even if it does have real zeros, we often have no way to find them. In this section we investigate methods of approximating any real zeros a polynomial function may have.

Much of our work in locating real zeros uses the following result, which follows from the fact that graphs of polynomial functions are continuous, with no gaps or sudden jumps.

THEOREM 8.11 If $P(x)$ is a polynomial function over the reals, a and b are real numbers such that $a < b$ and such that $P(a)$ and $P(b)$ are opposite in sign, then there exists a real number c, $a < c < b$, such that $P(c) = 0$.

The next theorem gives a good method for narrowing the search for real zeros.

THEOREM 8.12 If $P(x) = a_n x^n + a_{n-1}x^{n-1} + \cdots + a_1 x + a_0$ is a polynomial over the reals of degree $n \geq 1$, if $a_n > 0$, and if $P(x)$ is divided synthetically by $x - c$, then we have:

(a) if $c > 0$ and all numbers in the bottom row of the synthetic division are nonnegative, then c is an upper bound for the zeros of $P(x)$.

(b) if $c < 0$, and if the numbers in the bottom row of the synthetic division alternate in sign (with 0 considered positive or negative, as needed), then c is a lower bound for the zeros of $P(x)$.

Example 6 Approximate the real zeros of $P(x) = x^4 - 6x^3 + 8x^2 + 2x - 1$.

Let us begin to look for zeros by trying $c = -2$. To do this, we divide $P(x)$ by $x + 2$.

-2	1	-6	8	2	-1
		-2	16	-48	92
	1	-8	24	-46	91

Since the numbers in the bottom row alternate in sign, and since $-2 < 0$, we know by Theorem 8.12(b) that $x = -2$ is a lower bound for the zeros of $P(x)$. If we divide $P(x)$ by $x + 1$, we get

-1	1	-6	8	2	-1
		-1	7	-15	13
	1	-7	15	-13	12.

By this result, $x = -1$ is also a lower bound for any real zeros of $P(x)$. Note that $P(-1) = 12 > 0$, while $P(0) = -1 < 0$. Hence, there is at least one real number zero between $x = -1$ and $x = 0$.

Let us try $c = -.5$. If we divide $P(x)$ by $x + .5$, we have

$-.5$	1	-6	8	2	-1
		$-\ .5$	3.25	-5.625	1.8125
	1	-6.5	11.25	-3.625	.8125.

Since $P(-.5) > 0$ and $P(0) < 0$, the real number zero is between $-.5$ and 0.

Now try $c = -.4$.

$-.4$	1	-6	8	2	-1
		$-\ .4$	2.56	-4.224	.8896
	1	-6.4	10.56	-2.224	$-.1104$

Note that $P(-.5) > 0$, while $P(-.4) < 0$. Since $P(-.4)$ is much closer to zero than $P(-.5)$, we are probably safe in saying that, to one decimal place of accuracy, one real number zero of $P(x)$ is $x = -.4$. Further decimal places of accuracy could be obtained by continuing this process. In the same way, we can show that $P(x)$ has three more real zeros; to one decimal of accuracy these three other zeros are .3, 2.4, and 3.7.

Descartes' rule of signs, stated in the next theorem, gives a useful, practical test for determining the number of positive or negative real zeros of a given polynomial.

THEOREM 8.13 DESCARTES' RULE OF SIGNS

Let $P(x)$ be a polynomial over the reals.

(a) The number of positive real zeros of $P(x)$ is either equal to the number of variations in sign occurring in the coefficients of $P(x)$, or else is less than the number of variations by a positive even integer.

(b) The number of negative real zeros of $P(x)$ either equals the number of variations in sign of $P(-x)$, or else is less than the number of variations by a positive even integer.

For the purposes of this theorem, $(x - 1)^4$ is said to have four positive zeros, each equal to 1. As an example, $P(x) = x^4 - 6x^3 + 8x^2 + 2x - 1$ has three variations in sign:

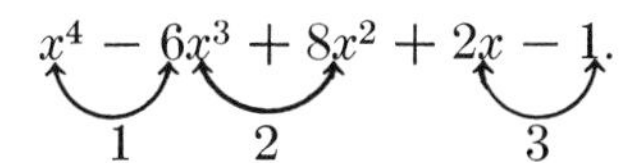

Thus, by Descartes' rule of signs, $P(x)$ has either 3 or $3 - 2 = 1$ positive real zeros. We found above that $P(x)$ has three positive real zeros. Since $P(x)$ is of degree 4, and has 3 positive real zeros, we know it must have 1 negative real zero, which also corresponds to the result above. We could also verify this with part (b) of Descartes' rule of signs. Note that

$$\begin{aligned} P(-x) &= (-x)^4 - 6(-x)^3 + 8(-x)^2 + 2(-x) - 1 \\ &= x^4 + 6x^3 + 8x^2 - 2x - 1 \end{aligned}$$

has only one variation in sign. Hence, $P(x)$ has only one negative real zero, which again corresponds to our result from above.

As a further example, note that

$$Q(x) = x^5 + 5x^4 + 3x^2 + 2x + 1$$

has no variations in sign and hence has no positive real zeros. Here

$$Q(-x) = -x^5 + 5x^4 + 3x^2 - 2x + 1,$$

which has three variations in sign. Hence, $Q(x)$ has either 3 or 1 negative real zeros.

8.6 Exercises

Show that each of the following polynomials has a real zero between the numbers given.

1. $x^3 + 3x^2 - 2x - 6$; 1 and 2
2. $x^3 + x^2 - 5x - 5$; 2 and 3
3. $2x^3 - 8x^2 + x + 16$; 2 and 2.5
4. $3x^3 + 7x^2 - 4$; $\frac{1}{2}$ and 1

Show that the real zeros of each of the following polynomials have the indicated bounds.

5. $x^4 - x^3 + 3x^2 - 8x + 8$; no real zeros greater than 2
6. $2x^5 - x^4 + 2x^3 - 2x^2 + 4x - 4$; no real zero greater than 1
7. $x^4 + x^3 - x^2 + 3$; no real zero less than -2
8. $x^5 + 2x^3 - 2x^2 + 5x + 5$; no real zero less than -1

Find all real zeros of each of the following polynomials. Approximate each zero as a decimal to the nearest tenth.

9. $x^3 + 3x^2 - 2x - 6$
10. $x^3 + x^2 - 5x - 5$
11. $x^3 - 4x^2 - 5x + 14$
12. $x^3 + 9x^2 + 34x + 13$
13. $x^3 + 6x - 13$
14. $x^3 - 3x^2 + 4x - 5$
15. $x^4 - 8x^3 + 17x^2 - 2x - 14$
16. $x^4 - 4x^3 - x^2 + 8x - 2$

8.7 Graphing Polynomial Functions

We have graphed many linear functions (straight lines) and we have graphed many quadratic functions (parabolas). Now we shall investigate the graphs of more general polynomial functions. We shall find that point plotting is more important here than with the other graphs mentioned.

As an example, let us graph the function $P(x) = 8x^3 - 12x^2 + 2x + 1$. By Descartes' rule of signs, $P(x)$ has 2 or 0 positive real zeros, and 1 negative real zero. We need to find some ordered pairs belonging to the graph. We can use synthetic division to evaluate, say, $P(3)$.

$$\begin{array}{r|rrrr} 3 & 8 & -12 & 2 & 1 \\ & & 24 & 36 & 114 \\ \hline & 8 & 12 & 38 & 115 \end{array}$$

Hence, $P(3) = 115$, and $(3, 115)$ belongs to the graph. We can also find $P(-1)$.

$$\begin{array}{r|rrrr} -1 & 8 & -12 & 2 & 1 \\ & & -8 & 20 & -22 \\ \hline & 8 & -20 & 22 & -21 \end{array}$$

From this result, $(-1, -21)$ belongs to the graph. Since we need several such points in order to sketch the graph, it is helpful to use the shortened form of synthetic division shown below, in which we find values of $P(x)$ corresponding to six different values of x. The numbers printed in bold face below are the numbers we obtained in the bottom row of the synthetic divisions above.

x				$P(x)$	ordered pair
	8	-12	2	1	
3	**8**	**12**	**38**	**115**	(3, 115)
2	8	4	10	21	(2, 21)
1	8	-4	-2	-1	$(1, -1)$
0	8	-12	2	1	(0, 1)
-1	**8**	**−20**	**22**	**−21**	$(-1, -21)$
-2	8	-28	58	-115	$(-2, -115)$

By Theorem 8.11, there is a zero between 0 and 1, and between -1 and 0, as well as between 1 and 2. To get the graph, we plot the points of the chart above, and then draw a continuous curve through them, as in Figure 8.1.

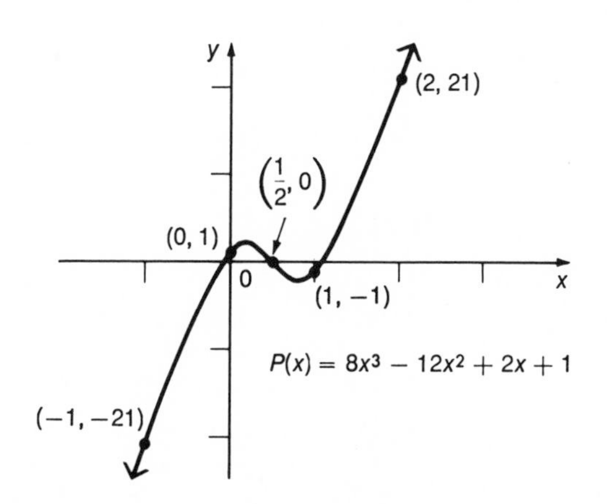

Figure 8.1

Example 7 Graph $P(x) = 3x^4 - 14x^3 + 54x - 3$.

To find points to plot we can again use the abbreviated form of synthetic division presented above.

x					$P(x)$	ordered pair
	3	-14	0	54	-3	
5	3	1	5	79	392	(5, 392)
4	3	-2	-8	22	85	(4, 85)
3	3	-5	-15	9	24	(3, 24)
2	3	-8	-16	22	41	(2, 41)
1	3	-11	-11	43	40	(1, 40)
0	3	-14	0	54	-3	(0, -3)
-1	3	-17	17	37	-40	(-1, -40)
-2	3	-20	40	-26	49	(-2, 49)
-3	3	-23	69	-153	456	(-3, 456)

Since the row in the chart for $x = 5$ contains all positive numbers, the polynomial has no zero greater than 5. Also, since the row for $x = -2$ has numbers which alternate in sign, there is no zero less than -2, by Theorem 8.12. We see here that the polynomial has zeros between 0 and 1, and between -2 and -1. If we plot the points found above and draw a continuous curve through them, we get the graph of Figure 8.2.

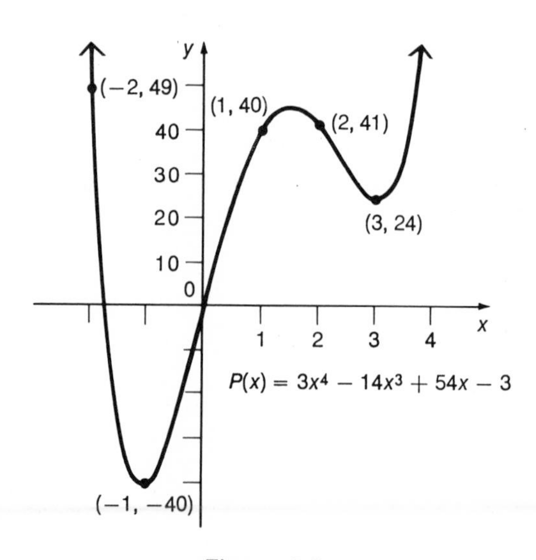

Figure 8.2

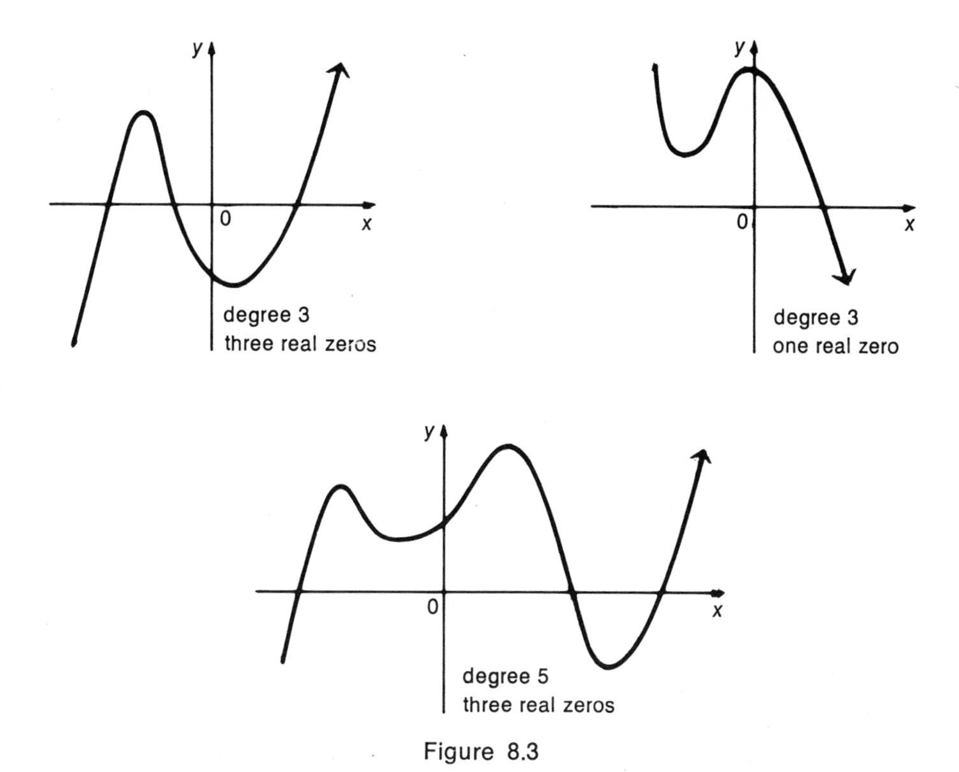

Figure 8.3

In general, the domain of a polynomial function is the set of all real numbers. The range of a polynomial function of odd degree is also the set of all real numbers. Some typical graphs of polynomial functions of odd degree are shown in Figure 8.3. These graphs illustrate that every polynomial function of odd degree has at least one real zero.

Polynomial functions of even degree have a range that takes the form $\{y \mid y \leq k\}$ or $\{y \mid y \geq k\}$ for some real number k. Figure 8.4 shows two typical graphs of polynomial functions of even degree.

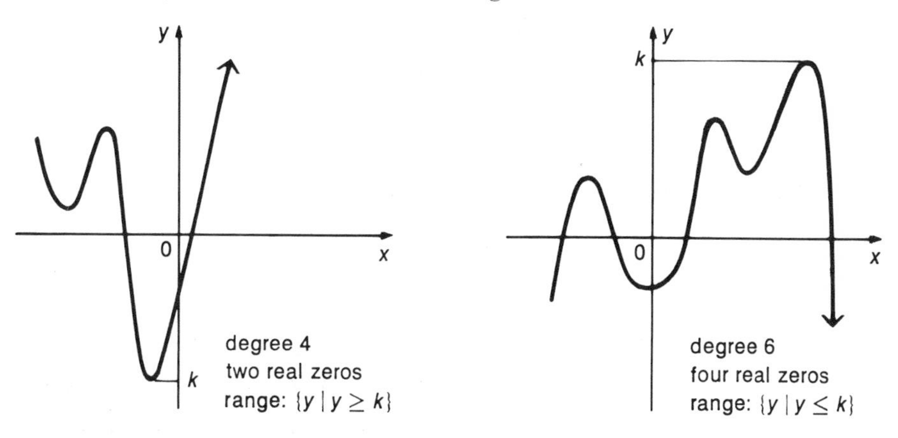

Figure 8.4

8.7 Exercises

Graph each of the following.

1. $P(x) = (x + 1)^3$
2. $P(x) = x^3 + 1$
3. $P(x) = x^3 - 7x - 6$
4. $P(x) = x^3 + x^2 - 4x - 4$
5. $P(x) = x^4 - 5x^2 + 6$
6. $P(x) = x^3 - 3x^2 - x + 3$
7. $P(x) = 6x^3 + 11x^2 - x - 6$
8. $P(x) = x^4 - 2x^2 - 8$
9. $P(x) = x^4 + x^3 - 2$
10. $P(x) = 6x^4 - x^3 - 23x^2 - 4x + 12$
11. $P(x) = 8x^4 - 2x^3 - 47x^2 - 52x - 15$

8.8 Rational Functions

A function of the form

$$\left\{(x, y) \,|\, y = \frac{P(x)}{Q(x)},\ P(x), Q(x) \text{ polynomials}, Q(x) \neq 0\right\}$$

is called a **rational function.** We shall graph several such functions in this section.

The function

$$y = \frac{2}{1 + x}$$

is undefined for $x = -1$. Hence, the graph of this function will not intersect the vertical line $x = -1$. Since x can equal any number except -1, we can let x approach -1 as closely as we wish. From the following table, note that as x gets closer and closer to -1, $1 + x$ gets closer and closer to 0, and $\frac{2}{(1 + x)}$ gets larger and larger. The vertical line $x = -1$ that is approached by the curve is called a **vertical asymptote.**

x	$1 + x$	$\frac{2}{1 + x}$
$-.5$	.5	4
$-.8$	.2	10
$-.9$	.1	20
$-.99$	.01	200

Note also that as $|x|$ gets larger and larger, $\frac{2}{1 + x}$ gets closer and closer to 0. Hence, the graph has the x-axis as a **horizontal asymptote.** By using these asymptotes and plotting the intercepts and a few points, we get the graph of Figure 8.5.

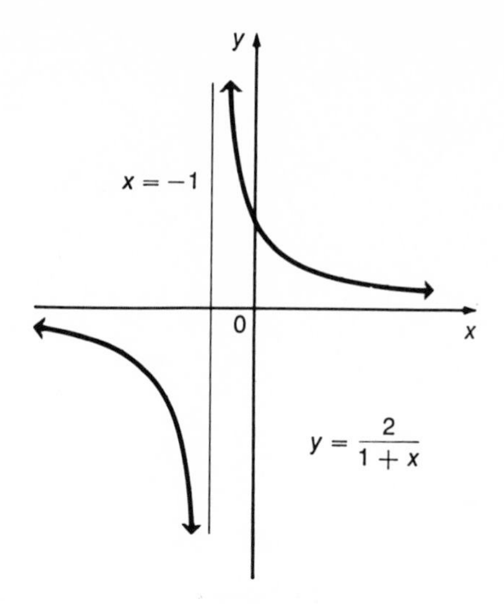

Figure 8.5

Example 8 Graph $y = \dfrac{1}{(x-1)(x+4)}$.

Note that there are vertical asymptotes at $x = 1$ and $x = -4$. As $|x|$ gets larger and larger, $|(x-1)(x+4)|$ also gets larger and larger. Thus, y gets closer and closer to 0, making the x-axis a horizontal asymptote. The vertical asymptotes divide the x-axis into three regions. It is often convenient to consider each region separately. If we find the intercepts and plot a few points, we get the result shown in Figure 8.6.

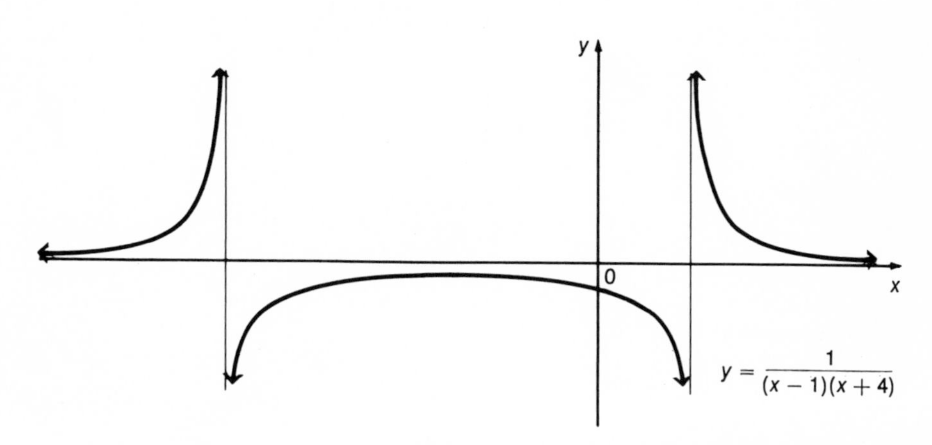

Figure 8.6

To graph

$$y = \frac{x^2 + 1}{x - 2},$$

we can divide $x - 2$ into $x^2 + 1$ using synthetic division.

$$\begin{array}{r|rrr} 2 & 1 & 0 & 1 \\ & & 2 & 4 \\ \hline & 1 & 2 & 5 \end{array}$$

This gives

$$y = x + 2 + \frac{5}{x - 2}.$$

As $|x|$ gets larger and larger, $\frac{5}{x - 2}$ gets closer and closer to 0. Hence, the graph approaches the **oblique asymptote** $y = x + 2$. Using this fact, and noticing that $x = 2$ is a vertical asymptote, we get the graph of Figure 8.7.

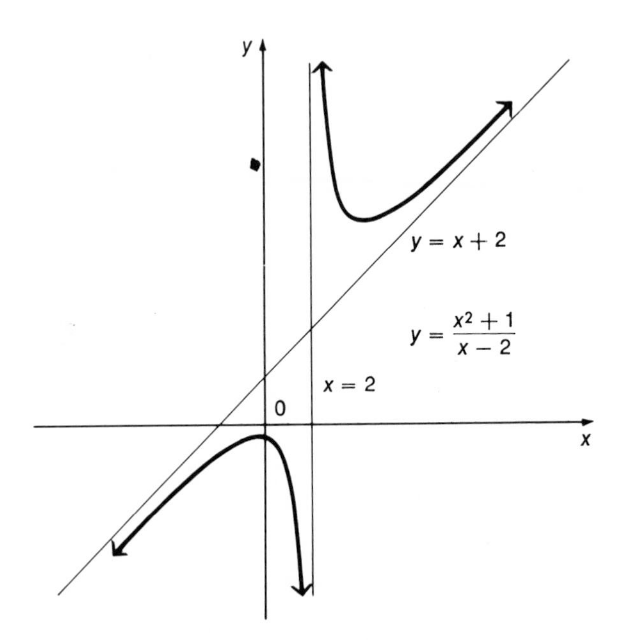

Figure 8.7

8.8 Exercises

Graph each of the following.

1. $y = \frac{1}{1 - x}$

2. $y = \frac{-2}{x}$

3. $y = \dfrac{2}{3 + 2x}$

4. $y = \dfrac{4}{5 + 3x}$

5. $y = \dfrac{1}{(x + 1)(x - 3)}$

6. $y = \dfrac{1}{(x - 2)(x + 4)}$

7. $y = \dfrac{3}{(x + 4)^2}$

8. $y = \dfrac{2}{(x - 3)^2}$

9. $y = \dfrac{3x}{x - 1}$

10. $y = \dfrac{4x}{1 - 3x}$

11. $y = \dfrac{x + 1}{x - 4}$

12. $y = \dfrac{x - 3}{x + 5}$

13. $y = \dfrac{x}{x^2 - 9}$

14. $y = \dfrac{x^2 + 1}{x + 3}$

15. $y = \dfrac{x^2 - 5}{x + 2}$

16. $y = \dfrac{x^2 - 3x + 2}{x - 3}$

COMBINATORICS

Most of the topics in this chapter are useful in the study of probability. Counting theory–the study of combinations and permutations–deals with methods for counting the number of ways in which certain objectives can be reached. We shall study only the basics of this interesting subject. The binomial theorem, which is important in many branches of mathematics, is a central idea in the study of probability theory.

9.1 Permutations

After making do with your old automobile for several years, you finally decide to replace it with a shiny new one. You drive over to Ned's New Car Emporium to choose the car of your dreams. Once there, you find that you can select from 5 models, each with 14 power options, in your choice of 8 exterior colors and 4 interior colors. How many different new cars are available to you? Problems of this sort are best solved by means of the counting principles which we shall discuss in this section and the next.

Suppose there are 3 roads from town A to town B and 2 roads from town B to town C. For each of the 3 roads from A to B, there are 2 different routes from B to C. Hence, to travel from A to C by way of B there are $3 \cdot 2 = 6$ different ways, as illustrated in Figure 9.1.

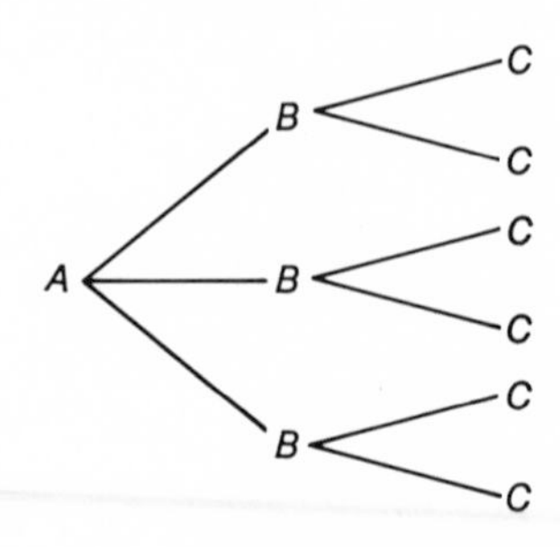

Figure 9.1

Generalizing from examples such as this, we have the **successive events axiom,**

▶ If one event can occur in m ways and then a second event can occur in n ways, then both events can occur in mn ways, provided the outcome of the first event does not influence the outcome of the second.

The successive events axiom can be extended to any number of events, provided that the outcome of no one event influences the outcome of another. Thus we can apply this axiom to determine how many choices we actually have in selecting a new car at Ned's. Applying the axiom we find that

$$5 \cdot 14 \cdot 8 \cdot 4 = 2240$$

different new cars are available from this one dealer.

Example 1 A restaurant offers a choice of 3 different salads, 5 main dishes, and 2 desserts. How many different meals can be selected?

Since the first event (selecting a salad) can occur in 3 ways, the second event (selecting a main dish) in 5 ways, and the third event (selecting a dessert) in 2 ways, there are $3 \cdot 5 \cdot 2$ or 30 different choices.

Example 2 A teacher has 5 different books which he wishes to arrange on his desk. How many different arrangements are possible?

With a little thought, we can use the successive events axiom here. There are 5 choices to be made, one for each space which will hold a book. To select a book for the first space, the teacher has 5 choices, for the second space, 4 choices (one book has already been put in the first space), for the third space, 3 choices, and so on. By the successive events axiom, we see that the number of arrangements is

$$5 \cdot 4 \cdot 3 \cdot 2 \cdot 1 = 120.$$

In using the successive events axiom, we shall frequently encounter such products as $5 \cdot 4 \cdot 3 \cdot 2 \cdot 1$, as we did in Example 2. For convenience in writing these products, we introduce the symbol $n!$ (read "n factorial"), which is defined as follows for any counting number n.

$$n! = n(n-1)(n-2)(n-3) \cdots (2)(1)$$

Thus, we may write $5 \cdot 4 \cdot 3 \cdot 2 \cdot 1$ as $5!$ Also, $3! = 3 \cdot 2 \cdot 1 = 6$. By this definition of $n!$, we see that $n[(n-1)!] = n!$ for all natural numbers $n \geq 2$. It is convenient to have this relation hold also for $n = 1$, so we define

▶ $$0! = 1.$$

Example 3 Suppose the teacher in Example 2 wishes to place only 3 of the 5 books on his desk. How many arrangements of 3 books are possible?

The teacher again has 5 ways to fill the first space, 4 ways to fill the second space, and 3 ways to fill the third. Since he wants to use only 3 books, there are only 3 spaces to be filled (3 events) instead of 5. Thus, there are

$$5 \cdot 4 \cdot 3 = 60 \text{ arrangements.}$$

The number 60 here is called the number of **permutations** of 5 things taken 3 at a time, written $60 = P(5, 3)$. Generalizing from these examples, we have the following theorems.

THEOREM 9.1 If $P(n, n)$ denotes the number of permutations of n elements taken n at a time, then

$$\boldsymbol{P(n, n) = n!.}$$

THEOREM 9.2 If $P(n, r)$ denotes the number of permutations of n elements taken r at a time, then

$$\boldsymbol{P(n, r) = \frac{n!}{(n - r)!}.}$$

As illustrated by Example 2, Theorem 9.1 is a direct consequence of the successive events axiom. To prove Theorem 9.2, we note that

$$\begin{aligned} P(n, r) &= n(n-1)(n-2) \cdots (n-r+1) \\ &= \frac{n(n-1)(n-2) \cdots (n-r+1)(n-r) \cdots (2)(1)}{(n-r)(n-r-1) \cdots (2)(1)} \\ &= \frac{n!}{(n-r)!}. \end{aligned}$$

Example 4 Find the number of permutations of 8 elements taken 3 at a time.

Using the successive events axiom, we note that there are 3 choices to be made, so that $P(8, 3) = 8 \cdot 7 \cdot 6 = 336$. However, we can instead use the formula given above for $P(n, r)$ as follows.

$$\begin{aligned} P(8, 3) &= \frac{8!}{5!} = \frac{8 \cdot 7 \cdot 6 \cdot 5 \cdot 4 \cdot 3 \cdot 2 \cdot 1}{5 \cdot 4 \cdot 3 \cdot 2 \cdot 1} \\ &= 8 \cdot 7 \cdot 6 \\ &= 336 \end{aligned}$$

Example 5 In how many ways can 6 students be seated in a row of 6 desks?

Here we have

$$P(6, 6) = 6! = 6 \cdot 5 \cdot 4 \cdot 3 \cdot 2 \cdot 1 = 720.$$

9.1 Exercises

1. Evaluate each of the following.
   (a) $P(7, 7)$
   (b) $P(5, 3)$
   (c) $P(6, 5)$
   (d) $P(4, 2)$
2. In how many ways can 5 pictures be arranged in a row? How many arrangements are possible if only 2 of them are used?
3. In how many ways can 6 people be seated in a row of 6 seats?
4. In how many ways can 7 out of 10 people be assigned to 7 seats?
5. In how many ways can 5 players be assigned to the 5 positions on a basketball team, assuming any player can play any position? In how many ways can 10 players be assigned to the 5 positions?
6. A couple has narrowed down their choice of names for a new baby to three first names and five middle names. How many different first and middle name arrangements are possible?
7. How many different homes are available if a builder offers a choice of 5 basic plans, 3 roof styles, and 2 exterior finishes?
8. An auto manufacturer produces 7 models, each available in 6 different colors, with 4 different upholstery fabrics, and 5 interior colors. How many varieties of the auto are available?
9. A concert is to consist of 5 works: 2 modern, 2 classical, and a piano concerto. In how many ways can the program be arranged?
10. If the program in Exercise 9 must be shortened to 3 works, chosen from the 5, how many arrangements are possible?
11. In Exercise 9, how many different programs are possible if the two modern works are to be played first, then the two classical, and then the concerto?
12. How many 4 letter radio station call letters can be made if the first letter must be K or W and no letter may be repeated? How many if repeats are allowed?
13. How many of the 4 letter call letters in Exercise 12 with no repeats end in K?
14. A business school gives courses in typing, shorthand, transcription, business English, technical writing, and accounting. How many ways can a student arrange his program if he takes 3 courses?
15. How many different license plate numbers can be formed using three letters followed by three digits if no repeats are allowed? How many if there are no repeats and either letters or numbers come first?
16. Show that $P(n, n - 1) = P(n, n)$.

9.2 Combinations

In the last section we discussed a method for finding the number of *ordered* subsets of size r which could be formed from a set of n elements ($n \geq r$). If we are not interested in the order within a subset, there will be fewer different subsets. For example, consider the set $\{a, b, c\}$. The number of ordered subsets of 2 elements which can be formed from the given set of 3 elements is

$$P(3, 2) = \frac{3!}{(3-2)!} = \frac{3!}{1!} = 3 \cdot 2 \cdot 1 = 6.$$

We can display these 6 subsets.

$$\begin{array}{ccc} \{a, b\} & \{a, c\} & \{b, c\} \\ \{b, a\} & \{c, a\} & \{c, b\} \end{array}$$

If we are not interested in the order of the elements, $\{a, b\}$ and $\{b, a\}$ are the same subset, $\{a, c\}$ and $\{c, a\}$ are the same, and $\{b, c\}$ and $\{c, b\}$ are the same. We see that there are only 3 subsets if we disregard order. These three sets are called the two-element **combinations** of the set $\{a, b, c\}$. There is only one three-element combination of $\{a, b, c\}$, the set itself, although there are $3!$, or 6, three-element permutations of $\{a, b, c\}$. To find the number of *ordered* subsets of r elements in a set of n elements, we use *permutations*. To find the number of *unordered* subsets, we use *combinations*. Since each combination of r elements, from a set of n elements, forms $r!$ permutations, we can find the number of combinations of n elements taken r at a time by dividing the number of permutations, $P(n, r)$, by $r!$ to get

$$\frac{P(n, r)}{r!} \text{ combinations.}$$

This expression can be rewritten as follows.

$$\frac{P(n, r)}{r!} = \frac{\dfrac{n!}{(n-r)!}}{r!}$$

$$= \frac{n!}{(n-r)!r!}$$

In summary, we have the following theorem.

THEOREM 9.3 If $C(n, r)$ denotes the number of combinations of n elements taken r at a time, then

$$C(n, r) = \frac{n!}{(n-r)!r!}.$$

We can now use Theorem 9.3 to find $C(3, 2)$ which we found to be equal to 3 in the discussion above.

$$C(3, 2) = \frac{3!}{(3-2)!2!} = \frac{3 \cdot 2 \cdot 1}{1 \cdot 2 \cdot 1} = \frac{3}{1} = 3$$

Example 6 How many committees of size three can be formed from a group of 8 people?

A committee is an unordered set, so we want $C(8, 3)$.

$$\begin{aligned} C(8, 3) &= \frac{8!}{5!3!} \\ &= \frac{8 \cdot 7 \cdot 6 \cdot 5 \cdot 4 \cdot 3 \cdot 2 \cdot 1}{5 \cdot 4 \cdot 3 \cdot 2 \cdot 1 \cdot 3 \cdot 2 \cdot 1} \\ &= \frac{8 \cdot 7 \cdot 6}{3 \cdot 2 \cdot 1} \\ &= 56 \end{aligned}$$

Example 7 From a group of 30 employees, 3 are to be selected to work on a special project.

(a) In how many different ways can the employees be selected?

(b) In how many ways can the group of three be selected if it has been decided that a particular man must work on the project?

(a) Here we wish to know how many three-element combinations can be formed from a set of 30 elements. (We want combinations, and not permutations, since order within the group of 3 is irrelevant.)

$$\begin{aligned} C(30, 3) &= \frac{30!}{27!3!} \\ &= \frac{30(29)(28)(27) \cdots (3)(2)(1)}{27(26)(25) \cdots (2)(1)(3)(2)(1)} \\ &= \frac{30 \cdot 29 \cdot 28}{3 \cdot 2 \cdot 1} \\ &= 4060 \end{aligned}$$

There are 4060 ways to select the project group.

(b) Since one man has already been selected for the project, the problem is reduced to selecting two more from the remaining 29 employees.

$$C(29, 2) = \frac{29!}{27!2!} = 406$$

In this case, the project group can be selected in 406 different ways.

Example 8 Find $C(n, n)$.

Using Theorem 9.3, with $r = n$, we have

$$C(n, n) = \frac{n!}{(n-n)!n!} = \frac{n!}{0!n!} = \frac{n!}{1 \cdot n!} = 1.$$

This is reasonable since there is only one way of selecting a group of n objects from a set of n objects when order is disregarded.

Example 9 Find $C(n, 1)$.

Again we can use Theorem 9.3.

$$\begin{aligned} C(n, 1) &= \frac{n!}{(n-1)!1!} \\ &= \frac{n(n-1)!}{(n-1)! \cdot 1} \\ &= n \end{aligned}$$

This result agrees with what we already know: there are n distinct one element subsets of an n element set.

Example 10 Find $C(n, n - 1)$.

Here we have

$$\begin{aligned} C(n, n-1) &= \frac{n!}{[n-(n-1)]!(n-1)!} \\ &= \frac{n!}{1!(n-1)!} \\ &= n. \end{aligned}$$

We can use the results of Examples 8–10 as theorems and apply them directly to special cases. Thus, from Example 8, $C(5, 5) = 1$, $C(8, 8) = 1$, and so on. Also, from Examples 9 and 10 we can write

$$C(7, 1) = C(7, 6) = 7,$$
$$C(10, 1) = C(10, 9) = 10,$$

and in general,

$$C(n, 1) = C(n, n-1) = n.$$

The examples above are all special cases of the general statement expressed in Theorem 9.4, which follows.

THEOREM 9.4 If $C(n, r)$ denotes the number of combinations of n elements taken r at a time, then

$$\boldsymbol{C(n, r) = C(n, n-r).}$$

The proof of this theorem is left for the exercises.

9.2 Exercises

1. Evaluate each of the following.
   (a) $C(6, 5)$
   (b) $C(4, 2)$
   (c) $C(8, 5)$
   (d) $C(10, 2)$
2. In how many ways can an employer select two new employees from a group of four applicants?
3. A club has 30 members. If a committee of 4 is selected in a random manner, how many committees are possible?
4. How many samples of 3 apples can be drawn from a crate of 25 apples?
5. A group of 3 students are to be selected randomly from a group of 12 students to participate in a special class. In how many ways can this be done? In how many ways can the group which will not participate be selected?
6. Hal's Hamburger Heaven sells hamburgers with cheese, relish, lettuce, tomato, mustard or ketchup. How many different hamburgers can be made using any three of the extras?
7. How many different two card hands can be dealt from a deck of 52 cards?
8. How many different 12 card pinochle hands can be dealt from a deck of 48 cards? (A pinochle deck has the 9, 10, J, Q, K, and A of each suit twice for a total of 48 cards.)
9. Five cards are marked with the numbers 1, 2, 3, 4, and 5, then shuffled, and two cards are then drawn. How many different two card combinations are possible?
10. If a bag contains 15 marbles, how many samples of two marbles can be drawn from it? How many samples of four marbles?
11. In Exercise 10, if the bag contains 3 yellow, 4 white, and 8 blue marbles, how many samples of two can be drawn in which both marbles are blue?
12. In Exercise 4, if it is known that there are 5 rotten apples in the crate:
   (a) How many samples of 3 could be drawn in which all 3 are rotten?
   (b) How many samples of 3 could be drawn in which there is one rotten apple and two good apples?
13. A city council is composed of 5 liberals and 4 conservatives. A delegation of 3 is to be selected randomly to attend a convention.
   (a) How many delegations are possible?
   (b) How many delegations could have all liberals?
   (c) How many delegations could have 2 liberals and 1 conservative?
   (d) If one member of the council serves as mayor, how many delegations are possible which include the mayor?

14. A group of 7 workers decide to send a delegation of 2 to the supervisor to discuss their grievances.
   (a) How many different delegations are possible?
   (b) If it is decided that a particular employee must be in the delegation, how many different delegations are possible?
   (c) If there are 2 women and 5 men in the group, how many delegations would include a woman?
15. Show that $C(n, r) = C(n, n - r)$.

9.3 Basic Properties of Probability

The study of probability is becoming increasingly popular partly because of its importance to applications and partly because of an inherent interest in the subject itself. It is human nature to be interested in chance, and probability deals with chance. In fact, the theory of probability was developed to describe games of chance. Modern usage, however, extends the application of probability theory to a much wider range of activities, including many practical applications.

We shall call the set S of all possible outcomes of a given experiment the **sample space** for the experiment. In this text, all sample spaces will be finite. For example, a sample space for the experiment of tossing a coin includes two outcomes, heads (H) and tails (T). We write this in set notation as

$$S = \{H, T\}.$$

Similarly, a sample space for the experiment of rolling a single die is

$$S = \{1, 2, 3, 4, 5, 6\}.$$

(A single die can have either 1, 2, 3, 4, 5, or 6 showing on top after it is rolled.)

Any subset of the sample space is called an **event.** In the experiment with the die, for example, "the number showing is a three" is an event, say E_1, such that $E_1 = \{3\}$. The **probability** of an event E, written $P(E)$, is the ratio of the number of outcomes in the sample space favorable to E compared to the total number of outcomes in the sample space. Thus, if there are s possible outcomes in the sample space S, each equally likely to occur, and n of these are favorable to event E, we have

▶ $$P(E) = \frac{n}{s}.$$

To use this definition to find $P(E_1)$ as given above, we note that the sample space for this experiment is $S = \{1, 2, 3, 4, 5, 6\}$, while the desired event is $E_1 = \{3\}$. Thus,

$$P(E_1) = \frac{1}{6},$$

since there is one element in E_1 and 6 elements in S.

Example 11 A single die is rolled. Write the following events in set notation and give the probability for each event.

(a) E_2; the number showing is even.
(b) E_3: the number showing is greater than 4.
(c) E_4: the number showing is less than 10.
(d) E_5: the number showing is 8.

Using the definitions above, we have the following.

(a) $E_2 = \{2, 4, 6\}$ and $P(E_2) = \frac{3}{6} = \frac{1}{2}$.

(b) $E_3 = \{5, 6\}$ and $P(E_3) = \frac{2}{6} = \frac{1}{3}$.

(c) $E_4 = \{1, 2, 3, 4, 5, 6\}$ and $P(E_4) = \frac{6}{6} = 1$.

(d) $E_5 = \varnothing$ and $P(E_5) = \frac{0}{6} = 0$.

Note that $E_4 = S$. The event E_4 is certain to occur every time the experiment is performed. We see that an event which is certain to occur, such as E_4, has a probability of 1. On the other hand, E_5 is $\varnothing$ and $P(E_5)$ is 0. The probability of an impossible event, such as E_5, is always 0, since none of the outcomes in the sample space favor the event.

The set of outcomes in the sample space which are *not* favorable to an event E is called the **complement** of E, written E'. For example, in the experiment of drawing a single card from a standard deck of 52 cards, let E be the event "the card is an ace." Then E' is the event "the card is not an ace." From the definition of E', we see that for any event E,

$$\boldsymbol{E \cup E' = S} \qquad \qquad \boldsymbol{E \cap E' = \varnothing}.$$

Example 12 In the experiment of drawing a card from a well-shuffled deck, find the probability of events E, (the card is an ace), and E'.

Since there are four aces in the deck of 52 cards, $P(E) = \frac{4}{52} = \frac{1}{13}$. Of the 52 cards, 48 are not aces, so that we have $P(E') = \frac{48}{52} = \frac{12}{13}$.

In Example 12, we note that $P(E) + P(E') = \frac{1}{13} + \frac{12}{13} = 1$. Recall that for any event E, $P(E) = \frac{n}{s}$ (where n is the number of outcomes favorable to E, and s is the total number of outcomes in the sample space). Since E' is the set of all outcomes not favorable to E, $P(E') = \frac{s - n}{s}$. Noting that

$$\frac{n}{s} + \frac{s-n}{s} = \frac{s}{s} = 1,$$

we see that in general, for any event E,

▶ $$P(E) + P(E') = 1.$$

This fact can be restated as

▶ $$P(E) = 1 - P(E') \quad \text{or} \quad P(E') = 1 - P(E).$$

These two equations suggest an alternate way to compute the probability of an event. For example, if it is known that $P(E) = \frac{1}{10}$, then we have

$$P(E') = 1 - \frac{1}{10} = \frac{9}{10}.$$

Example 13 Two dice are rolled. Find
(a) the probability that the sum of the numbers showing is 3,
(b) the probability that the sum is greater than 3.

The sample space for this experiment has 36 outcomes. These can be displayed as follows.

		Die 2					
		1	2	3	4	5	6
Die 1	1	1, 1	1, 2	1, 3	1, 4	1, 5	1, 6
	2	2, 1	2, 2	2, 3	2, 4	2, 5	2, 6
	3	3, 1	3, 2	3, 3	3, 4	3, 5	3, 6
	4	4, 1	4, 2	4, 3	4, 4	4, 5	4, 6
	5	5, 1	5, 2	5, 3	5, 4	5, 5	5, 6
	6	6, 1	6, 2	6, 3	6, 4	6, 5	6, 6

(a) To find the probability of a sum of 3, we need to find the number of outcomes favorable to a sum of 3. From the sample space, we see that there are 2 such outcomes: 1, 2, and 2, 1. Since the number of outcomes in the sample space is 36, the probability we seek is 2/36 or 1/18.
(b) The sample space contains 33 outcomes in which the sum is greater than 3. Thus the probability of a sum greater than 3 is 33/36 or 11/12. Another approach to this problem is to use the relationship

$$P(E) = 1 - P(E').$$

For the event "the sum is greater than 3," the complementary event is "the sum is 3 or less," which is satisfied by 3 of the 36 outcomes in the sample space. Here, $P(E') = 3/36$, or 1/12, and the required probability, then, is

$$1 - \frac{1}{12} = \frac{11}{12}.$$

Sometimes probability statements are expressed in terms of odds, a comparison of $P(E)$ with $P(E')$. The **odds** in favor of an event E are expressed as the ratio of $P(E)$ to $P(E')$ or as the fraction $P(E)/P(E')$. For example, if the probability of rain can be established as 1/3, the odds that it will rain are

$$P(\text{rain}) \text{ to } P(\text{no rain}) = \frac{1}{3} \text{ to } \frac{2}{3}$$

$$= \frac{\frac{1}{3}}{\frac{2}{3}}$$

$$= \frac{1}{2} \text{ (or 1 to 2).}$$

On the other hand, the odds that it will not rain are 2 to 1 (or 2/3 to 1/3). If we know that the odds in favor of an event are, say, 3 to 5, then we can see that the probability of the event is 3/8, while the probability of the complement of the event is 5/8. In general, if the odds favoring event E are m to n, then

▶ $$P(E) = \frac{m}{m+n} \quad \text{and} \quad P(E') = \frac{n}{m+n}.$$

The probabilities that we have considered to this point have been determined by inherent properties of the experiment in question. Such probabilities are called *a priori* and are based on theory as opposed to observation. Empirical probabilities, those based only on observation, are more likely to occur in practical applications. An example is the life tables used by life insurance companies to determine premiums. A portion of such a table is shown below.

From the table, we see that the probability of dying before the age of 10 is .033, while a 70-year old person has a .259 probability of dying before age 80.

Life table

The table gives the probability that a person alive at the beginning of an age interval will die before reaching the beginning of the next age interval.

period	probability of death	period	probability of death
0–9	.033	50–59	.112
10–19	.007	60–69	.245
20–29	.013	70–79	.259
30–39	.019	80–89	.201
40–49	.047	90–99	.064

When using a life table, it is important to note for which group of individuals the table has been prepared. A table valid for people in the United States would not be applicable to people in other countries such as India. Also, a table prepared on the basis of men would not be valid for women, who usually live about 5 years longer than men.

9.3 Exercises

Write sample spaces in which each outcome is equally likely to occur for the following experiments.

1. A two-headed coin is tossed once.
2. Two fair coins are tossed.
3. Three fair coins are tossed.
4. Slips of paper marked with the numbers 1, 2, 3, 4, and 5 are placed in a box. After mixing well, two slips are drawn.
5. An unprepared student takes a three question true-false quiz in which he guesses the answer for all three questions.
6. A die is rolled and then a coin is tossed.

Write the following events in set notation and give the probability of each event.

7. In the experiment of Exercise 2:
   (a) both coins show the same face.
   (b) at least one coin turns up heads.
8. In Exercise 1:
   (a) the result of the toss is heads.
   (b) the result of the toss is tails.
9. In Exercise 4:
   (a) both slips are marked with even numbers.
   (b) both slips are marked with odd numbers.
   (c) both slips are marked with the same number.
   (d) one slip is marked with an odd number, the other with an even number.
10. In Exercise 5:
    (a) the student gets all three answers correct.
    (b) he gets all three answers wrong.
    (c) he gets exactly two answers correct.
    (d) he gets at least one answer correct.
11. A marble is drawn at random from a box containing 3 yellow, 4 white, and 8 blue marbles. Find the following probabilities.
    (a) a yellow marble is drawn.
    (b) a blue marble is drawn.
    (c) a black marble is drawn.

(d) what are the odds in favor of drawing a yellow marble?
(e) what are the odds against drawing a blue marble?

12. A baseball player with a batting average of .300 comes to bat. What are the odds in favor of his getting a hit?
13. In Exercise 4, what are the odds that the sum of the numbers on the two slips of paper is 5?
14. If the odds that it will rain are 4 to 5, what is the probability of rain?
15. If the odds that a candidate will win an election are 3 to 2, what is the probability that the candidate will lose?
16. A card is drawn from a well-shuffled deck of 52 cards. Find the probability that the card is
    (a) a 9,
    (b) black,
    (c) a black 9,
    (d) a heart,
    (e) the 9 of hearts,
    (f) a face card. (The K, Q, J of any suit.)
17. Two dice are rolled. Find the probability that
    (a) the sum of the points is 5.
    (b) the sum of the points is 12.
    (c) the sum of the points is less than 12. (Hint: use $P(E) = 1 - P(E')$.)
    (d) the sum of the points is an even number.
    (e) both dice show the same number of points.
18. Use the life table to find the probability of dying in the 4th decade; the 7th decade.

9.4 Probabilities of Alternate and Successive Events

We now consider the probability of a compound event which involves an alternative such as (E or F) where E and F are simple events. For example, in the experiment of rolling a die, suppose H is the event "the result is a 3," while K is the event "the result is an even number." What is the probability of (H or K), "the result is a 3 or an even number?" The three event sets and their probabilities are

$$H = \{3\}; \qquad P(H) = \frac{1}{6}.$$

$$K = \{2, 4, 6\}; \qquad P(K) = \frac{3}{6} = \frac{1}{2}.$$

$$(H \text{ or } K) = \{2, 3, 4, 6\}; \qquad P(H \text{ or } K) = \frac{4}{6} = \frac{2}{3}.$$

Here $P(H) + P(K) = P(H \text{ or } K)$. Before we assume that this relationship is true in general, let us consider another event for this experiment, "the result is a 2," event G. For G and (K or G) we have

$$G = \{2\}; \qquad P(G) = \frac{1}{6}.$$

$$(K \text{ or } G) = \{2, 4, 6\}; \qquad P(K \text{ or } G) = \frac{3}{6} = \frac{1}{2}.$$

In this case $P(K) + P(G) \neq P(K \text{ or } G)$.

The difference in the two examples above comes from the fact that events H and K cannot occur simultaneously. Such events are called **mutually exclusive events.** We see that $H \cap K = \varnothing$, which is true in general for any two mutually exclusive events. Events K and G, however, can occur simultaneously. Both are satisfied if the result of the roll is a 2, the element in their intersection ($K \cap G = \{2\}$). Generalizing the above discussion, we have the following theorem.

THEOREM 9.5 For any events E and F,

$$\boldsymbol{P(E \text{ or } F) = P(E \cup F) = P(E) + P(F) - P(E \cap F).}$$

Example 14 For the experiment consisting of one roll of a pair of dice, find the probability that the sum of the points showing is at most 4.

We can rewrite "at most 4" as "2 or 3 or 4." (A sum smaller than 2 is meaningless here.) Then, by Theorem 9.5, we can write

$$\begin{aligned} P(\text{at most } 4) &= P(2 \text{ or } 3 \text{ or } 4) \\ &= P(2) + P(3) + P(4), \end{aligned} \tag{1}$$

since the events represented by "2," "3," and "4" are mutually exclusive. By the definition of probability, $P(2) = 1/36$, $P(3) = 2/36$, and $P(4) = 3/36$. Substituting into equation (1) above we have

$$P(\text{at most } 4) = \frac{1}{36} + \frac{2}{36} + \frac{3}{36} = \frac{6}{36} = \frac{1}{6}.$$

Example 15 One card is drawn from a well-shuffled deck of 52 cards. What is the probability that it is an ace or a spade? a three or a king?

The events "drawing an ace" and "drawing a spade" are not mutually exclusive since it is possible to draw the ace of spades, an outcome favorable to both events.

By Theorem 9.5, we know

$$
\begin{aligned}
P(\text{ace or spade}) &= P(\text{ace}) + P(\text{spade}) - P(\text{ace and spade}) \\
&= \frac{4}{52} + \frac{13}{52} - \frac{1}{52} \\
&= \frac{16}{52} \\
&= \frac{4}{13}.
\end{aligned}
$$

"Drawing a 3" and "drawing a king" are mutually exclusive events because it is impossible to draw one card which is both a 3 and a king. From Theorem 9.5 we have

$$P(3 \text{ or } K) = P(3) + P(K) - P(3 \text{ and } K) = \frac{4}{52} + \frac{4}{52} - 0 = \frac{8}{52} = \frac{2}{13}.$$

An experiment can consist of more than one event. Suppose we wish to find the probability of the two-event outcome, getting a 3 on a roll of a die and then getting heads on the toss of a coin. A successful outcome here requires success in both parts of the experiment. The sample space has 12 outcomes as shown below.

$$S = \begin{Bmatrix} 1, H & 2, H & 3, H & 4, H & 5, H & 6, H \\ 1, T & 2, T & 3, T & 4, T & 5, T & 6, T \end{Bmatrix}$$

Since only one of the twelve outcomes favors the desired event, the probability we seek is 1/12. By the successive events axiom, probability of a composite event which consists of successive parts (or events) can also be found by multiplying the probabilities of the parts. Thus, in the experiment above, the probability of the first event, $P(3)$, multiplied by the probability of the second event, $P(H)$, gives $P(3 \text{ and } H)$.

$$P(3) = \frac{1}{6}, \qquad P(H) = \frac{1}{2}, \qquad P(3 \text{ and } H) = \frac{1}{12} = \frac{1}{6} \cdot \frac{1}{2}$$

From this example, it might appear that we can generalize to a product theorem. Before we do, let us consider another example.

Example 16 Two marbles are drawn (without replacement) from a box containing 1 red, 3 white, and 2 green marbles. Find the probability that the two marbles drawn are both white.

The sample space for this experiment consists of all possible sets of two marbles. There are $C(6, 2)$ ways in which a sample of two marbles can be drawn from the 6 marbles, and $C(3, 2)$ ways in which the sample can include 2 white marbles. (There are 3 white marbles in the box.) We can compute the probability using the definition as follows.

$$P(W \text{ and } W) = \frac{C(3, 2)}{C(6, 2)}$$

$$= \frac{\frac{3!}{2!1!}}{\frac{6!}{2!4!}}$$

$$= \frac{3!2!4!}{2!1!6!}$$

$$= \frac{3 \cdot 2}{6 \cdot 5}$$

$$= \frac{1}{5}.$$

To compute the probability by multiplication, as we did above, we find that the probability of drawing a white marble on one draw is 3/6 or 1/2, so that

$$P(W) \cdot P(W) = \frac{1}{2} \cdot \frac{1}{2} = \frac{1}{4}.$$

However, $P(W \text{ and } W) = 1/5$ from the work above. The difficulty in this case is that the probability of drawing white for the second marble is *not* 1/2 but 2/5, since drawing a white marble on the first draw reduces the remaining marbles to 5, of which 2 are white. Thus, to use multiplication to compute $P(W \text{ and } W)$, we must compute the probability that the second marble is white taking into account the first draw of a white marble. This gives

$$P(W \text{ and } W) = P(W \text{ first}) \cdot P(W \text{ second}) = \frac{1}{2} \cdot \frac{2}{5} = \frac{1}{5},$$

which agrees with the probability we computed using the definition.

In the experiment discussed above, if the first marble were replaced before the second was drawn, the probability that both draws resulted in a white marble becomes

$$P(W \text{ first}) \cdot P(W \text{ second}) = \frac{1}{2} \cdot \frac{1}{2} = \frac{1}{4} = P(W \text{ and } W).$$

Thus, it is important to consider the relationship between the events in a two-part experiment.

As Example 16 illustrates, when computing the probability of the second event in a two-part experiment, it is important to consider whether or not the prior occurrence of the first event has affected the outcome of the second event. That is, we must find the probability of the second event *given that the first event has already occurred*. This kind of probability is called **conditional probability.** The conditional probability of event F given that event E has occurred is written $P(F | E)$. We can now state a product theorem as follows.

THEOREM 9.6 If E and F are two successive events, the probability of the composite event (E and F) is given by

$$P(E \text{ and } F) = P(E) \cdot P(F|E).$$

If $P(F|E)$ equals $P(F)$, the product theorem becomes

$$P(E \text{ and } F) = P(E) \cdot P(F).$$

When this is the case, E and F are said to be **independent events.** Intuitively, this means that the occurrence of E does not affect the occurrence of F. For example, the two events "heads on the first toss of a coin" (H_1) and "heads on the second toss of a coin" (H_2) are independent, which can be shown as follows. We know that $P(H_1) = 1/2$ and $P(H_2) = 1/2$. Also, $P(H_1 \text{ and } H_2) = 1/4$, because only one of the four outcomes in the sample space

$$\{H_1H_2,\ H_1T_2,\ T_1H_2,\ T_1T_2\},$$

satisfies the condition (H_1 and H_2). Thus,

$$P(H_1) \cdot P(H_2) = P(H_1 \text{ and } H_2)$$

as required by the definition of independent events. Events which are not independent are called **dependent events.**

Example 17 Find the probability of getting heads on five successive tosses of a coin.

For each toss of the coin, $P(H) = 1/2$. Each of the tosses is independent of the others, since what happened on the first toss has no affect on the outcome of the second toss, and so on. Using the product theorem, we have

$$P(5 \text{ heads}) = \frac{1}{2} \cdot \frac{1}{2} \cdot \frac{1}{2} \cdot \frac{1}{2} \cdot \frac{1}{2} = \left(\frac{1}{2}\right)^5 = \frac{1}{32}.$$

9.4 Exercises

1. Mrs. Elliott invites 10 relatives to a party: her mother, two uncles, three brothers, and four cousins. If the chances of any one guest arriving first are equally likely, find the following probabilities.
   (a) The first guest is an uncle or a brother.
   (b) The first guest is a brother or cousin.
   (c) The first guest is a brother or her mother.
2. One card is drawn from a standard deck of 52 cards. What is the probability that the card is
   (a) a 9 or a 10?
   (b) red or a 3?
   (c) a heart or black?
   (d) less than a four? (Consider aces as ones.)

3. Two dice are rolled. Find the probability that
   (a) the sum of the points is at least 10.
   (b) the sum of the points is either 7 or at least 10.
   (c) the sum of the points is 2 or the dice both show the same number.
4. A die is rolled three times. What is the probability that
   (a) all three results are even?
   (b) the first result is a 2 and the second is an even number and the third is any number?
5. If the probability of having a boy is 1/2, what is the probability in a three-child family of having
   (a) two boys and then a girl?
   (b) two boys and one girl in any order?
   (c) all three children of the same sex?
6. If two marbles are drawn without replacement from a jar with four black and three white marbles, what is the probability that
   (a) both are white?
   (b) both are black?
   (c) the first is black and the second is white?
   (d) one is black and the other is white?
7. Two cards are drawn from a deck of 52 cards, without replacement. What is the probability that
   (a) both cards are hearts?
   (b) both cards are black?
   (c) both cards are twos?
   (d) the first card is black and the second is a heart?
   (e) the first card is black and the second is a club?
8. If a marble is drawn from a bag containing 3 yellow, 4 white, and 8 blue marbles, what is the probability that
   (a) the marble is either yellow or white?
   (b) it is either yellow or blue?
   (c) it is either red or white?
9. A smooth talking young man has a 1/3 probability of talking a policeman out of giving him a speeding ticket when stopped. The probability that he is stopped during a given weekend is 1/2. Find the probability that
   (a) he will receive no tickets on a given weekend.
   (b) he will receive no tickets on three consecutive weekends.
10. If two fair coins are tossed, find the probability that
    (a) both coins show heads or both show tails.
    (b) only one coin comes up heads or both show heads.
    (c) at least one coin comes up heads or both coins show tails.
11. In a club with 20 senior and 10 junior members, a committee of three is to be randomly selected. Find the probability that

(a) the committee is composed entirely of senior members or entirely of junior members.
(b) the committee has at least one senior member.

12. Slips of paper marked with the digits 1, 2, 3, 4, and 5 are placed in a box and mixed well. If two slips are drawn (without replacement), find the probability that
(a) the first is even and the second is odd.
(b) both are even.
(c) the first is a three and the second is a number greater than three.
(d) both are marked three.

13. If two cards are drawn from a deck of 52 cards, what is the probability that they are both queens or both red?

14. Use the life table on page 221 to find the following probabilities.
(a) the probability of dying by age 20. (Hint: dying by age 20 means dying in either the first or the second decade.)
(b) the probability of dying by age 50.
(c) the probability of living to age 50.
(d) the probability of living to age 80.
(e) For approximately what age is there a 1/2 probability of dying before that age and a 1/2 probability of surviving beyond it?

9.5 The Binomial Theorem

If we write the expression $(x + y)^n$ for positive integer values of n, we get a family of expressions which are important in the study of mathematics generally, and, in particular, in the study of probability theory. For example

$$
\begin{aligned}
(x + y)^1 &= x + y \\
(x + y)^2 &= x^2 + 2xy + y^2 \\
(x + y)^3 &= x^3 + 3x^2y + 3xy^2 + y^3 \\
(x + y)^4 &= x^4 + 4x^3y + 6x^2y^2 + 4xy^3 + y^4 \\
(x + y)^5 &= x^5 + 5x^4y + 10x^3y^2 + 10x^2y^3 + 5xy^4 + y^5.
\end{aligned}
$$

From inspection, we see that these expressions follow a pattern. Let us try to identify the pattern so that we can write a general expression for $(x + y)^n$.

First, observe that each expression begins with x raised to the same power as the binomial itself. That is, the expansion of $(x + y)^1$ has a first term of x^1, $(x + y)^2$ has a first term of x^2, $(x + y)^3$ has a first term of x^3, and so on. Also, the last term in each expansion is y to the same power as the binomial. We can see that the expression of $(x + y)^n$ should begin with the term x^n and the end with the term y^n.

Further, we note that the exponents on x decrease by one in each term after the first, while the exponents on y, beginning with y in the second term, increase by one in each succeeding term. Thus, the *variables,* in the expansion of $(x + y)^n$ have the following pattern.

$$x^n,\ x^{n-1}y,\ x^{n-2}y^2,\ x^{n-3}y^3,\ \cdots,\ xy^{n-1},\ y^n$$

From this pattern, it can be seen that the sum of the exponents on x and y in each term is n. For example, in the third term above, the variable is $x^{n-2}y^2$, and the sum of the exponents, $n - 2$ and 2, is n.

Now let us try to find a pattern for the *coefficients* in the terms of the expansions shown above. In the product

$$(x + y)^5 = (x + y)(x + y)(x + y)(x + y)(x + y), \tag{2}$$

the term x occurs 5 times, once in each factor. To get the first term of the expansion, we form the product of these 5 x's to get x^5. The product x^5 can occur in just one way, by taking an x from each factor in equation (2), so the coefficient of x^5 is 1. We can get the term with x^4y in more than one way. For example, we can take the x's from the first four factors and the y from the last factor, or we may take the x's from the last four factors and the y from the first, and so on. Since there are five factors in equation (2), from which we wish to select four x's for the term x^4y, there are $C(5, 4)$ ways in which this can be done. Thus the term x^4y occurs in $C(5, 4)$ or 5 ways and has coefficient 5. Continuing in this manner, we can use combinations to find the coefficients for each of the terms of the expansion as follows.

$$(x + y)^5 = x^5 + C(5, 4)x^4y + C(5, 3)x^3y^2 + C(5, 2)x^2y^3 + C(5, 1)xy^4 + y^5$$

The coefficient, 1, of the first and last terms could be written $C(5, 5)$ or $C(5, 0)$ to complete the pattern.

In general, then, the coefficient for a term of $(x + y)^n$ in which the variable is $x^{n-r}y^r$ is $C(n, n - r)$. Thus, we have the **binomial theorem,** which gives the **general binomial expansion.**

THEOREM 9.7 For any positive integer n,

$$\begin{aligned}(x + y)^n = x^n &+ C(n, n - 1)x^{n-1}y \\ &+ C(n, n - 2)x^{n-2}y^2 + C(n, n - 3)x^{n-3}y^3 \\ &+ \cdots + C(n, 1)xy^{n-1} + y^n.\end{aligned}$$

A formal proof of this theorem is given at the end of the next section.

Example 18 Write out the binomial expansion of $(x + y)^9$.

We can use the binomial theorem to write

$$\begin{aligned}(x + y)^9 &= x^9 + C(9, 8)x^8y + C(9, 7)x^7y^2 + C(9, 6)x^6y^3 \\ &\quad + C(9, 5)x^5y^4 + C(9, 4)x^4y^5 + C(9, 3)x^3y^6 \\ &\quad + C(9, 2)x^2y^7 + C(9, 1)xy^8 + y^9 \\ &= x^9 + \frac{9!}{8!1!}x^8y + \frac{9!}{7!2!}x^7y^2 + \frac{9!}{6!3!}x^6y^3 + \frac{9!}{5!4!}x^5y^4 \\ &\quad + \frac{9!}{4!5!}x^4y^5 + \frac{9!}{3!6!}x^3y^6 + \frac{9!}{2!7!}x^2y^7 + \frac{9!}{1!8!}xy^8 + y^9 \\ &= x^9 + 9x^8y + 36x^7y^2 + 84x^6y^3 + 126x^5y^4 + 126x^4y^5 \\ &\quad + 84x^3y^6 + 36x^2y^7 + 9xy^8 + y^9.\end{aligned}$$

Example 19 Expand $\left(a - \frac{b}{2}\right)^5$.

Again we can use the binomial theorem.

$$\begin{aligned}\left(a - \frac{b}{2}\right)^5 &= a^5 + C(5, 4)a^4\left(-\frac{b}{2}\right) + C(5, 3)a^3\left(-\frac{b}{2}\right)^2 \\ &\quad + C(5, 2)a^2\left(-\frac{b}{2}\right)^3 + C(5, 1)a\left(-\frac{b}{2}\right)^4 + \left(-\frac{b}{2}\right)^5 \\ &= a^5 + 5a^4\left(-\frac{b}{2}\right) + 10a^3\left(-\frac{b}{2}\right)^2 + 10a^2\left(-\frac{b}{2}\right)^3 \\ &\quad + 5a\left(-\frac{b}{2}\right)^4 + \left(-\frac{b}{2}\right)^5 \\ &= a^5 - \frac{5}{2}a^4b + \frac{5}{2}a^3b^2 - \frac{5}{4}a^2b^3 + \frac{5}{16}ab^4 - \frac{1}{32}b^5.\end{aligned}$$

We can also write any single term of a binomial expansion. For example, if we want to write the tenth term of $(x + y)^n$, $(n \geq 9)$, we see that in the tenth term y is raised to the ninth power (since y has the power of 1 in the second term, the power of 2 in the third term, and so on.) Since the exponents on x and y in any term must have a sum of n, the exponent on x in the tenth term is $n - 9$. From this we have

$$C(n, n - 9)x^{n-9}y^9 = \frac{n!}{(n - 9)!9!}x^{n-9}y^9$$

for the tenth term of the expansion. Stated generally, we have Theorem 9.8.

THEOREM 9.8 The rth term of the binomial expansion of $(x + y)^n$, where $n \geq r - 1$, is

$$\boldsymbol{C(n, n - (r - 1))x^{n-(r-1)}\,y^{r-1}}.$$

Example 20 Find the sixth term of $(a + 2b)^{10}$.

In the sixth term we see that $2b$ has an exponent of 5, while a has an exponent of $10 - 5$ or 5. The sixth term thus becomes

$$\begin{aligned} C(10, 5)a^5(2b)^5 &= \frac{10!}{5!5!}a^5(2b)^5 \\ &= 252a^5(32b^5) \\ &= 8064a^5b^5. \end{aligned}$$

Example 21 Use the binomial theorem to find $(1.01)^8$ correct to three decimal places.

We can write 1.01 as $1 + .01$. Then, using the binomial theorem we have

$$\begin{aligned} (1.01)^8 &= (1 + .01)^8 \\ &= 1^8 + C(8, 7)(1)^7(.01) + C(8, 6)(1)^6(.01)^2 + \cdots + (.01)^8 \\ &= 1 + 8(1)(.01) + 28(1)(.0001) + \cdots + (.01)^8 \\ &= 1 + .08 + .0028 + \cdots + (.01)^8 \\ &\approx 1.083. \end{aligned}$$

Since we need only three decimal places there was no need to evaluate more terms of the expansion, although additional terms would give the answer to more decimal places of accuracy.

9.5 Exercises

Write out the binomial expansion for the following and simplify the terms.

1. $(x + y)^6$
2. $(m + n)^4$
3. $(p - q)^5$
4. $(r^2 - s)^7$
5. $(2x + t^3)^4$
6. $(3x - 2y)^6$
7. $\left(\frac{m}{2} - 3n\right)^5$
8. $\left(2p + \frac{q}{3}\right)^4$

For each of the following, write the indicated term of the binomial expansion.

9. 5th term of $(m - 2p)^{12}$
10. 7th term of $(3x + y)^{15}$
11. 17th term of $\left(p^2 + \frac{q}{2}\right)^{20}$
12. 8th term of $(x^3 + 2y)^{14}$

Use the binomial expansion to evaluate each of the following to four decimal places.

13. $(1.10)^{10}$
14. $(0.99)^{15}$
15. $(1.99)^8$
16. $(3.02)^6$

9.6 The Binomial Theorem and Probability

Many probability problems are concerned with an experiment in which an event is repeated many times. For example, we may wish to find the probability of getting 7 heads in 8 tosses of a coin, or hitting a target 6 times out of 6, or finding 1 defective item in a sample of 15 items. These problems all have certain things in common.

(1) The same experiment is repeated several times.
(2) The probability of the desired outcome remains the same for each trial.
(3) The repeated trials are independent.

Probability problems of this kind are called **repeated trials** problems. In each case, some outcome is designated a success and any other outcome is considered a failure. Thus, if the probability of a success in a single trial is p, the probability of failure will be $1 - p$.

Let us consider the solution of a problem of this type. Suppose we want to find the probability of getting 6 ones on 6 rolls of a die. The probability of getting a one on 1 roll is $1/6$, while the probability of any other result is $5/6$. Using the product theorem, and noting that the throws are independent events we have

$$\begin{aligned} P(\text{6 ones on 6 rolls of a die}) &= P(1)P(1)P(1)P(1)P(1)P(1) \\ &= \left(\frac{1}{6}\right)^6 \\ &\approx .00002. \end{aligned}$$

Now, let us find the probability of getting exactly 5 ones in 6 rolls of the die. The desired outcome for this experiment can occur in more than one way as shown below where S represents success (getting a one) and F represents failure (any other result).

$$\begin{array}{cccccc} S & S & S & S & S & F \\ S & S & S & S & F & S \\ S & S & S & F & S & S \\ S & S & F & S & S & S \\ S & F & S & S & S & S \\ F & S & S & S & S & S \end{array}$$

The probability of each of these six outcomes is

$$\left(\frac{1}{6}\right)^5\left(\frac{5}{6}\right).$$

Since the six outcomes represent mutually exclusive alternative events, we add the six probabilities to get

$$P(5 \text{ ones in 6 rolls of a die}) = 6\left(\frac{1}{6}\right)^5\left(\frac{5}{6}\right) = \frac{5}{6^5} \approx .0006.$$

In the same way, we can compute the probability of 4 ones in 6 rolls of a die. The probability of 4 successes and 2 failures will be

$$\left(\frac{1}{6}\right)^4\left(\frac{5}{6}\right)^2$$

Again, the desired outcomes can occur in more than one way. To find the number of alternative outcomes, we can use combinations. We want to find the number of ways in which we can combine 4 successes and 2 failures–that is, we want $C(6, 4)$ (or $C(6, 2)$). Since $C(6, 4) = 6!/(4!2!) = 15$, we have

$$\begin{aligned} P(4 \text{ ones in 6 rolls of a die}) &= 15\left(\frac{1}{6}\right)^4\left(\frac{5}{6}\right)^2 \\ &= \frac{15 \cdot 25}{6^6} \\ &\approx .008. \end{aligned}$$

In general, for experiments of this kind where the same experiment is repeated with the probability of success in a single trial the same for all trials, we have the following theorem.

THEOREM 9.9 If p is the probability of success in a single trial, the probability of x successes and $n - x$ failures in n independent repeated trials of an experiment is

$$\boldsymbol{C(n, x)p^x(1 - p)^{n-x}}.$$

In Theorem 9.9, if we let $1 - p = q$ and $x = r$, we have

$$C(n, r)p^rq^{n-r},$$

the expression for one of the terms of the binomial expansion of $(p + q)^n$. For example, in the problem discussed above, if we expand $\left(\frac{1}{6} + \frac{5}{6}\right)^6$, we have

$$\begin{aligned} \left(\frac{1}{6} + \frac{5}{6}\right)^6 = \left(\frac{1}{6}\right)^6 &+ C(6, 5)\left(\frac{1}{6}\right)^5\left(\frac{5}{6}\right) + C(6, 4)\left(\frac{1}{6}\right)^4\left(\frac{5}{6}\right)^2 \\ &+ C(6, 3)\left(\frac{1}{6}\right)^3\left(\frac{5}{6}\right)^3 + C(6, 2)\left(\frac{1}{6}\right)^2\left(\frac{5}{6}\right)^4 \\ &+ C(6, 1)\left(\frac{1}{6}\right)\left(\frac{5}{6}\right)^5 + \left(\frac{5}{6}\right)^6. \end{aligned}$$

The terms of this binomial expansion give respectively the probabilities of 6 ones, 5 ones, 4 ones, 3 ones, 2 ones, 1 one, and 0 ones in 6 rolls of a die. This result can be generalized as a theorem.

THEOREM 9.10 The rth term of the binomial expansion of $(p + q)^n$ (where $q = 1 - p$) gives the probability of getting $n - (r - 1)$ successes and $r - 1$ failures in n repeated trials.

Example 22 Find the probability of getting 7 heads in 8 tosses of a coin.

The probability of success (getting a head) in a single toss is 1/2. Thus the probability of failure (getting a tail) is 1/2. We have

$$\begin{aligned} P(7 \text{ heads in 8 tosses of a coin}) &= C(8, 7)\left(\frac{1}{2}\right)^7\left(\frac{1}{2}\right) \\ &= 8\left(\frac{1}{2}\right)^8 \\ &\approx .0312. \end{aligned}$$

Example 23 Assuming that all selections of items for a sample are independent trials, find the probability of one defective item in a sample of 15 items from a production line, if the probability that an item will be defective is .01.

The probability of success (a defective item) is .01, while the probability of failure (a good item), is .99. Thus,

$$\begin{aligned} P(1 \text{ defective item in 15 items}) &= C(15, 1)(.01)(.99)^{14} \\ &= 15(.01)(.99)^{14} \\ &\approx .130. \end{aligned}$$

If it seems wrong to call a defective item a success, one could call it a failure and find the probability of 14 successes in 15 trials. In that case we have $p = .99$ and $q = .01$.

$$\begin{aligned} P(14 \text{ good items in 15 items}) &= C(15, 14)(.99)^{14}(.01) \\ &= 15(.99)^{14}(.01) \\ &\approx .130. \end{aligned}$$

9.6 Exercises

1. Find the probability that a family with seven children will have exactly five boys. (Assume the probability of having a boy is 1/2.)
2. Find the probability of hitting a bull's eye eight out of ten times if the probability of getting a bull's eye in a single trial is .6.

3. A die is rolled twelve times. Find the probability of
   (a) a 1 twelve times.
   (b) six 1's out of twelve tries.
   (c) exactly one 1 in the twelve tries.
4. A coin is tossed five times. Find the probability of
   (a) all heads.
   (b) exactly three heads.
   (c) at least three heads.
5. A factory tests a sample of twenty bulbs for defectives. The probability that a particular bulb will be defective has been established by past experience to be .05.
   (a) What is the probability that there are no defectives in the sample?
   (b) What is the probability that the number of defectives in the sample is at most two?
6. In a ten question multiple choice test with five choices for each question, a student who was not prepared guessed on each item. Find the probability that he gets
   (a) six questions correct.
   (b) at least eight correct.
   (c) less than eight correct.

9.7 Mathematical Induction

Most of the reasoning used in this course has been **deductive reasoning.** By deductive reasoning, we mean the kind of reasoning which procedes from a hypothesis to a conclusion, forming a chain of reasoning using previously established facts as shown below.

To prove: Given that A is true, show that C is true.
Hypothesis: A is true.
Previously known fact: If A is true, then B is true.
Conclusion: B is true.
Previously known fact: If B is true, then C is true.
Conclusion: C is true.

There is another kind of reasoning called **inductive reasoning.** Inductive reasoning is used whenever a generalization is made based on a number of observations. This kind of reasoning is often used by scientists. On the basis of the outcome of experiments in the laboratory, a scientist may theorize that a certain fact is generally true. Further repetitions of the experiment may tend to strengthen the claim or may refute it. However, no matter how many times experimental results confirm the theory, it is never absolutely certain. There is always the chance that it may be wrong.

In mathematics, this type of reasoning is used to generalize from examples. Any conclusion arrived at in this manner must then be proved by deductive methods. One such method, called **mathematical induction,** depends on the **Axiom of Mathematical Induction:**

▶ If a given open sentence with the variable n is true for $n = 1$, and if the truth of the sentence for $n = k$, k any natural number, implies its truth for $n = k + 1$, then the sentence is true for every natural number.

The method of mathematical induction is particularly useful in proving theorems which claim that some property is true for all natural numbers. A typical example of the kind of result best proved by mathematical induction is given below.

Prove The sum of the first n natural numbers is $\frac{n(n+1)}{2}$.

To prove this, we must show that the statement is true when n is replaced by any of the natural numbers 1, 2, 3, ... That is, we must show

$$1 + 2 + 3 + \cdots + n = \frac{n(n+1)}{2},$$

for all natural number values of n.

Using inductive reasoning we could easily establish the truth of the statement for the first few natural numbers.

$$1 = \frac{1(1+1)}{2}$$

$$1 + 2 = \frac{2(2+1)}{2}$$

$$1 + 2 + 3 = \frac{3(3+1)}{2}$$

$$1 + 2 + 3 + 4 = \frac{4(4+1)}{2},$$

and so on. No matter how many such examples we show to be true, we still have not *proved* the result to be true for *all* natural numbers. However, the result can be proved by mathematical induction.

In a proof by mathematical induction, two steps are required.

(1) Show that the statement is true for $n = 1$.
(2) Assume that the statement is true for some natural number k. Then show that this implies that the statement is true for the next natural number, $k + 1$.

We show in step 2 that *if* the statement is true for the natural number k, *then* it will also be true for the next natural number, $k + 1$. In step 1 we show that the statement is true for the natural number 1. Then, by step 2, it is true for 2, and hence for 3, and so on, for any value of n.

Now, let us use mathematical induction to prove the statement discussed above.

Prove $1 + 2 + 3 + \cdots + n = \dfrac{n(n+1)}{2}$ for any natural number n.

Proof

STEP 1. We must show that $1 = \dfrac{1(1+1)}{2}$.

Since
$$\frac{1(1+1)}{2} = \frac{1(2)}{2} = 1,$$

the statement is true for $n = 1$.

STEP 2. Assume the statement is true for some natural number k, so that

$$1 + 2 + 3 + \cdots + k = \frac{k(k+1)}{2}.$$

Now we must show that this implies

$$1 + 2 + 3 + \cdots + (k+1) = \frac{(k+1)[(k+1)+1]}{2}.$$

Our hypothesis is

$$1 + 2 + 3 + \cdots + k = \frac{k(k+1)}{2}.$$

By adding the quantity $(k + 1)$ to both sides of the equation, we have

$$1 + 2 + 3 + \cdots + (k+1) = \frac{k(k+1)}{2} + (k+1).$$

Then, factoring on the right, we have

$$= (k+1)\left(\frac{k}{2} + 1\right)$$

$$= (k+1)\left(\frac{k+2}{2}\right)$$

$$1 + 2 + 3 + \cdots + (k+1) = \frac{(k+1)[(k+1)+1]}{2}.$$

The final result is the statement we wished to establish for $n = (k + 1)$. Thus, we have shown that if the statement is true for $n = k$, it is also true for $n = (k + 1)$ and hence for any natural number value of n.

Example 24 Use the method of mathematical induction to prove that for all natural numbers n, $n < 2^n$.

Proof

STEP 1. $$1 < 2^1$$

STEP 2. Assuming $k < 2^k$, we want to show that $(k + 1) < 2^{(k+1)}$. If we multiply both sides of the hypothesis, $k < 2^k$, by 2, we have

$$2k < 2 \cdot 2^k$$
$$2k < 2^{k+1}.$$

Also, $$2k = k + k,$$
and $$k + 1 \leq k + k \text{ (since } 1 \leq k)$$
Then, $$k + 1 \leq 2k < 2^{(k+1)},$$
or $$k + 1 < 2^{(k+1)}.$$

Thus, the statement is true for $(k + 1)$ whenever it is true for k, and the proof is completed.

The binomial theorem of the previous sections

$$(x + y)^n = x^n + \frac{n!}{(n-1)!1!}x^{n-1}y + \frac{n!}{(n-2)!2!}x^{n-2}y^2 + \cdots + y^n,$$

can be proved by the method of mathematical induction. To begin, we show that the theorem holds for $n = 1$.

$$(x + y)^1 = x^1 + y^1 = x + y$$

Now, assume that the theorem is true for $n = k$. We wish to show that the theorem will then be true for $n = k + 1$, or that

$$(x + y)^{k+1} = x^{k+1} + \frac{(k+1)!}{k!1!}x^k y + \frac{(k+1)!}{(k-1)!2!}x^{k-1}y^2 + \cdots + y^{k+1}.$$

From the assumption that the theorem holds for $n = k$, we have

$$(x + y)^k = x^k + \frac{k!}{(k-1)!1!}x^{k-1}y + \frac{k!}{(k-2)!2!}x^{k-2}y^2 + \cdots + y^k.$$

We can multiply both sides of this equation by $(x + y)$.

$$(x + y)^k(x + y) = \left(x^k + \frac{k!}{(k-1)!1!}x^{k-1}y + \cdots + y^k\right)(x + y)$$

$$(x + y)^{k+1} = x^{k+1} + \frac{k!}{(k-1)!1!}x^k y + \frac{k!}{(k-2)!2!}x^{k-1}y^2$$
$$+ \cdots + xy^k + x^k y + \frac{k!}{(k-1)!1!}x^{k-1}y^2$$
$$+ y^{k+1}$$

Collect like terms on the right-hand side.

$$= x^{k+1} + \left[\frac{k!}{(k-1)!1!} + 1\right]x^k y$$

$$+ \left[\frac{1}{2} + \frac{1}{k-1}\right]\left[\frac{k!}{(k-2)!1!}x^{k-1}y^2\right] + \cdots + y^{k+1}$$

$$= x^{k+1} + \frac{(k+1)!}{k!1!}x^k y + \frac{(k+1)!}{(k-1)!2!}x^{k-1}y^2 + \cdots + y^{k+1}$$

This gives the desired expression for $n = k + 1$. (In the second term, the coefficient of $x^k y$ was simplified as follows.)

$$\frac{k!}{(k-1)!1!} + 1 = \frac{k! + (k-1)!}{(k-1)!1!}$$

$$= \frac{k(k-1)! + (k-1)!}{(k-1)!1!}$$

$$= \frac{(k+1)(k-1)!}{(k-1)!1!}$$

$$= \frac{(k+1)(k)(k-1)!}{k(k-1)!1!}$$

$$= \frac{(k+1)!}{k!1!}$$

Thus, by the axiom of mathematical induction, the theorem is true for any natural number n.

9.7 Exercises

Use the method of mathematical induction to prove the following statements. Assume n is a positive integer.

1. $2 + 4 + 6 + \cdots + 2n = n(n+1)$.
2. $1 + 3 + 5 + \cdots + (2n-1) = n^2$.
3. $3 + 6 + 9 + \cdots + 3n = \frac{3n(n+1)}{2}$.
4. $5 + 10 + 15 + \cdots + 5n = \frac{5n(n+1)}{2}$
5. $2 + 4 + 8 + \cdots + 2^n = 2^{n+1} - 2$.
6. $1^2 + 2^2 + 3^2 + \cdots + n^2 = \frac{n(n+1)(2n+1)}{6}$.
7. $1^3 + 2^3 + 3^3 + \cdots + n^3 = \frac{n^2(n+1)^2}{4}$.
8. $(a^m)^n = a^{mn}$, (assume m is a constant).

9. $\frac{1}{2} + \frac{1}{2^2} + \frac{1}{2^3} + \cdots + \frac{1}{2^n} = 1 - \frac{1}{2^n}.$

10. $(a + d) + (a + 2d) + (a + 3d) + \cdots + (a + nd) = \frac{n}{2}[2a + (n + 1)d],$
(assume a and d are constant.)

11. $ar + ar^2 + ar^3 + \cdots + ar^n = \frac{ar(1 - r^n)}{1 - r}$, (assume a and r are constant.)

12. $(1 + a)^n \geq 1 + a^n$, $(a \geq 0)$, (assume a is a constant.)

13. What is wrong with the following proof by mathematical induction?
Prove: Any natural number equals the next natural number.
To begin, we assume the statement true for some natural number k:

$$k = k + 1.$$

We must now show that the statement is true for $n = k + 1$. If we add 1 to both sides, we have

$$k + 1 = k + 1 + 1$$
$$k + 1 = k + 2.$$

Hence, if the statement is true for $n = k$, it is also true for $n = k + 1$. Thus, the theorem is proved.

APPENDIX

Table 1 Squares and Square Roots

n	n^2	$\sqrt{n}$	$\sqrt{10n}$	n	n^2	$\sqrt{n}$	$\sqrt{10n}$
1	1	1.000	3.162	51	2601	7.141	22.583
2	4	1.414	4.472	52	2704	7.211	22.804
3	9	1.732	5.477	53	2809	7.280	23.022
4	16	2.000	6.325	54	2916	7.348	23.238
5	25	2.236	7.071	55	3025	7.416	23.452
6	36	2.449	7.746	56	3136	7.483	23.664
7	49	2.646	8.367	57	3249	7.550	23.875
8	64	2.828	8.944	58	3364	7.616	24.083
9	81	3.000	9.487	59	3481	7.681	24.290
10	100	3.162	10.000	60	3600	7.746	24.495
11	121	3.317	10.488	61	3721	7.810	24.698
12	144	3.464	10.954	62	3844	7.874	24.900
13	169	3.606	11.402	63	3969	7.937	25.100
14	196	3.742	11.832	64	4096	8.000	25.298
15	225	3.873	12.247	65	4225	8.062	25.495
16	256	4.000	12.649	66	4356	8.124	25.690
17	289	4.123	13.038	67	4489	8.185	25.884
18	324	4.243	13.416	68	4624	8.246	26.077
19	361	4.359	13.784	69	4761	8.307	26.268
20	400	4.472	14.142	70	4900	8.367	26.458
21	441	4.583	14.491	71	5041	8.426	26.646
22	484	4.690	14.832	72	5184	8.485	26.833
23	529	4.796	15.166	73	5329	8.544	27.019
24	576	4,899	15.492	74	5476	8.602	27.203
25	625	5.000	15.811	75	5625	8.660	27.386
26	676	5.099	16.125	76	5776	8.718	27.568
27	729	5.196	16.432	77	5929	8.775	27.749
28	784	5.292	16.733	78	6084	8.832	27.928
29	841	5.385	17.029	79	6241	8.888	28.107
30	900	5.477	17.321	80	6400	8.944	28.284
31	961	5.568	17.607	81	6561	9.000	28.460
32	1024	5.657	17.889	82	6724	9.055	28.636
33	1089	5.745	18.166	83	6889	9.110	28.810
34	1156	5.831	18.439	84	7056	9.165	28.983
35	1225	5.916	18.708	85	7225	9.220	29.155
36	1296	6.000	18.974	86	7396	9.274	29.326
37	1369	6.083	19.235	87	7569	9.327	29.496
38	1444	6.164	19.494	88	7744	9.381	29.665
39	1521	6.245	19.748	89	7921	9.434	29.833
40	1600	6.325	20.000	90	8100	9.487	30.000
41	1681	6.403	20.248	91	8281	9.539	30.166
42	1764	6.481	20.494	92	8464	9.592	30.332
43	1849	6.557	20.736	93	8649	9.644	30.496
44	1936	6.633	20.976	94	8836	9.695	30.659
45	2025	6.708	21.213	95	9025	9.747	30.822
46	2116	6.782	21.448	96	9216	9.798	30.984
47	2209	6.856	21.679	97	9409	9.849	31.145
48	2304	6.928	21.909	98	9604	9.899	31.305
49	2401	7.000	22.136	99	9801	9.950	31.464
50	2500	7.071	22.361	100	10000	10.000	31.623

Table 2 Common Logarithms

n	0	1	2	3	4	5	6	7	8	9
1.0	.0000	.0043	.0086	.0128	.0170	.0212	.0253	.0294	.0334	.0374
1.1	.0414	.0453	.0492	.0531	.0569	.0607	.0645	.0682	.0719	.0755
1.2	.0792	.0828	.0864	.0899	.0934	.0969	.1004	.1038	.1072	.1106
1.3	.1139	.1173	.1206	.1239	.1271	.1303	.1335	.1367	.1399	.1430
1.4	.1461	.1492	.1523	.1553	.1584	.1614	.1644	.1673	.1703	.1732
1.5	.1761	.1790	.1818	.1847	.1875	.1903	.1931	.1959	.1987	.2014
1.6	.2041	.2068	.2095	.2122	.2148	.2175	.2201	.2227	.2253	.2279
1.7	.2304	.2330	.2355	.2380	.2405	.2430	.2455	.2480	.2504	.2529
1.8	.2553	.2577	.2601	.2625	.2648	.2672	.2695	.2718	.2742	.2765
1.9	.2788	.2810	.2833	.2856	.2878	.2900	.2923	.2945	.2967	.2989
2.0	.3010	.3032	.3054	.3075	.3096	.3118	.3139	.3160	.3181	.3201
2.1	.3222	.3243	.3263	.3284	.3304	.3324	.3345	.3365	.3385	.3404
2.2	.3424	.3444	.3464	.3483	.3502	.3522	.3541	.3560	.3579	.3598
2.3	.3617	.3636	.3655	.3674	.3692	.3711	.3729	.3747	.3766	.3784
2.4	.3802	.3820	.3838	.3856	.3874	.3892	.3909	.3927	.3945	.3962
2.5	.3979	.3997	.4014	.4031	.4048	.4065	.4082	.4099	.4116	.4133
2.6	.4150	.4166	.4183	.4200	.4216	.4232	.4249	.4265	.4281	.4298
2.7	.4314	.4330	.4346	.4362	.4378	.4393	.4409	.4425	.4440	.4456
2.8	.4472	.4487	.4502	.4518	.4533	.4548	.4564	.4579	.4594	.4609
2.9	.4624	.4639	.4654	.4669	.4683	.4698	.4713	.4728	.4742	.4757
3.0	.4771	.4786	.4800	.4814	.4829	.4843	.4857	.4871	.4886	.4900
3.1	.4914	.4928	.4942	.4955	.4969	.4983	.4997	.5011	.5024	.5038
3.2	.5051	.5065	.5079	.5092	.5105	.5119	.5132	.5145	.5159	.5172
3.3	.5185	.5198	.5211	.5224	.5237	.5250	.5263	.5276	.5289	.5302
3.4	.5315	.5328	.5340	.5353	.5366	.5378	.5391	.5403	.5416	.5428
3.5	.5441	.5453	.5465	.5478	.5490	.5502	.5514	.5527	.5539	.5551
3.6	.5563	.5575	.5587	.5599	.5611	.5623	.5635	.5647	.5658	.5670
3.7	.5682	.5694	.5705	.5717	.5729	.5740	.5752	.5763	.5775	.5786
3.8	.5798	.5809	.5821	.5832	.5843	.5855	.5866	.5877	.5888	.5899
3.9	.5911	.5922	.5933	.5944	.5955	.5966	.5977	.5988	.5999	.6010
4.0	.6021	.6031	.6042	.6053	.6064	.6075	.6085	.6096	.6107	.6117
4.1	.6128	.6138	.6149	.6160	.6170	.6180	.6191	.6201	.6212	.6222
4.2	.6232	.6243	.6253	.6263	.6274	.6284	.6294	.6304	.6314	.6325
4.3	.6335	.6345	.6355	.6365	.6375	.6385	.6395	.6405	.6415	.6425
4.4	.6435	.6444	.6454	.6464	.6474	.6484	.6493	.6503	.6513	.6522
4.5	.6532	.6542	.6551	.6561	.6571	.6580	.6590	.6599	.6609	.6618
4.6	.6628	.6637	.6646	.6656	.6665	.6675	.6684	.6693	.6702	.6712
4.7	.6721	.6730	.6739	.6749	.6758	.6767	.6776	.6785	.6794	.6803
4.8	.6812	.6821	.6830	.6839	.6848	.6857	.6866	.6875	.6884	.6893
4.9	.6902	.6911	.6920	.6928	.6937	.6946	.6955	.6964	.6972	.6981
5.0	.6990	.6998	.7007	.7016	.7024	.7033	.7042	.7050	.7059	.7067
5.1	.7076	.7084	.7093	.7101	.7110	.7118	.7126	.7135	.7143	.7152
5.2	.7160	.7168	.7177	.7185	.7193	.7202	.7210	.7218	.7226	.7235
5.3	.7243	.7251	.7259	.7267	.7275	.7284	.7292	.7300	.7308	.7316
5.4	.7324	.7332	.7340	.7348	.7356	.7364	.7372	.7380	.7388	.7396
n	0	1	2	3	4	5	6	7	8	9

Table 2 (continued)

n	0	1	2	3	4	5	6	7	8	9
5.5	.7404	.7412	.7419	.7427	.7435	.7443	.7451	.7459	.7466	.7474
5.6	.7482	.7490	.7497	.7505	.7513	.7520	.7528	.7536	.7543	.7551
5.7	.7559	.7566	.7574	.7582	.7589	.7597	.7604	.7612	.7619	.7627
5.8	.7634	.7642	.7649	.7657	.7664	.7672	.7679	.7686	.7694	.7701
5.9	.7709	.7716	.7723	.7731	.7738	.7745	.7752	.7760	.7767	.7774
6.0	.7782	.7789	.7796	.7803	.7810	.7818	.7825	.7832	.7839	.7846
6.1	.7853	.7860	.7868	.7875	.7882	.7889	.7896	.7903	.7910	.7917
6.2	.7924	.7931	.7938	.7945	.7952	.7959	.7966	.7973	.7980	.7987
6.3	.7993	.8000	.8007	.8014	.8021	.8028	.8035	.8041	.8048	.8055
6.4	.8062	.8069	.8075	.8082	.8089	.8096	.8102	.8109	.8116	.8122
6.5	.8129	.8136	.8142	.8149	.8156	.8162	.8169	.8176	.8182	.8189
6.6	.8195	.8202	.8209	.8215	.8222	.8228	.8235	.8241	.8248	.8254
6.7	.8261	.8267	.8274	.8280	.8287	.8293	.8299	.8306	.8312	.8319
6.8	.8325	.8331	.8338	.8344	.8351	.8357	.8363	.8370	.8376	.8382
6.9	.8388	.8395	.8401	.8407	.8414	.8420	.8426	.8432	.8439	.8445
7.0	.8451	.8457	.8463	.8470	.8476	.8482	.8488	.8494	.8500	.8506
7.1	.8513	.8519	.8525	.8531	.8537	.8543	.8549	.8555	.8561	.8567
7.2	.8573	.8579	.8585	.8591	.8597	.8603	.8609	.8615	.8621	.8627
7.3	.8633	.8639	.8645	.8651	.8657	.8663	.8669	.8675	.8681	.8686
7.4	.8692	.8698	.8704	.8710	.8716	.8722	.8727	.8733	.8739	.8745
7.5	.8751	.8756	.8762	.8768	.8774	.8779	.8785	.8791	.8797	.8802
7.6	.8808	.8814	.8820	.8825	.8831	.8837	.8842	.8848	.8854	.8859
7.7	.8865	.8871	.8876	.8882	.8887	.8893	.8899	.8904	.8910	.8915
7.8	.8921	.8927	.8932	.8938	.8943	.8949	.8954	.8960	.8965	.8971
7.9	.8976	.8982	.8987	.8993	.8998	.9004	.9009	.9015	.9020	.9025
8.0	.9031	.9036	.9042	.9047	.9053	.9058	.9063	.9069	.9074	.9079
8.1	.9085	.9090	.9096	.9101	.9106	.9112	.9117	.9122	.9128	.9133
8.2	.9138	.9143	.9149	.9154	.9159	.9165	.9170	.9175	.9180	.9186
8.3	.9191	.9196	.9201	.9206	.9212	.9217	.9222	.9227	.9232	.9238
8.4	.9243	.9248	.9253	.9258	.9263	.9269	.9274	.9279	.9284	.9289
8.5	.9294	.9299	.9304	.9309	.9315	.9320	.9325	.9330	.9335	.9340
8.6	.9345	.9350	.9355	.9360	.9365	.9370	.9375	.9380	.9385	.9390
8.7	.9395	.9400	.9405	.9410	.9415	.9420	.9425	.9430	.9435	.9440
8.8	.9445	.9450	.9455	.9460	.9465	.9469	.9474	.9479	.9484	.9489
8.9	.9494	.9499	.9504	.9509	.9513	.9518	.9523	.9528	.9533	.9538
9.0	.9542	.9547	.9552	.9557	.9562	.9566	.9571	.9576	.9581	.9586
9.1	.9590	.9595	.9600	.9605	.9609	.9614	.9619	.9624	.9628	.9633
9.2	.9638	.9643	.9647	.9652	.9657	.9661	.9666	.9671	.9675	.9680
9.3	.9685	.9689	.9694	.9699	.9703	.9708	.9713	.9717	.9722	.9727
9.4	.9731	.9736	.9741	.9745	.9750	.9754	.9759	.9763	.9768	.9773
9.5	.9777	.9782	.9786	.9791	.9795	.9800	.9805	.9809	.9814	.9818
9.6	.9823	.9827	.9832	.9836	.9841	.9845	.9850	.9854	.9859	.9863
9.7	.9868	.9872	.9877	.9881	.9886	.9890	.9894	.9899	.9903	.9908
9.8	.9912	.9917	.9921	.9926	.9930	.9934	.9939	.9943	.9948	.9952
9.9	.9956	.9961	.9965	.9969	.9974	.9978	.9983	.9987	.9991	.9996
n	0	1	2	3	4	5	6	7	8	9

ANSWERS TO SELECTED EXERCISES

1.1 Exercises (page 3)

1. $\{1, 2, 3, 4\}$ **3.** $\{-2, -1, 0, 1, 2\}$ **5.** $\{7, 9, 11, 13, \ldots\}$ **7.** $\{-1, 0, 1, 2, 3, 4, 6, 8\}$ **9.** $\{x|x \text{ is an integer}\} = U$ **11.** $\{-1\}$ **13.** P **15.** $\{-3, -2, -1\}$ **17.** $\{-1, 0, 2, 4, 6, 8\}$ **19.** $\{-1\}$ **21.** $\subset$ **23.** $\cup$ **25.** $\cap$ **27.** $\cup$ **29.** $=$ **31.** X and Y are disjoint **33.** $X \subset Y$ **35.** $X = \varnothing$ **37.** $X \subset Y$ **39.** no special requirements

1.2 Exercises (page 7)

1. true **3.** false $(0 - 1 = -1)$ **5.** true **7.** false $(0 \notin N)$ **9.** false (1/0 does not exist) **11.** commutative axiom **13.** distributive axiom **15.** associative axiom **17.** inverse axiom **19.** substitution axiom **21.** subset of, equals **23.** Set axioms include commutative axioms, associative axioms, identity axioms, closure axioms and distributive axioms. **25.** definition of subtraction; identity axiom of addition; substitution **27.** commutative axiom; associative axiom; inverse axiom; identity axiom of multiplication; substitution **29.** associative axiom; associative axiom; commutative axiom; associative axiom; associative axiom; substitution

1.3 Exercises (page 12)

1. commutative and associative axioms; additive inverse axiom; identity axiom; definition; definition; additive inverse is unique **3.** addition property of equality; associative axiom; additive inverse axiom; identity axiom; definition of subtraction **5.** definition of division; Exercise 4 above; definition of division **7.** reflexive property; commutative and associative axioms; Exercise 6 above, part II

1.4 Exercises (page 17)

1. false **3.** false **5.** true **7.** transitive axiom **9.** Theorem 1.8(b)
11. Theorem 1.8(a)

	Greatest Lower bound	*Least Upper bound*
13.	none	6
15.	1	4
17.	none	none

19. definition of $<$; Theorem 1.2; distributive axiom; given; closure axiom; definition of $<$
21. given; Theorem 1.8(c); Exercise 9 of Section 1.3; substitution

PART II

$(-1)(-a) > (-1)(-b)$	Exercise 20
$1 \cdot a > 1 \cdot b$	Theorem 1.7(d)
$a > b$	identity

23. definition of $<$; definition of $<$; addition property of equality and substitution; associative and commutative axioms; closure axiom; definition of $<$

1.5 Exercises (page 22)

1. $=$ **3.** $\leq$ **5.** $=$ **7.** $=$ **9.** $\leq, \leq$ **11.** $\geq, \leq$ **13.** $=$ **15.** $=$
17. $\leq$ **19.** $=$ **21.** $=$

23.

25.

3 6

27.

1 3

29.

2 5

31. $\varnothing$ **33.** $\{1, -1\}$ **35.** $\{-4, -1\}$ **37.** $\{3, -9\}$

2.1 Exercises (page 27)

1. 30 **3.** 3 **5.** $-3/2$ **7.** $2m + 1$ **9.** $2m^2 + 4m$ **11.** $70 + 24h + 2h^2$
13. $8x^2 - 16x + 6$ **15.** $2x^2 + 4xr + 2r^2 - 4x - 4r$ **17.** 1 **19.** 448
21. $x^2 - x + 3$ **23.** $9y^2 - 4y + 4$ **25.** $-14q^2 + 11q - 14$
27. $28r^2 + r - 2$ **29.** $18p^2 - 27p - 35$ **31.** $25r^2 - 4$ **33.** $4k^2 - 12k + 9$
35. $12k^4 + 21k^3 - 5k^2 + 3k + 2$ **37.** $8z^3 - 12z^2 + 6z - 1$ **39.** $8m^3 + 1$
41. $216m^9$ **43.** $\frac{-96}{m^7}$ **45.** y^{1+2m} **47.** $\frac{-2}{m^{n-2}}$
49. If $n = 0$, we have $a^m \cdot a^0 = a^m \cdot 1 = a^m$. Also, $a^{m+n} = a^{m+0} = a^m$.
Hence, $a^m a^n = a^{m+n}$.

2.2 Exercises (page 31)

1. $2m^2 - 4m + 8$ **3.** $3y + 2$ **5.** $5p - 1$ **7.** $3x + 2 + \frac{-2}{4x - 5}$

9. $4z - 3 + \frac{-5}{2z + 5}$ **11.** $x^2 - 3x + 2$ **13.** $5x^2 - 3x + 4 + \frac{4}{3x + 4}$

15. $2z^3 - z^2 - \frac{1}{2}z - \frac{5}{4} + \frac{(-\frac{13}{4})z + \frac{17}{4}}{2z^2 + z + 1}$ **17.** $m^3 - m^2 - 6m$

19. $3x^2 + 4x + \frac{3}{x - 5}$ **21.** $4m^2 - 4m + 1 + \frac{-3}{m + 1}$

23. $x^4 + x^3 + 2x - 1 + \frac{3}{x + 2}$ **25.** $\frac{1}{3}x^2 - \frac{1}{9}x + \frac{1}{x - \frac{1}{3}}$

27. $x^3 + x^2 + x + 1$ **29.** $x^3 + x^2 + x + 1 + \frac{2}{x - 1}$

2.3 Exercises (page 34)

1. $4m(3n - 2)$ **3.** $2px(3x - 4x^2 - 6)$ **5.** $(4p - 1)(p + 1)$
7. $(2z - 5)(z + 6)$ **9.** $(3q + 4)(2q - 3)$ **11.** $3(2r - 1)(2r + 5)$
13. $(6r - 5s)(3r + 2s)$ **15.** $(8pq + 11mn)(8pq - 11mn)$
17. $(9m^2 + 4n^4)(3m + 2n^2)(3m - 2n^2)$ **19.** $x^2(3 - x)^2$
21. $(m + 1)(m^2 - m + 1)(m - 1)(m^2 + m + 1)$
23. $(2m^2 - 3n)(4m^4 + 6m^2n + 9n^2)$ **25.** $(2x - y)y$
27. $(3x - y)(19y^2 - 24xy + 9x^2)$ **29.** $(x + y + 5z)(x + y - 3z)$
31. $[3(a + b) + 8c][2(a + b) - 5c]$ **33.** $4pq$ **35.** $(m^n + 4)(m^n - 4)$
37. $(x^n - y^{2n})(x^{2n} + x^ny^{2n} + y^{4n})$ **39.** $(2x^n + 3y^n)(x^n - 13y^n)$

2.4 Exercises (page 37)

1. $\frac{4}{x - 1}$ **3.** $a + 2$ **5.** 1 **7.** $\frac{m - 4}{m + 4}$ **9.** $\frac{x^2 - 1}{x^2}$ **11.** $\frac{x + y}{x - y}$

13. $\frac{x^2 - xy + y^2}{x^2 + xy + y^2}$ **15.** 1 **17.** $\frac{2y + 1}{y(y + 1)}$ **19.** $\frac{3}{2(a + b)}$ **21.** $\frac{-2}{(a + 1)(a - 1)}$

23. $\frac{2m^2 + 2}{(m - 1)(m + 1)}$ **25.** $\frac{4}{a - 2}$ **27.** $\frac{5}{(a - 2)(a - 3)(a + 2)}$

29. $\frac{x - 11}{(x + 4)(x - 4)(x - 3)}$ **31.** $\frac{p + 5}{p(p + 1)}$ **33.** $\frac{-1}{x(x + h)}$ **35.** $\frac{x + 1}{x - 1}$

37. $\frac{(2 - b)(1 + b)}{b(1 - b)}$ **39.** $\frac{2x + 1}{x + 1}$ **41.** $\frac{3}{8}$

2.5 Exercises (page 41)

1. $\frac{1}{2^3} = \frac{1}{8}$ **3.** $\frac{1}{(-4)^3} = \frac{-1}{64}$ **5.** 8 **7.** $-\frac{1}{8}$ **9.** $2^6 = 64$ **11.** $\frac{x^2}{2y}$

13. p^4r^5 **15.** $\frac{x^6}{4y^4}$ **17.** $\frac{9m^5}{n^3}$ **19.** $a + b$ **21.** -1 **23.** $\frac{1}{ab}$

25. $\frac{y(xy - 9)}{x^2y^2 - 9}$

2.6 Exercises (page 45)

1. 2 **3.** 4 **5.** $3^{-2} = \frac{1}{9}$ **7.** $(\frac{2}{3})^{-3} = \frac{27}{8}$ **9.** $5^{-2} = \frac{1}{25}$ **11.** $4p^2$
13. $9x^4$ **15.** $\frac{4z^2}{x^5y}$ **17.** $\frac{r^2 - 2r + 1}{r}$ **19.** 1 **21.** $x^{11/12}$ **23.** $\frac{16}{y^{11/12}}$
25. $m^{3/2}$ **27.** r^{6+p} **29.** $\frac{x+1}{x^{1/2}}$ **31.** $12^{1/2}$ **33.** $x^{2/3}$ **35.** $y^{4/3}$
37. $y^{9/2}$ **39.** $m^{5/6}$ **41.** (a) Since $1^2 < 2$, $1 \in A$. (b) If $x = 2$, we have $x^2 = 4$. If $x^2 < 2$, then $x^2 < 4$, and 2 is an upper bound. (c) A has a least upper bound (d) $\sqrt{2}$

2.7 Exercises (page 47)

1. $3\sqrt[3]{3}$ **3.** $\frac{2\sqrt[4]{2}}{3}$ **5.** $72\sqrt{2}$ **7.** $3\sqrt{5}$ **9.** $9\sqrt{3}$ **11.** $\sqrt{2}$ **13.** $2\sqrt[3]{2}$
15. $2\sqrt{3}$ **17.** $7x^2z^4\sqrt{2x}$ **19.** $3py^2\sqrt[3]{3p^2x^2}$ **21.** $y\sqrt[3]{x^{2/3}}$ **23.** $\frac{\sqrt{6x}}{3x}$
25. $\frac{3\sqrt{2y}}{2y}$ **27.** $\frac{2\sqrt[3]{x}}{x}$ **29.** $\frac{x^2y\sqrt{xy}}{z}$ **31.** $\frac{a^2b\sqrt[4]{bc}}{c}$ **33.** $\frac{xy^2\sqrt[3]{9x^2zw}}{z^2w}$
35. $\frac{r\sqrt[4]{rs^3t^2}}{t}$ **37.** $\frac{\sqrt[4]{x^2y^3}}{y^2}$ **39.** $\frac{m\sqrt[3]{n^2}}{n}$ **41.** $ab\sqrt{ab}(b - 2a^2 + b^3)$
43. $\frac{x\sqrt{x+1}}{x+1}$ **45.** $\frac{2\sqrt{1+r}}{1+r}$ **47.** $-2x - 2\sqrt{x(x+1)} - 1$ **49.** $\frac{3}{2\sqrt{3}}$
51. $\frac{-1}{2(1-\sqrt{2})}$ **53.** $\frac{x}{\sqrt{x}+x}$ **55.** $\frac{-1}{2x - 2\sqrt{x(x+1)} + 1}$

3.1 Exercises (page 53)

1. Domain: all reals; Range: all reals; Function **3.** Domain: all reals; Range: all reals; not a Function **5.** Domain: all reals; Range: nonnegative reals; Function **7.** Domain: $\{x \mid -4 \le x \le 4\}$; Range: $\{y \mid 0 \le y \le 4\}$; Function **9.** Domain: all reals; Range: $\{y \mid y = -2\}$: Function **11.** Domain: $\{x \mid x \neq 0\}$; Range: $\{y \mid y \neq 0\}$; Function **13.** -1 **15.** -10 **17.** 4 **19.** $3a - 1$ **21.** 11 **23.** 5 **25.** 2 **27.** 16 **29.** 1 **31.** 3 **33.** even **35.** even **37.** neither

3.2 Exercises (page 56)

1.

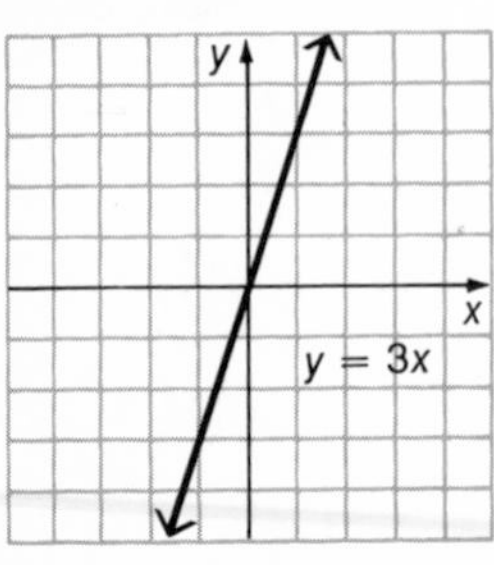

3.

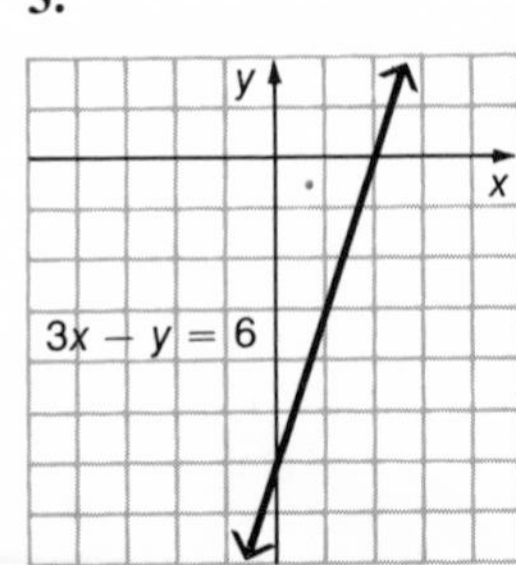

5.

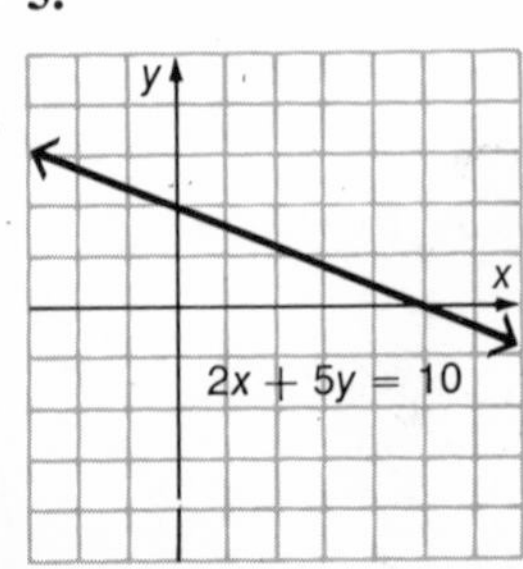

7.

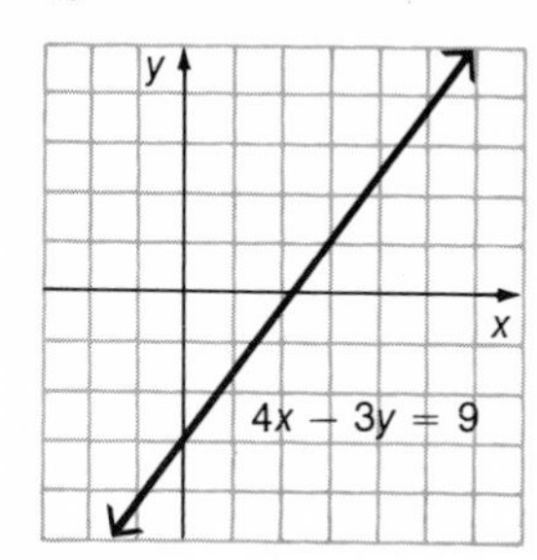

9.

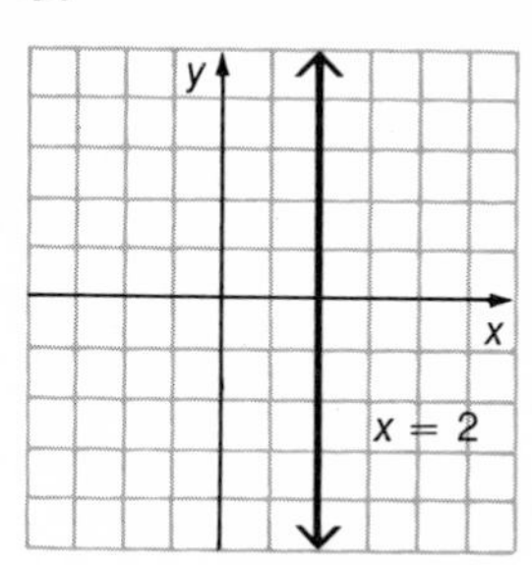

11.

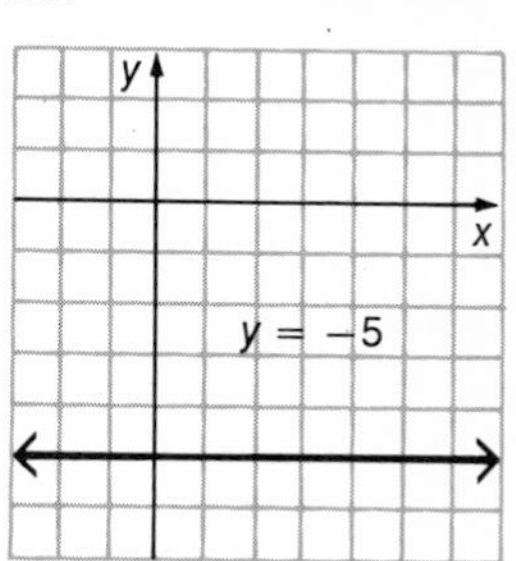

13. (4, 2) **15.** (5, −2) **17.** Lines do not intersect.

3.3 Exercises (page 60)

1. $m = 2$ **3.** $m = 4$ **5.** $m = -\frac{3}{4}$ **7.** $m = \frac{1}{5}$ **9.** $m = 7$ **11.** $m = 0$
13. $m = 0$ **15.** $y - 2x = 5$ **17.** $2y + 3x = -7$ **19.** $4y + 3x = 6$
21. $y = 2$ **23.** $x = -8$ **25.** $x + 3y = 10$ **27.** $4y - x = 13$
29. $4y + 3x = 12$ **31.** $2x - 3y = 6$ **33.** $x = -5$

35.

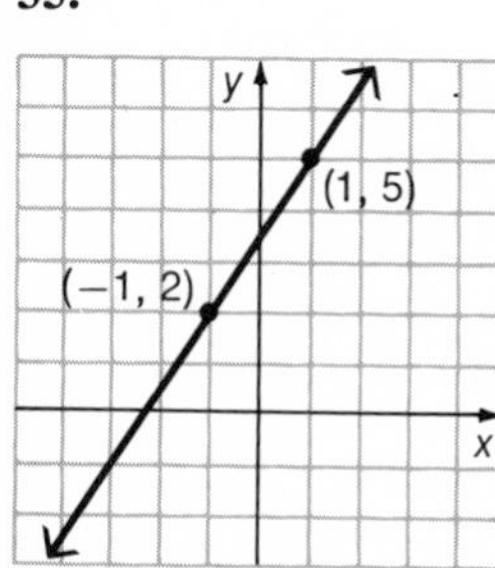

37.

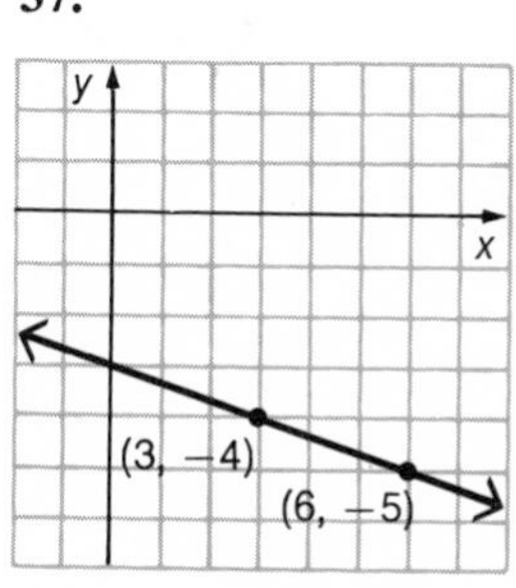

39.

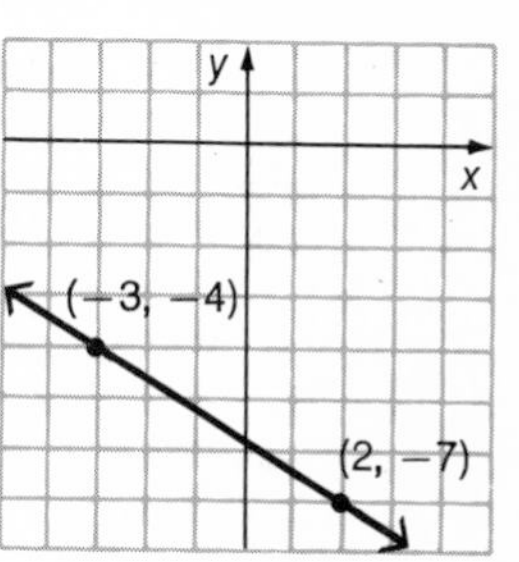

41. $x + 3y = 11$ **43.** $x - y = 7$ **45.** no

3.4 Exercises (page 64)

1. $x = 4$ **3.** all real numbers **5.** $x = 3$ **7.** $x = 12$ **9.** $m = -\frac{12}{5}$
11. $\varnothing$ **13.** $x = -\frac{5}{13}$ **15.** $x \leq 4$ **17.** $p \geq -1$ **19.** $a = 1, a = 3$
21. $m = 1, m = -\frac{1}{3}$ **23.** $x = \frac{8}{3}, x = \frac{2}{3}$ **25.** $-2 < a < 10$
27. $m < \frac{1}{3}$ or $m > 1$ **29.** $b \leq -\frac{3}{2}$ or $b \geq \frac{1}{4}$ **31.** $V = k/P$
33. $g = \dfrac{V - V_0}{t}$ **35.** $R = \dfrac{r_1 r_2}{r_1 + r_2}$ **37.** $\frac{18}{5}$ hours **39.** $\frac{37}{8}$ hours

3.5 Exercises (page 67)

1.

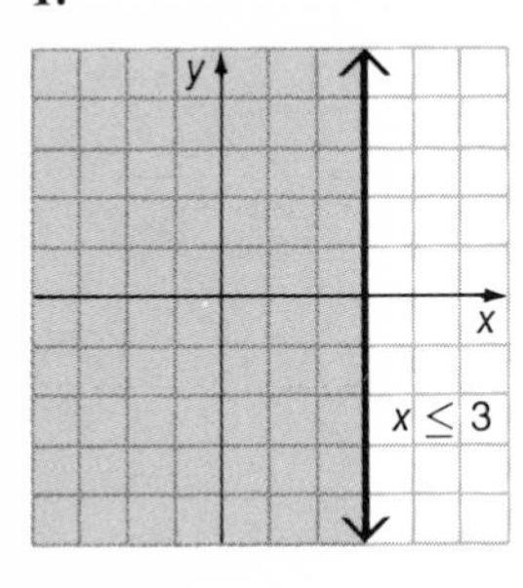

3.

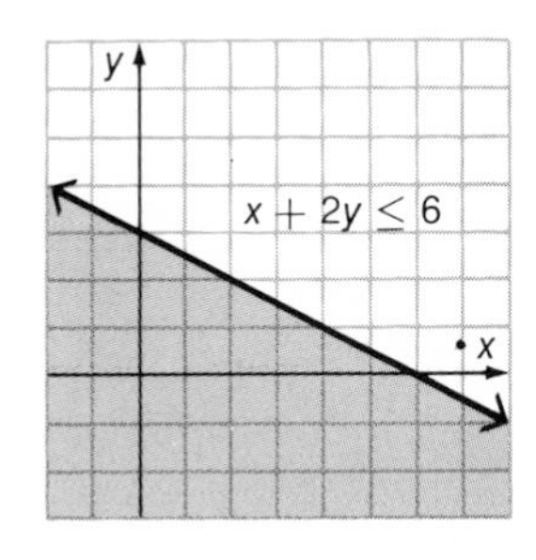

5.

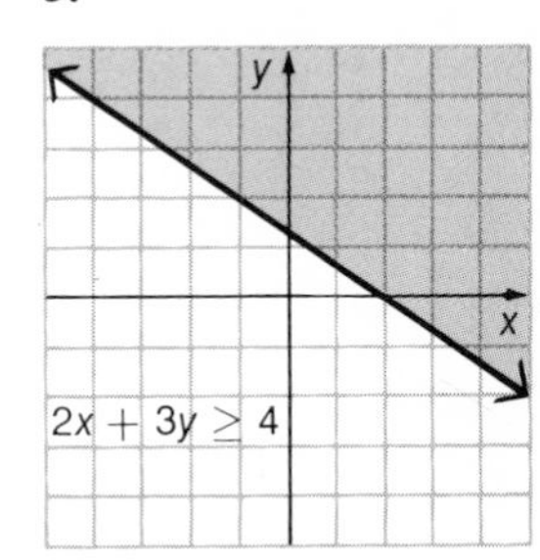

7.

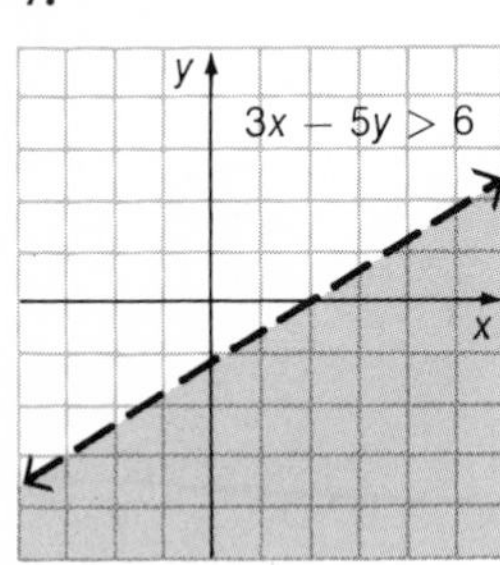

9.

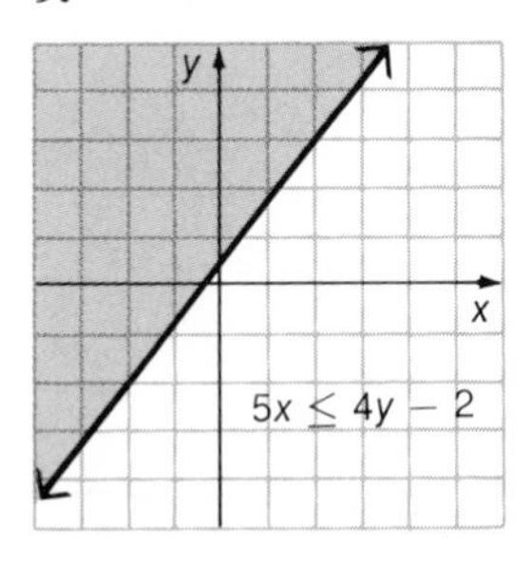

11.

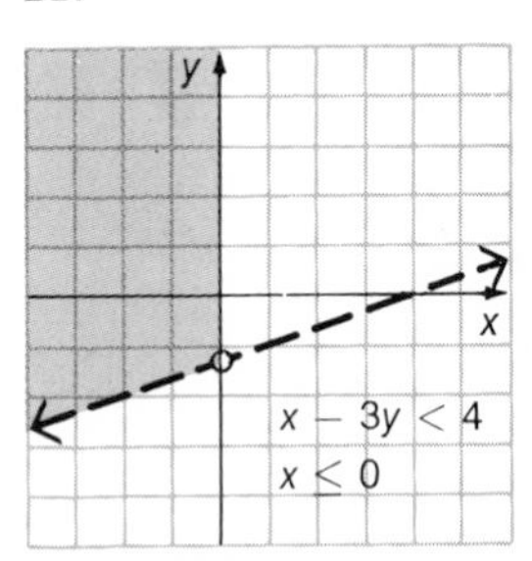

13.

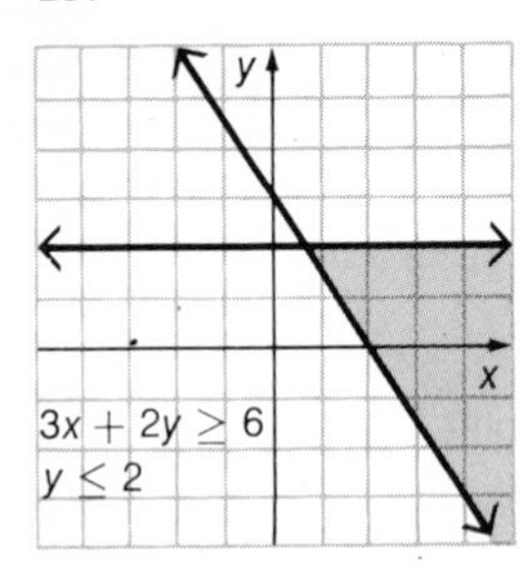

15.

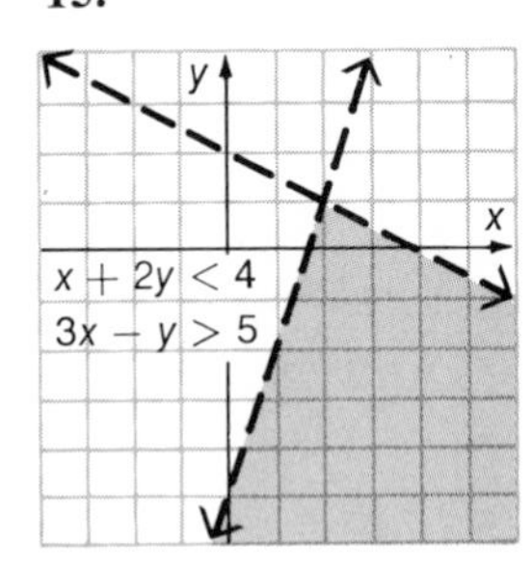

17.

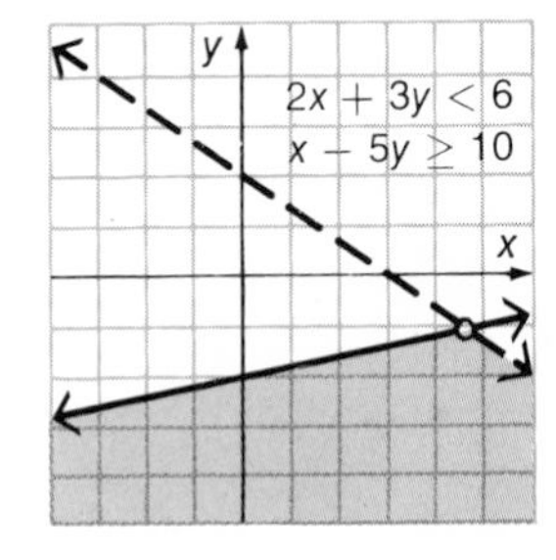

3.6 Exercises (page 71)

1.

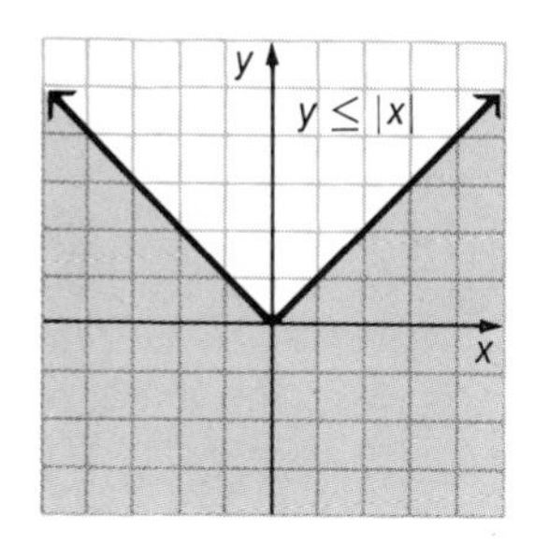

3.

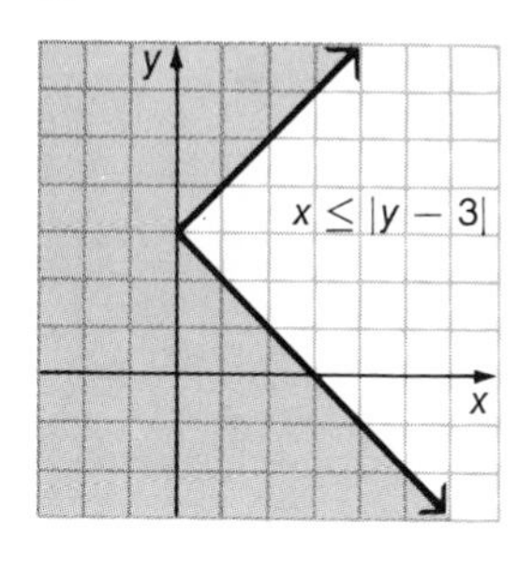

5.

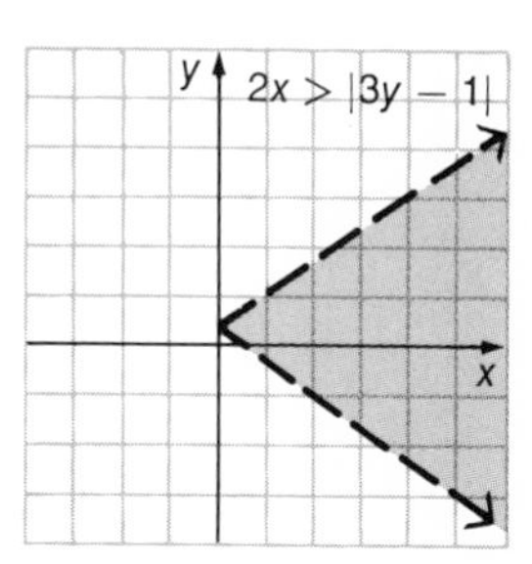

7.

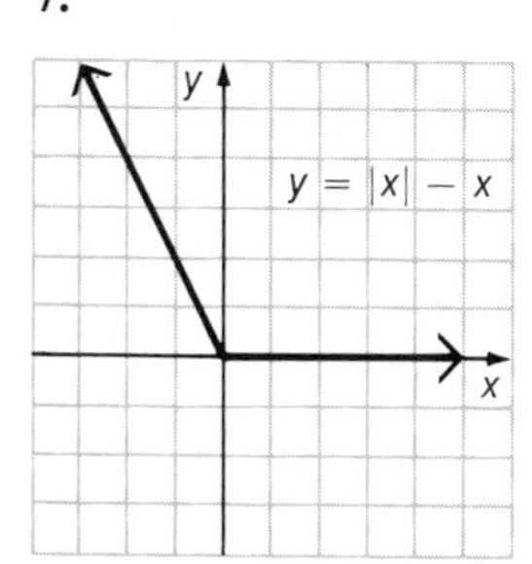

9.

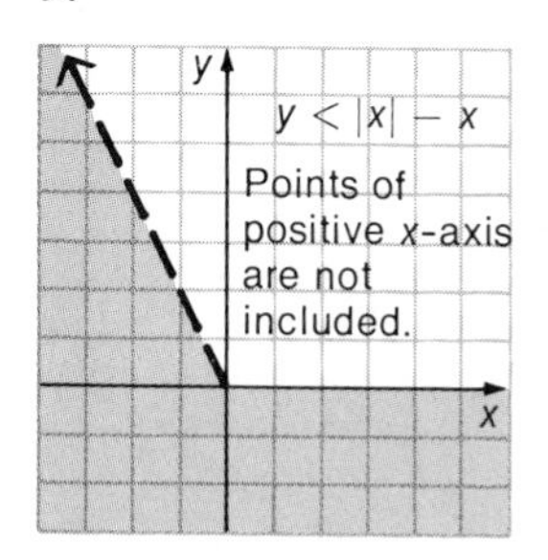

11.

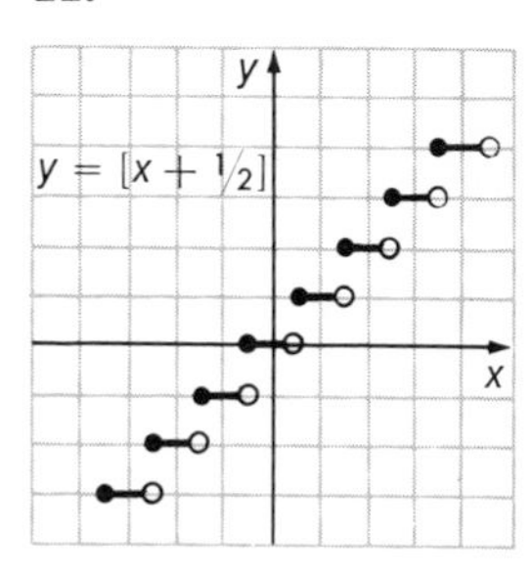

13.

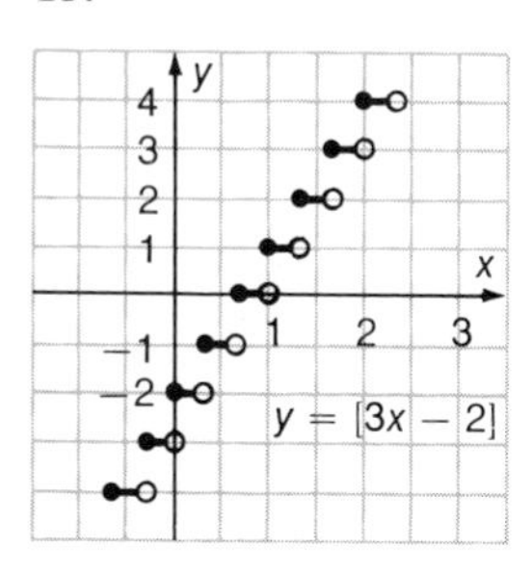

15.

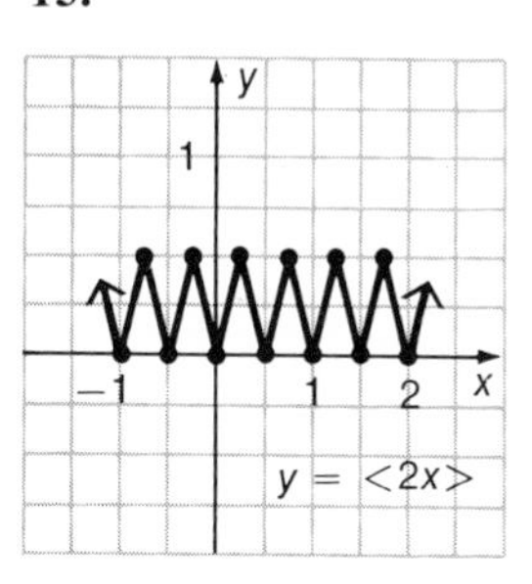

17.

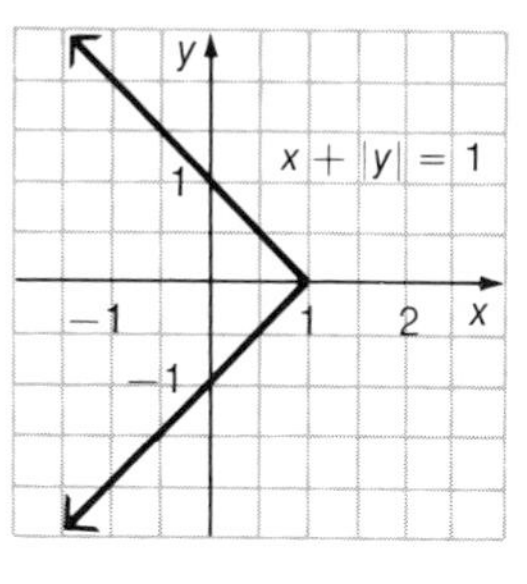

19.

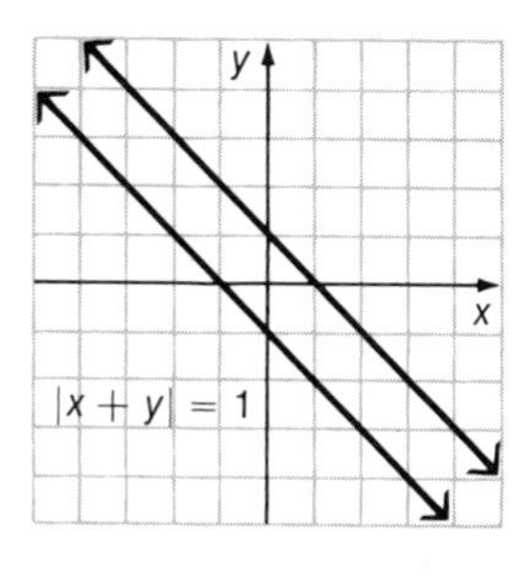

21.

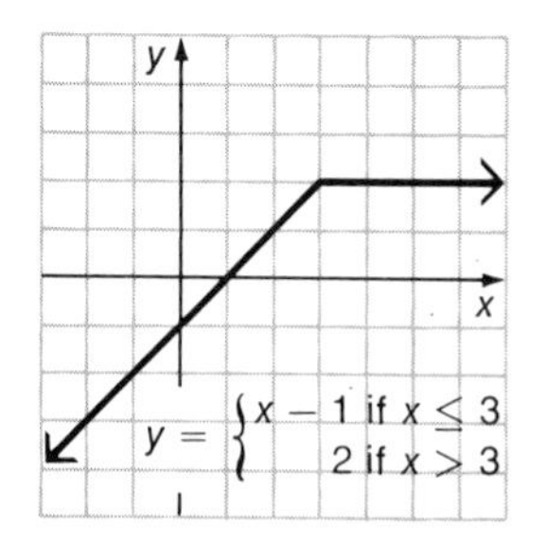

23.

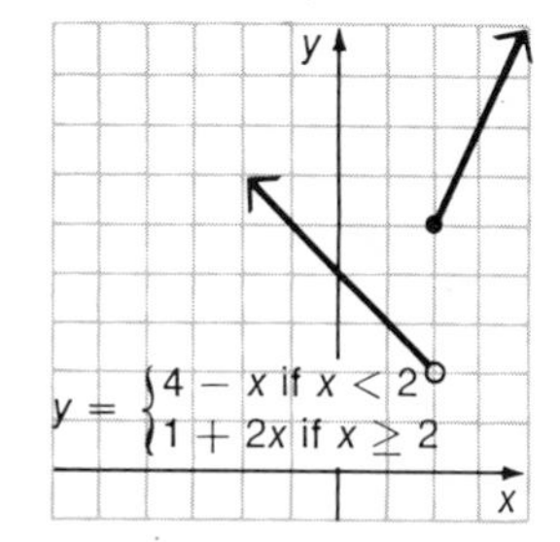

3.7 Exercises (page 74)

1. 19 **3.** 10 **5.** 7 **7.** -6 **9.** -9 **11.** -3 **13.** $x + 21$
15. $4m$ **17.** 215 **19.** 30 **21.** 125 **23.** 230 **25.** $\frac{205}{2}$ **27.** $\frac{155}{2}$
29. 6 **31.** 75 **33.** 231 **35.** 1281 **37.** 4680

4.1 Exercises (page 79)

1.

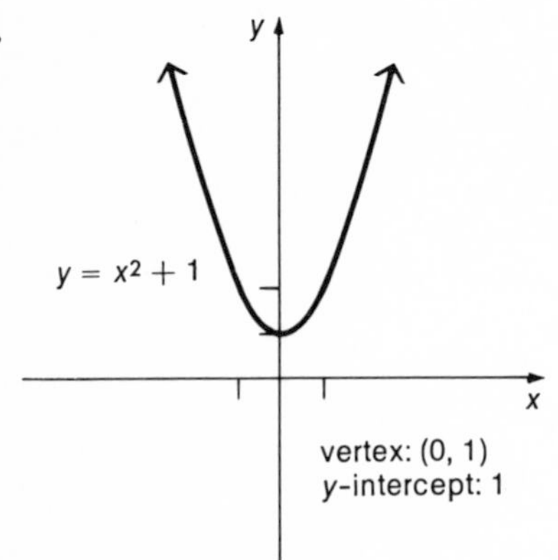

3.

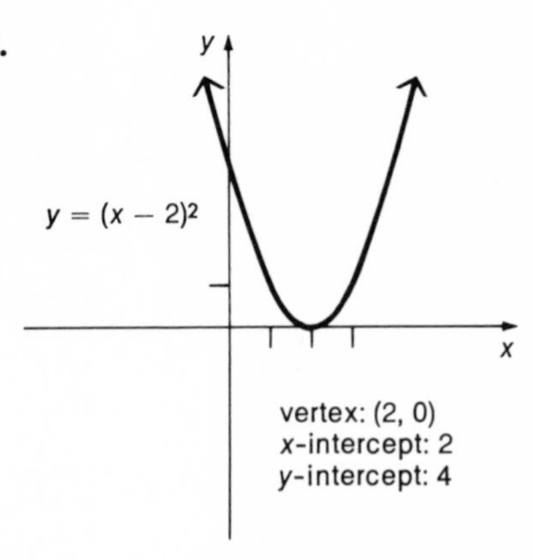

5.

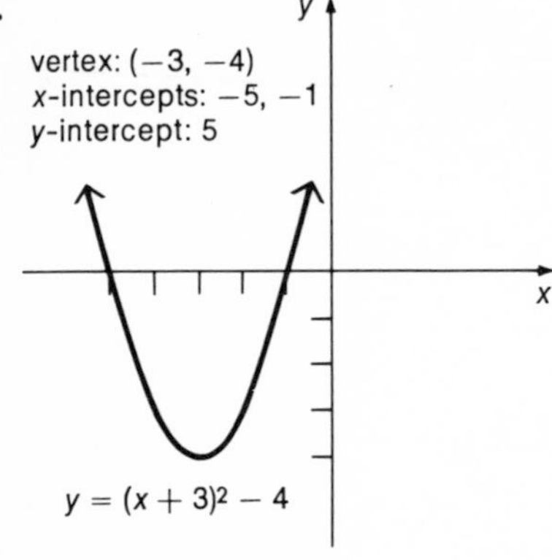

7.

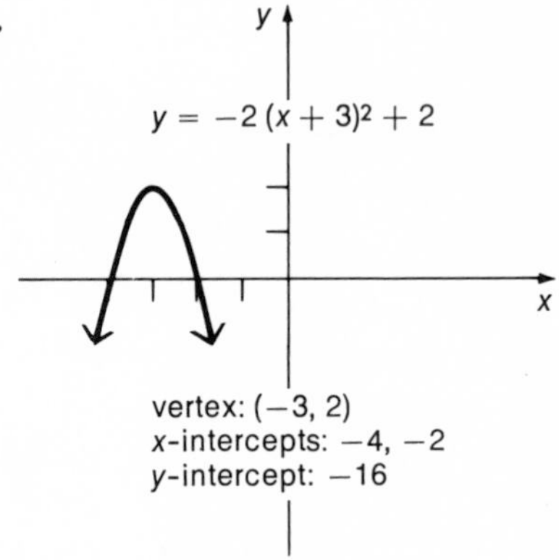

9.

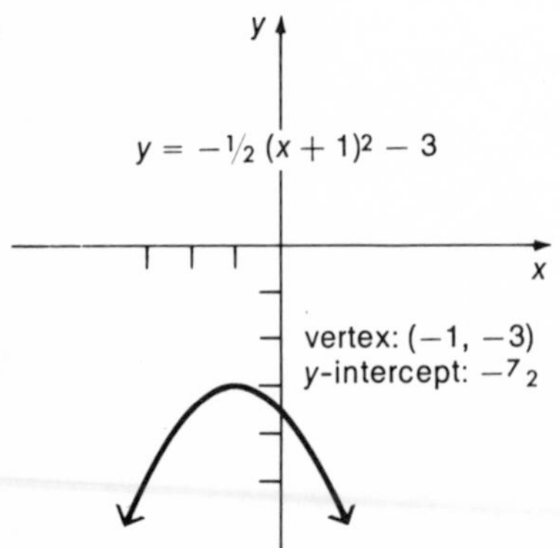

11.

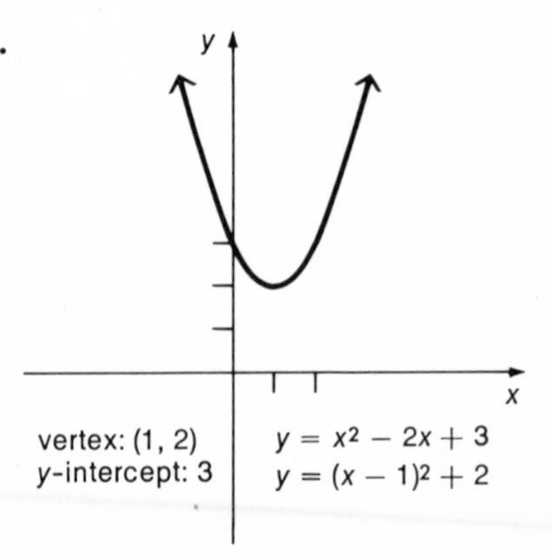

13.

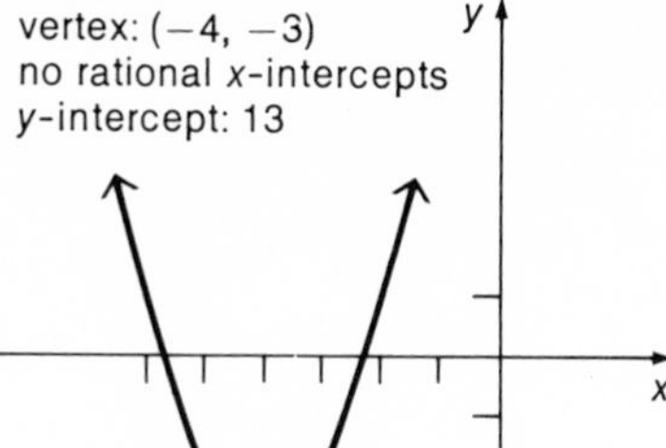

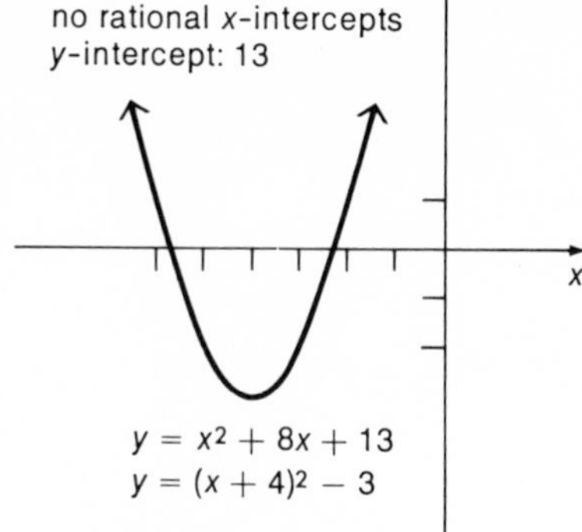

15.

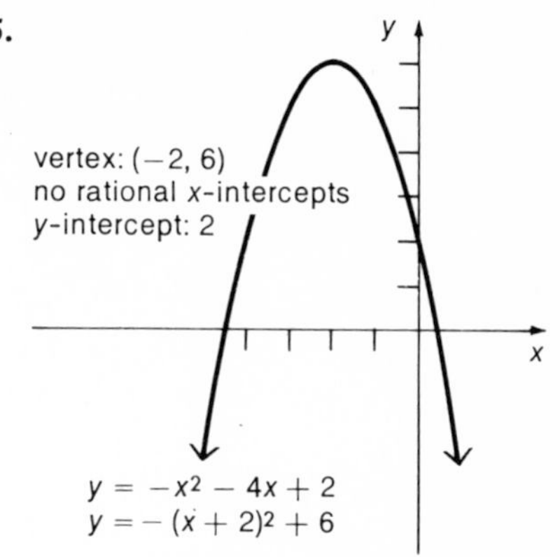

17.

y

$y = 2x^2 - 4x + 5$
$y = 2(x - 1)^2 + 3$

x

vertex: (1, 3)
y-intercept: 5

19.

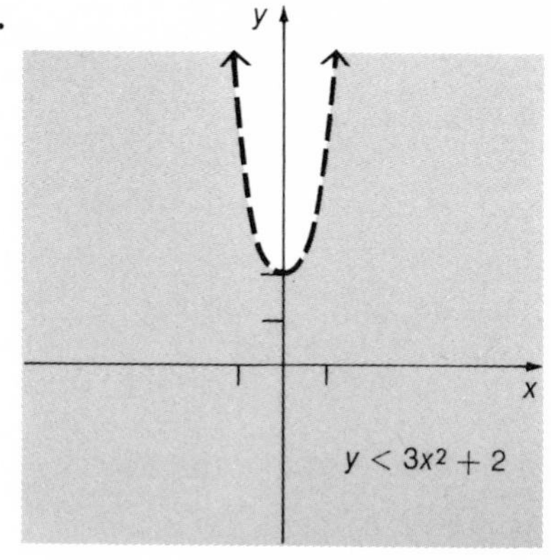

21.

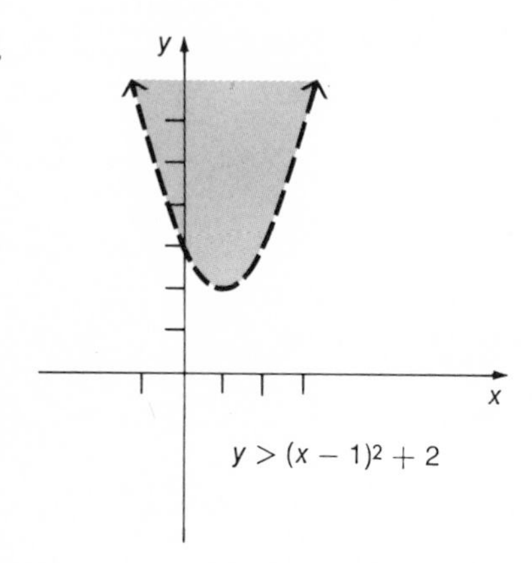

23. 80×160 **25.** 10 and 10

4.2 Exercises (page 82)

1. 0, 2, or -3 **3.** 2 or 3 **5.** $\frac{10}{3}$ or $-\frac{5}{2}$ **7.** $-\frac{3}{2}$ or $-\frac{1}{4}$

9. $\frac{1 + \sqrt{5}}{2}$ or $\frac{1 - \sqrt{5}}{2}$ **11.** $\frac{-1 + \sqrt{7}}{2}$ or $\frac{-1 - \sqrt{7}}{2}$

13. $\frac{-7+\sqrt{73}}{6}$ or $\frac{-7-\sqrt{73}}{6}$ **15.** $3+\sqrt{2}$ or $3-\sqrt{2}$ **17.** $-\frac{3}{4}$ or $\frac{1}{3}$

19. $\frac{5+5\sqrt{2}}{3}$ or $\frac{5-5\sqrt{2}}{3}$ **21.** $-\frac{1}{4}$ or 3 **23.** $\frac{3}{2}$ or 1

25. $\frac{\sqrt{2}+\sqrt{6}}{2}$ or $\frac{\sqrt{2}-\sqrt{6}}{2}$ **27.** $\sqrt{2}$ or $\sqrt{2}/2$

29. $\frac{-n+\sqrt{n^2-8m}}{2m}$ or $\frac{-n-\sqrt{n^2-8m}}{2m}$ **31.** 0; one real zero

33. 1; 2 real zeros **35.** 84; 2 real zeros **37.** -23; no real zeros

4.3 Exercises (page 85)

We give only the real solutions.

1. 1 **3.** $\sqrt{3}$ or $-\sqrt{3}$ **5.** $\frac{\sqrt{2}}{2}$ or $\frac{-\sqrt{2}}{2}$ **7.** $\frac{\sqrt{10}}{2}$, $\frac{-\sqrt{10}}{2}$, 1, or -1

9. 4 or 6 **11.** $\frac{-5+\sqrt{21}}{2}$ or $\frac{-5-\sqrt{21}}{2}$

13. $\frac{-6+2\sqrt{3}}{3}$ or $\frac{-6-2\sqrt{3}}{3}$ or $\frac{-4+\sqrt{2}}{2}$ or $\frac{-4-\sqrt{2}}{2}$

15. 1 or $\frac{-5+\sqrt{21}}{2}$ or $\frac{-5-\sqrt{21}}{2}$ **17.** -63 or 28 **19.** 2 or 3

21. $-\frac{5}{2}$ **23.** 9 **25.** 2

4.4 Exercises (page 89)

1. $-3 \leq x \leq 3$ **3.** $0 < y < 10$ **5.** $r \leq -2-\sqrt{3}$ or $r \geq -2+\sqrt{3}$
7. $-2 \leq x \leq 3$ **9.** $k < \frac{1}{2}$ or $k > 4$ **11.** $x \neq 0$ **13.** $x < 6$ or $x \geq \frac{15}{2}$
15. $k < 2$ or $k > 5$ **17.** $x \leq -2$ or $0 \leq x \leq 2$

4.5 Exercises (page 91)

1. $\sqrt{26}$ **3.** $\sqrt{53}$ **5.** the distance between (1,1) and (2, -2) is $\sqrt{10}$. The distance between (2, -2) and (3, -5) is $\sqrt{10}$. The distance between (1, 1) and (3, -5) is $2\sqrt{10}$. Therefore the three points lie on the same straight line.

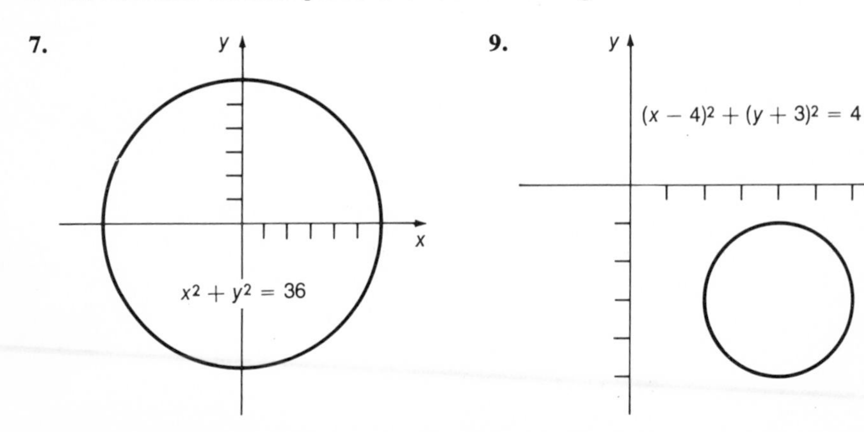

11.

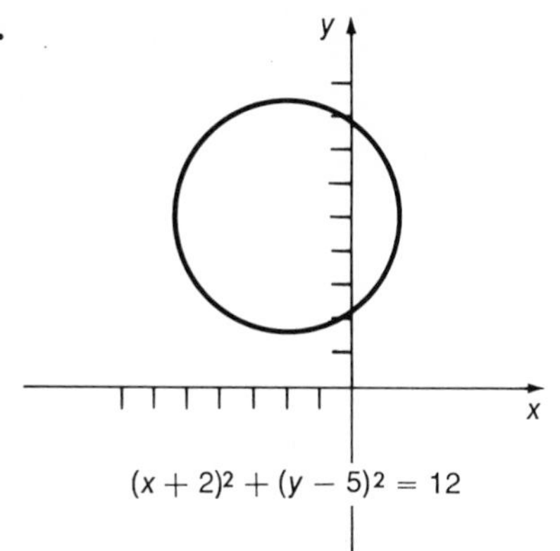

13.

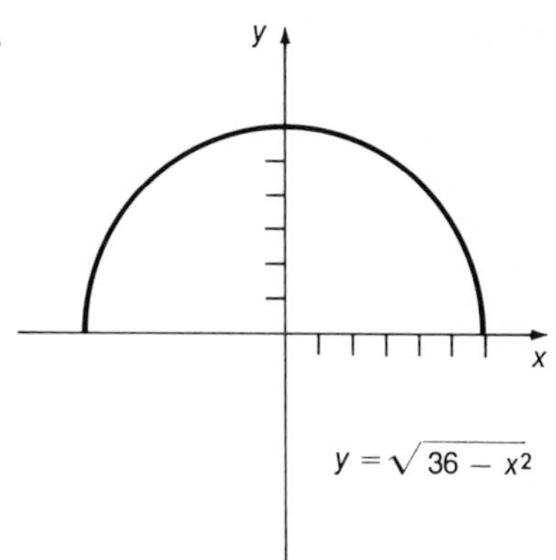

15.

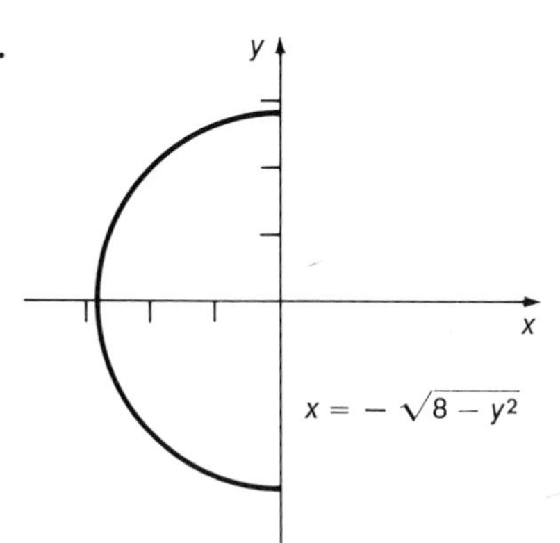

17.

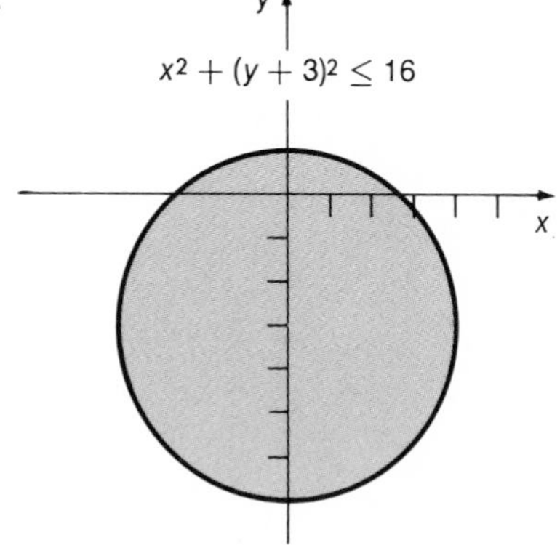

19.

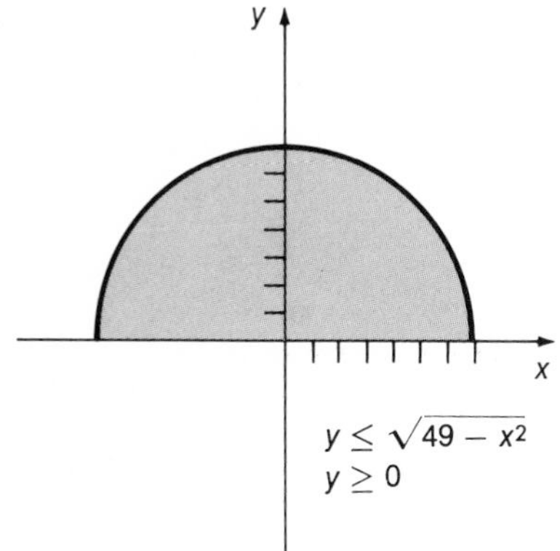

	center	radius
21.	$(-3, -4)$	4
23.	$(6, -5)$	6
25.	$(-4, 7)$	0

4.6 Exercises (page 98)

1.

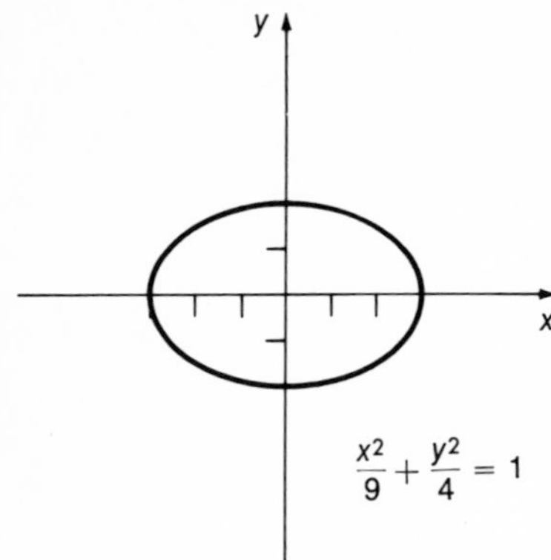

3.

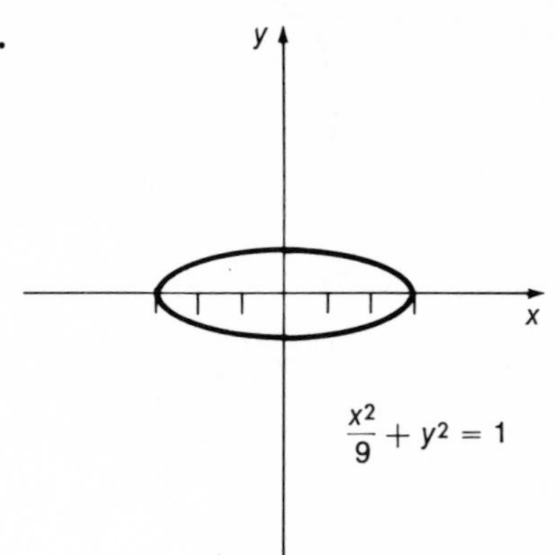

5.

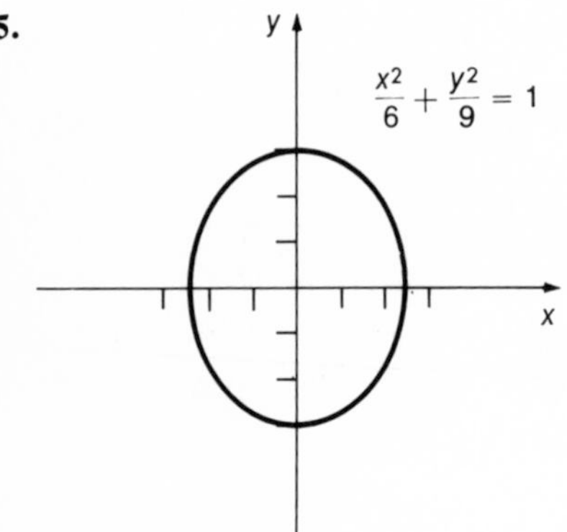

7.

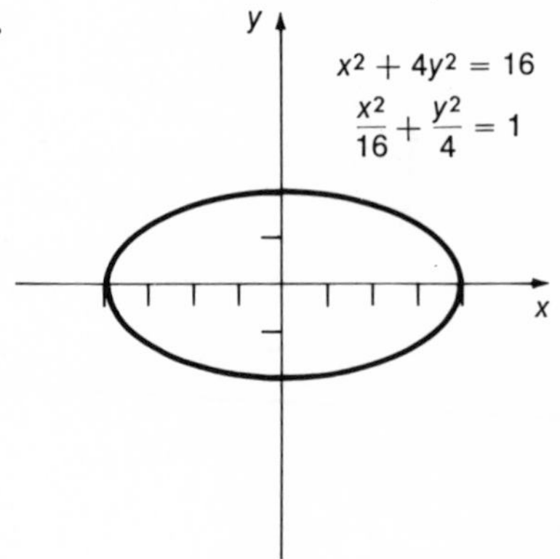

9.

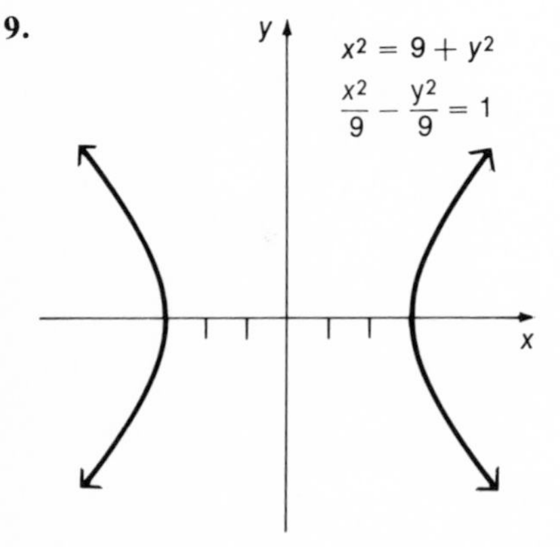

11.

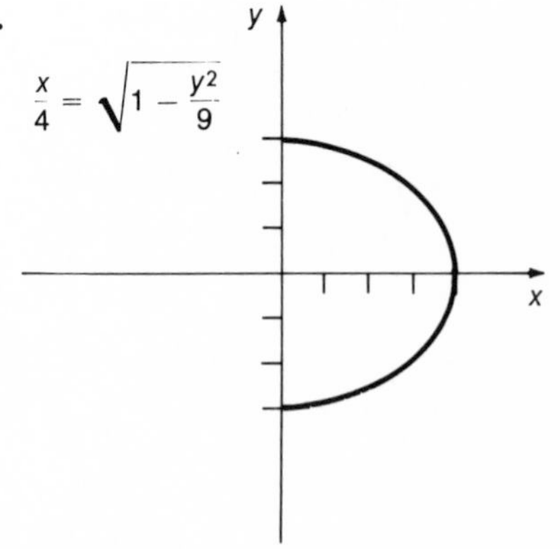

13.

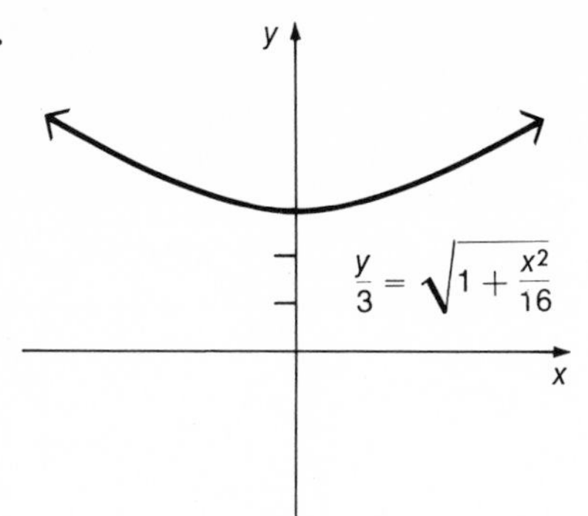

15.

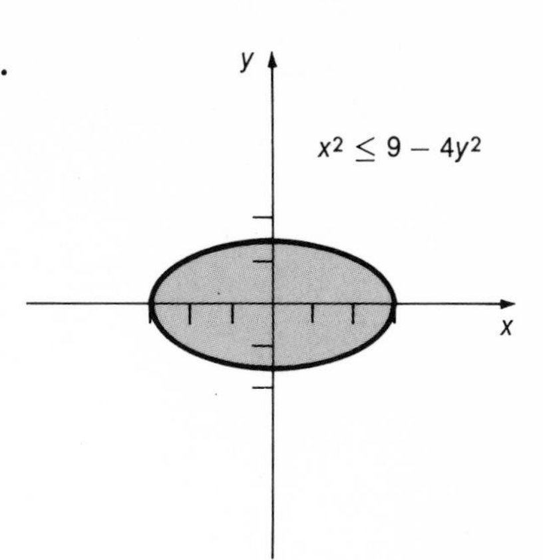

17.

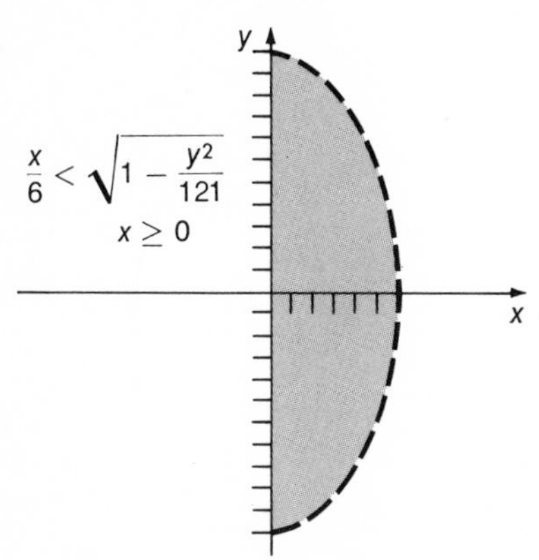

19. $y = \frac{5}{3}x$ or $y = -\frac{5}{3}x$ **21.** $y = \frac{1}{2}x$ or $y = -\frac{1}{2}x$

4.7 Exercises (page 101)

1. $\frac{70}{3}$ **3.** 2304 feet **5.** .0444 ohms **7.** 250 pounds

5.1 Exercises (page 106)

1.

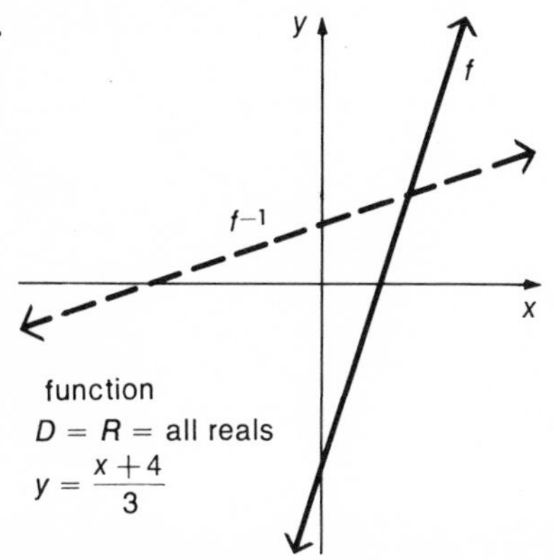

3.

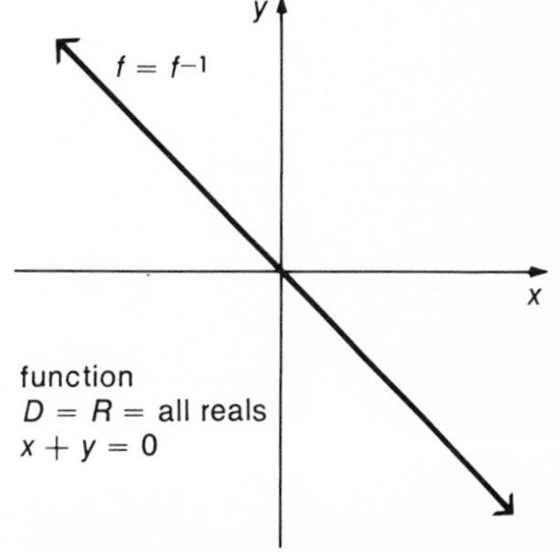

5.

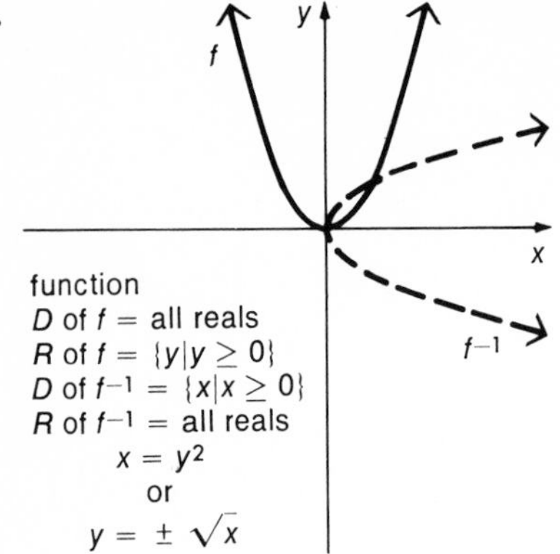

y
f
x
function
D of f = all reals
R of f = {y|y ≥ 0}
D of f⁻¹ = {x|x ≥ 0}
R of f⁻¹ = all reals
x = y²
or
y = ± √x
f⁻¹

7.

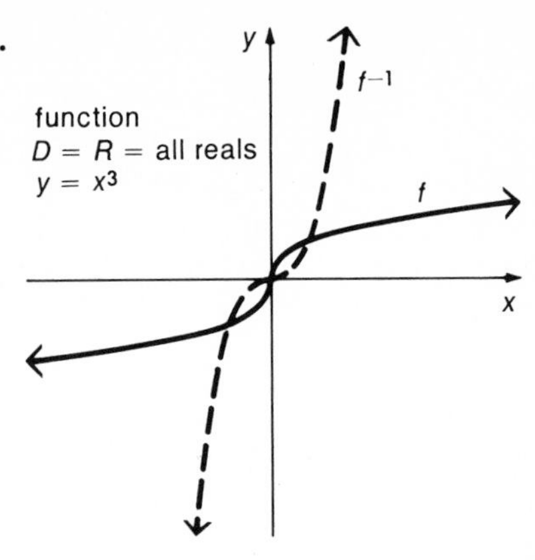

y
f⁻¹
function
D = R = all reals
y = x³
f
x

9.

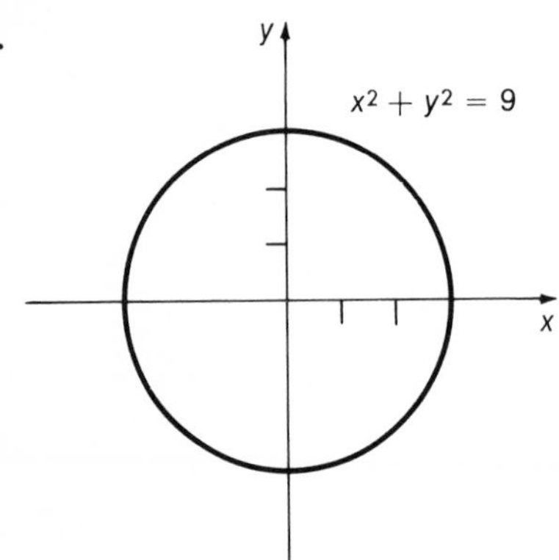

y
x² + y² = 9
x

11.

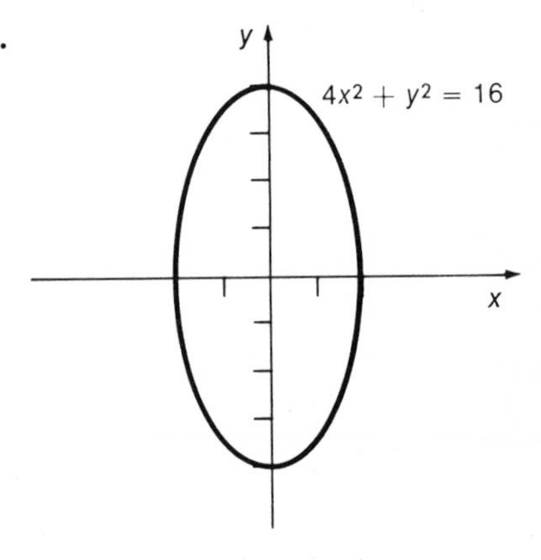

y
4x² + y² = 16
x

13.

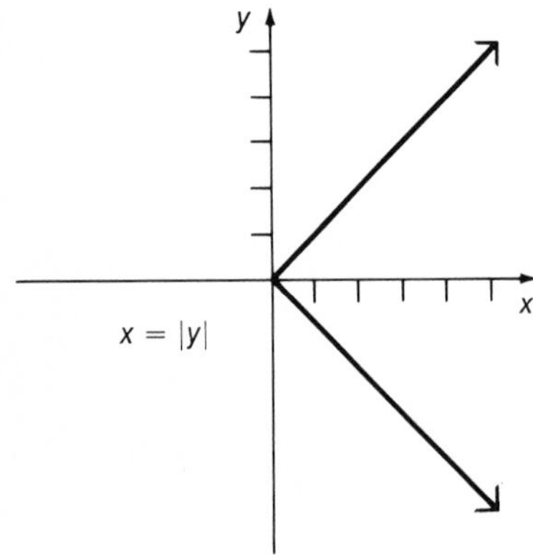

y
x
x = |y|

15.

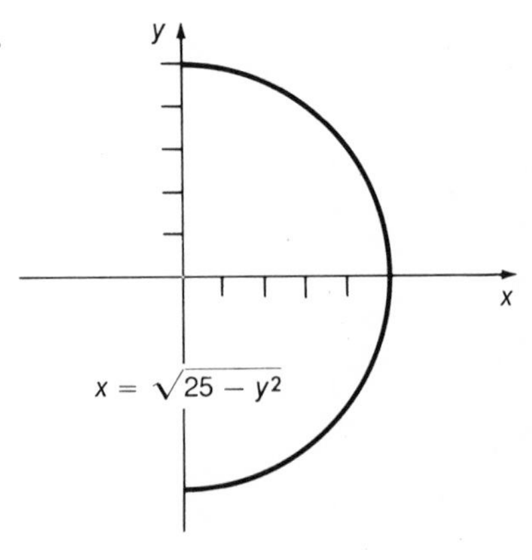

y
x
x = √(25 − y²)

17.

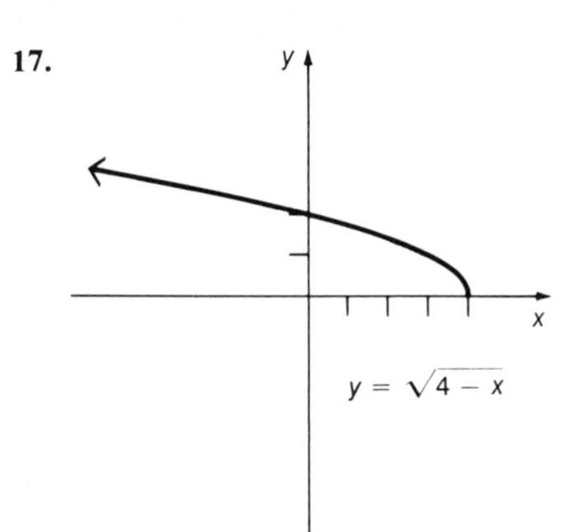

5.2 Exercises (page 109)

1.

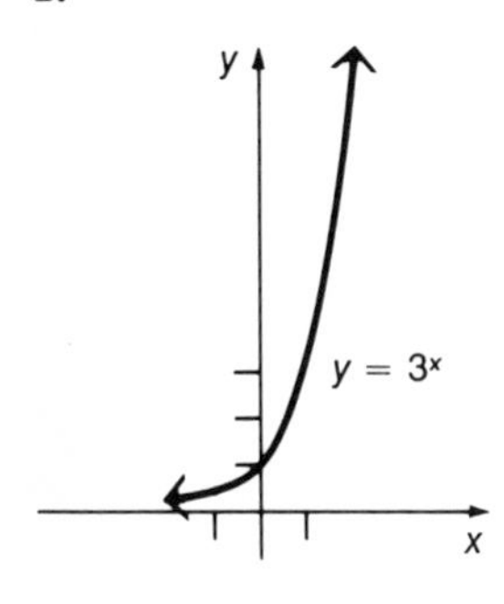

3.

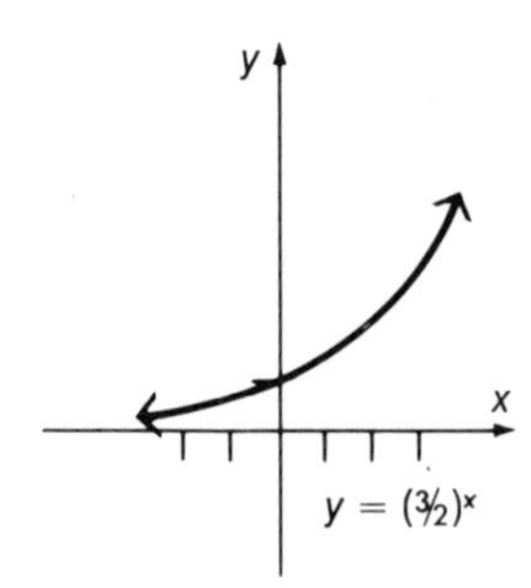

5.

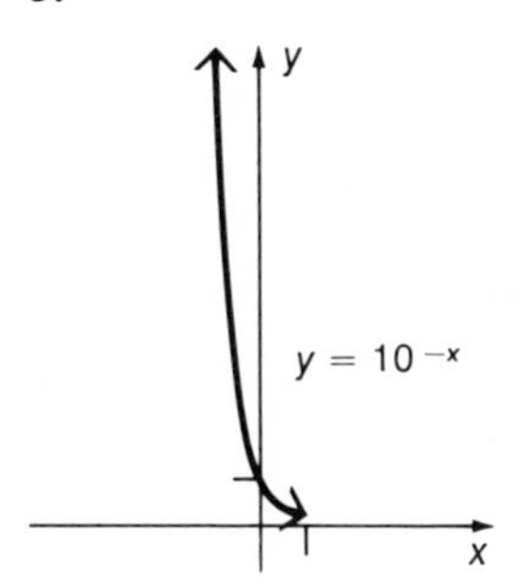

7.

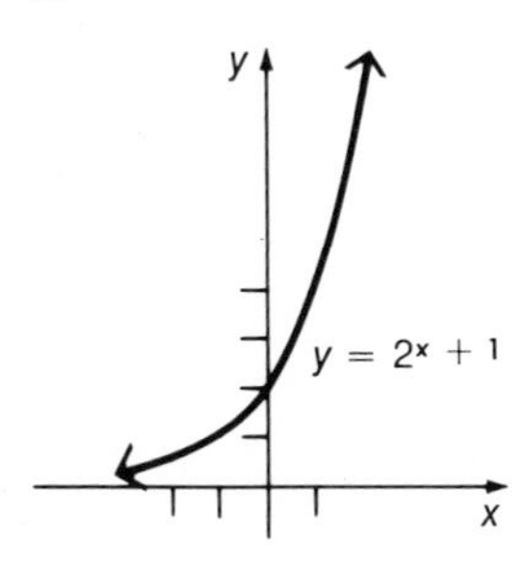

9.

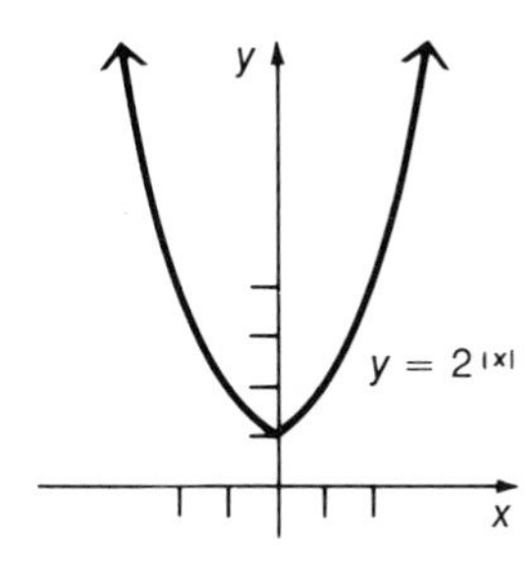

11. $y = 2^x + 2^{-x}$

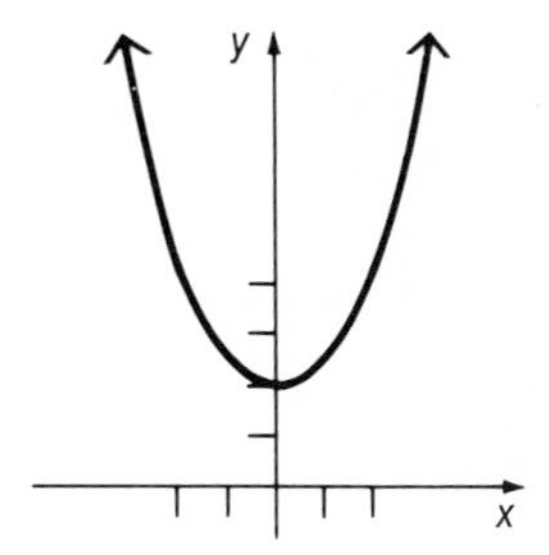

13. (a) 1.4 (b) 2.1

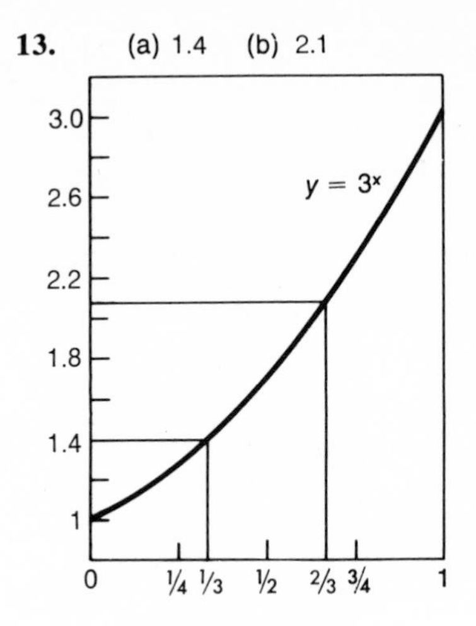

15. A series of points, one for each rational number on the number line.

5.3 Exercises (page 112)

1. $\log_3 81 = 4$ **3.** $\log_{10} 10000 = 4$ **5.** $\log_{1/2} 16 = -4$ **7.** $\log_{10} .0001 = -4$
9. $6^2 = 36$ **11.** $(\sqrt{3})^8 = 81$ **13.** $m^n = k$ **15.** $\log_a \left(\frac{xy}{m}\right)$
17. $\log_m \left(\frac{a^2}{b^6}\right)$ **19.** $\log_x a^{3/2}b^{-4}$

5.4 Exercises (page 116)

1. 2 **3.** 6 **5.** −4 **7.** −5 **9.** 2.9420 **11.** 4.1072 **13.** 0.8825
15. 6.9509 − 10 **17.** 0.5884 **19.** 4.8341 **21.** 6.5825 − 10 **23.** 34.4
25. .0747 **27.** 3124 **29.** 636.7 **31.** 81.6 **33.** 713 **35.** 2.01
37. .735 **39.** 36.3 **41.** 2.77 **43.** 31.2 pounds per cubic feet
45. (a) 350 years (b) 3990 years (c) 2300 years

5.5 Exercises (page 120)

1. 1.63 **3.** 1.16 **5.** $\varnothing$ **7.** 2.26 **9.** 1.08 **11.** −0.20 **13.** 11
15. 5 **17.** 4 **19.** 8 **21.** 10^{-2} or 1/100 **23.** $\varnothing$ **25.** $\varnothing$ **27.** 17.5

5.6 Exercises (page 124)

1. 4.83 **3.** −.11 **5.** 4.59 **7.** 1.43 **9.** .59 **11.** .90 **13.** −1.59
15. .96 **17.** 495 **19.** 960,800 **21.** 17.3 years **23.** We get $\frac{R}{r} = 1$.
25. $t = \frac{(5600)(\ln R/r)}{\ln 2}$ **27.** 378 feet **29.** 18.9

5.7 Exercises (page 128)

1. $a_5 = 48; S_5 = 93$ **3.** $a_5 = \frac{3}{4}; S_5 = \frac{33}{4}$ **5.** $a_5 = \frac{16}{9}; S_5 = \frac{55}{9}$
7. $a_5 = 64; S_5 = 124$ **9.** $a_5 = 9; S_5 = \frac{121}{9}$ **11.** $a_8 = \frac{64}{243}$ **13.** 8 or -8
15. 4, 8, 16 or $-4, 8, -16$ **17.** 30 **19.** $\frac{255}{256}$ **21.** 372 **23.** $\frac{650}{81}$

5.8 Exercises (page 133)

1. $\frac{3}{2}$ **3.** 2 **5.** $\frac{1}{5}$ **7.** cannot be found by the formula of this section
9. $\frac{1}{3}$ **11.** the sum does not exist **13.** $\frac{1}{20}$ **15.** $\frac{1}{4}$ **17.** $\frac{5}{9}$ **19.** $\frac{31}{99}$
21. $\frac{311}{900}$ **23.** 70 feet

6.1 Exercises (page 136)

1. False—not all corresponding elements are equal. **3.** True **5.** True
7. $\begin{pmatrix} 9 & 12 & 0 & 2 \\ 1 & -1 & 2 & -4 \end{pmatrix}$
9. Not possible—only matrices of the same dimension can be added.
11. $\begin{pmatrix} -2 & -3 & 3 \\ 4 & 3 & -1 \end{pmatrix}$
13. $X + T = \begin{pmatrix} x + r & y + s \\ z + t & w + u \end{pmatrix}$, and so is $T + X$.
15. $X + (-X) = \begin{pmatrix} x + (-x) & y + (-y) \\ z + (-z) & w + (-w) \end{pmatrix} = 0.$
17. All of the statements would still be true as long as 0, P, T, and X are of the same order.

6.2 Exercises (page 140)

1. True **3.** False—A could be 2×4 and B could be 4×2. Both BA and AB would then exist.
5. False—$\begin{pmatrix} 2 & 3 \\ 0 & 0 \end{pmatrix}\begin{pmatrix} -3 & 0 \\ 2 & 0 \end{pmatrix} = \begin{pmatrix} 0 & 0 \\ 0 & 0 \end{pmatrix}$. **7.** $\begin{pmatrix} 14 & 18 \\ 7 & 14 \end{pmatrix}$
9. $\begin{pmatrix} -6 & 6 & 6 \\ -5 & 5 & 5 \\ -4 & 4 & 4 \end{pmatrix}$
11. (3) **13.** It is not possible to find this product.
19. (a) $\begin{pmatrix} 17\frac{3}{4} & 23\frac{3}{4} \\ 9 & 12\frac{3}{4} \\ 33 & 42 \\ 6 & 8 \end{pmatrix}$
(b) $(220 \quad 890 \quad 105 \quad 125 \quad 70)$ (c) $(4165 \quad 5605)$

6.3 Exercises (page 147)

1. $\begin{pmatrix} -1/2 & 3/2 \\ 1 & -2 \end{pmatrix}$ **3.** $\begin{pmatrix} 1 & 0 \\ 0 & 1/2 \end{pmatrix}$
5. Does not exist.
7. $\begin{pmatrix} -1/13 & 7/26 \\ 2/13 & -1/26 \end{pmatrix}$ **9.** $\begin{pmatrix} 3 \\ 5 \end{pmatrix}$ **11.** $\begin{pmatrix} -5 \\ -18 \end{pmatrix}$

6.4 Exercises (page 151)

1. -36 **3.** 7 **5.** 0 **7.** 17 **9.** 166 **11.** 0 **13.** 0 **15.** 1

6.5 Exercises (page 156)

12. (a) -32 (b) -6 **13.** $\frac{5}{2}$ **15.** 7 **17.** 8

7.1 Exercises (page 163)

1. $x = 1, y = 3$ **3.** $p = 4, q = -2$ **5.** $x = 2, y = -2$
7. $x = 12, y = 6$ **9.** $x = 2, y = 2$ **11.** $x = 1, y = 2, z = 3$
13. $m = -1, n = 2, p = 1$ **15.** $a = 4, b = 1, c = 2$ **17.** $a = -3, b = -1$
19. 3000 at 8%, 6000 at 9%, and 1000 at 5%

7.2 Exercises (page 168)

1. $x = 2, y = 0$ **3.** $x = \frac{9}{7}, y = \frac{2}{7}$ **5.** $x = \frac{1}{2}, y = \frac{3}{8}$ **7.** no solution
9. $x = \frac{197}{91}, y = -\frac{118}{91}, z = -\frac{23}{91}$ **11.** $x = \frac{31}{5}, y = \frac{19}{10}, z = -\frac{29}{10}$

7.3 Exercises (page 172)

1. $x = 2, y = 0$ **3.** $x = \frac{7}{2}, y = -1$ **5.** $x = 5, y = 0$
7. $x = -1, y = 23, z = 16$ **9.** $x = 3, y = 2, z = -4$
11. $x = 2, y = 1, z = -1$ **13.** $x = -1, y = 2, z = 5, w = 1$

7.4 Exercises (page 174)

1. $x = \frac{1}{2}, y = \frac{1}{2}$ **3.** $x = \frac{53}{29}, y = \frac{-7}{29}$ **5.** $x = \frac{55}{41}, y = \frac{-19}{82}$ **7.** $x = \frac{55}{7}, y = \frac{15}{7}$
9. $x = \frac{7}{2}, y = -1$

7.5 Exercises (page 179)

1. $(-6/5, 17/5), (2, -3)$ **3.** $(-3/5, 7/5), (-1, 1)$
5. $(3, 6), (-3, 6), (3, -6), (-3, -6)$ **7.** any x and y such that $x^2 + y^2 = 10$
9. no solution **11.** $(3, 5), (-3, -5)$ **13.** $(3, 2), (-3, -2), (4, 3/2), (-4, -3/2)$
15. $(\sqrt{5}, 0), (-\sqrt{5}, 0), (\sqrt{5}, \sqrt{5}), (-\sqrt{5}, -\sqrt{5})$ **17.** $(1, 3), (31/14, -9/14)$
19. $\left(\frac{1 + \sqrt{13}}{2}, \frac{-1 + \sqrt{13}}{2}\right), \left(\frac{-1 - \sqrt{21}}{2}, \frac{3 + \sqrt{21}}{2}\right)$

7.6 Exercises (page 182)

1.

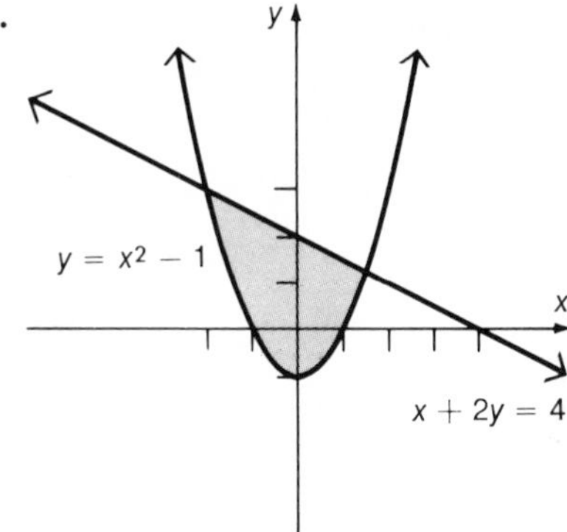

3.

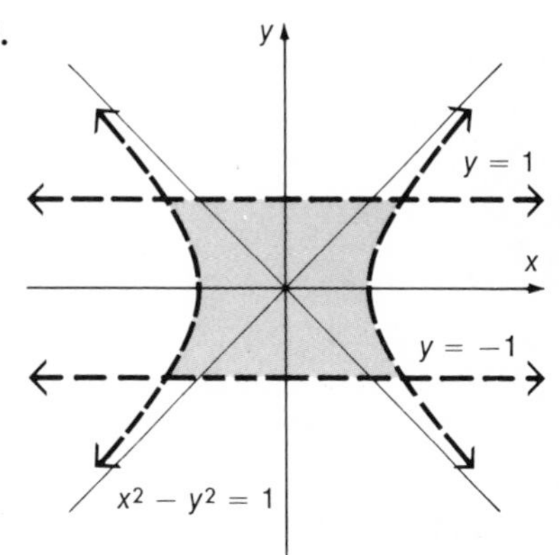

5.

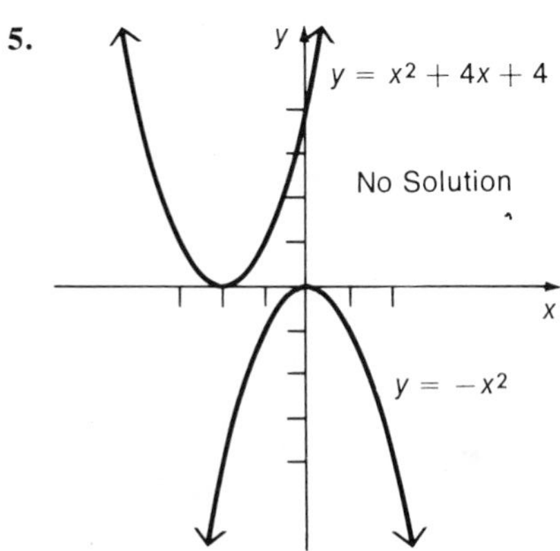

7.

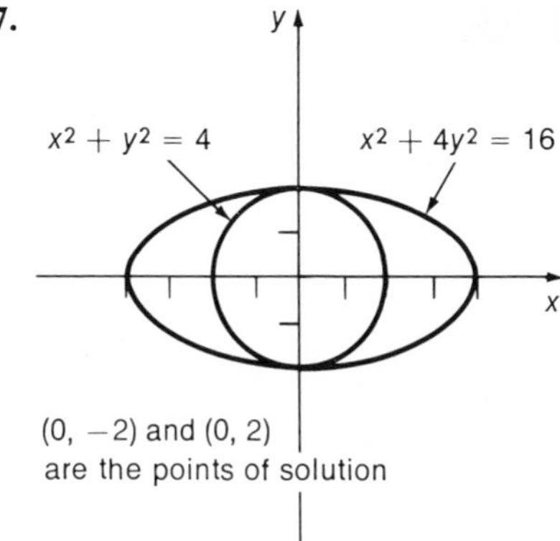

9.

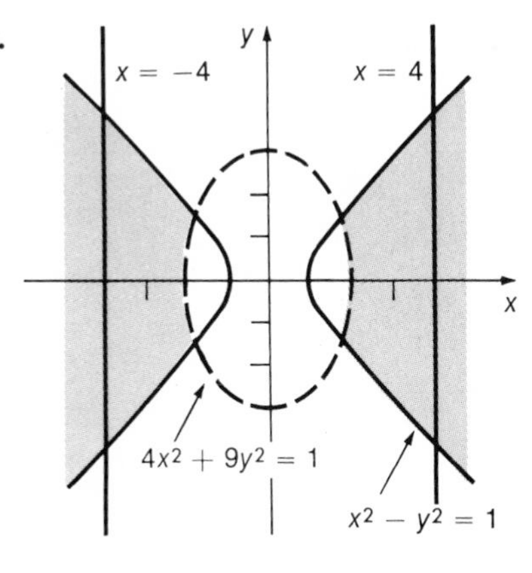

11.

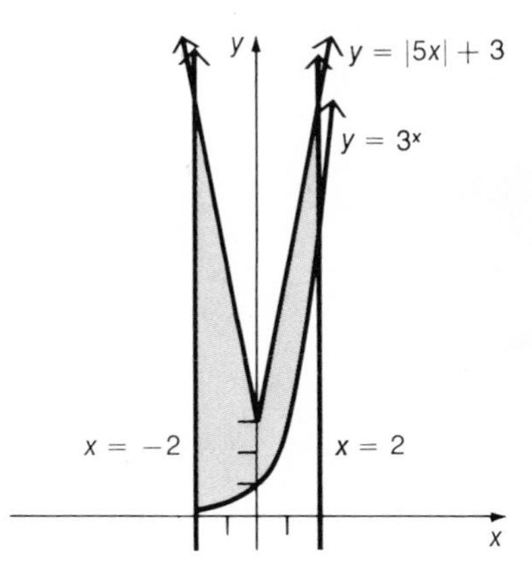

8.1 Exercises (page 186)

1. $7 - i$ **3.** $3 - 6i$ **5.** $2 + i$ **7.** $8 - i$ **9.** $-14 + 2i$ **11.** $31 - 5i$
13. $3 + 4i$ **15.** 5 **17.** $25i$ **19.** $24 - 7i$ **21.** i **23.** $\frac{7}{25} - \frac{24}{25}i$
25. $\frac{13}{20} - \frac{1}{20}i$ **27.** $\frac{32}{37} - \frac{7}{37}i$ **29.** $-1 - 2i$ **31.** -1 **33.** 1 **35.** i
37. $3i$ **39.** $3i\sqrt{2}$ **41.** $5i\sqrt{6}$

8.2 Exercises (page 190)

1. $3i, -3i$ **3.** $\frac{5}{2}i, \frac{-5}{2}i$ **5.** $2 + 2i, 2 - 2i$ **7.** $\frac{1}{2} + i, \frac{1}{2} - i$
9. $-1 + \sqrt{2}i, -1 - \sqrt{2}i$ **11.** $\frac{1 + \sqrt{7}i}{4}, \frac{1 - \sqrt{7}i}{4}$
13. $\frac{1 + \sqrt{23}i}{6}, \frac{1 - \sqrt{23}i}{6}$ **15.** $\frac{-i + \sqrt{5}i}{2}, \frac{-i - \sqrt{5}i}{2}$
17. $\frac{1 + \sqrt{2}}{i}, \frac{1 - \sqrt{2}}{i}$ **19.** $\frac{1 + \sqrt{7}i}{2 + 2i}, \frac{1 - \sqrt{7}i}{2 + 2i}$ **21.** $x^2 + 1 = 0$
23. $x^2 - 4x + 13 = 0$ **25.** $x^2 + 3 = 0$ **27.** $x^2 - 2x + 3 = 0$
29. $x^2 + (2i - 3)x + (5 - i) = 0$ **31.** $x^2 + (-5 - 2i)x + (14 + 8i) = 0$

8.3 Exercises (page 192)

1. yes **3.** yes **5.** no **7.** yes **9.** yes **11.** yes **13.** no **15.** 2
17. $-6 - i$ **19.** $-2 - 3i$ **21.** -20 **23.** 4

8.4 Exercises (page 196)

1. $x^2 - 10x + 26$ **3.** $x^3 - 4x^2 + 6x - 4$ **5.** $x^3 - 3x^2 + x + 1$
7. $x^4 - 6x^3 + 10x^2 + 2x - 15$ **9.** $x^3 - 8x^2 + 22x - 20$
11. $x^4 - 4x^3 + 5x^2 - 2x - 2$ **13.** $-1 + i, -1 - i$ **15.** $3, -2 - 2i$
17. $-\frac{1}{2} + \frac{\sqrt{5}}{2}i, -\frac{1}{2} - \frac{\sqrt{5}}{2}i$ **19.** $i, 2i, -2i$ **21.** $3, -2, 1 - 3i$

8.5 Exercises (page 198)

1. $-1, -2, 5$ **3.** $2, -3, -5$ **5.** no rational solutions **7.** $1, -2, -3, -5$
9. $\frac{3}{2}, -\frac{1}{3}, -4$ **11.** $\frac{1}{2}, -\frac{2}{3}, -\frac{3}{2}$ **13.** $\frac{1}{2}$ **15.** no rational solutions

8.6 Exercises (page 202)

9. $-3, -1.4, 1.4$ **11.** $-2, 1.6, 4.4$ **13.** 1.6 **15.** $-.7, 1.6, 2.7, 4.4$

8.7 Exercises (page 206)

1.

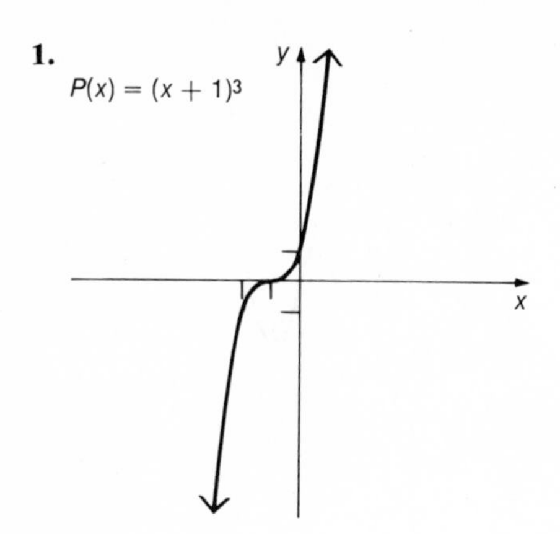

3. 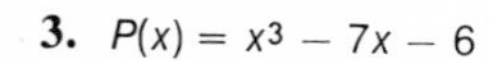

$P(x) = x^3 - 7x - 6$

	1	0	−7	−6
−3	1	−3	2	−12
−2	1	−2	−3	0
$-3/2$	1	$-3/2$	$-19/4$	$9/8$
−1	1	−1	−6	0
1	1	1	−6	−12
2	1	2	−3	−12
3	1	3	2	0

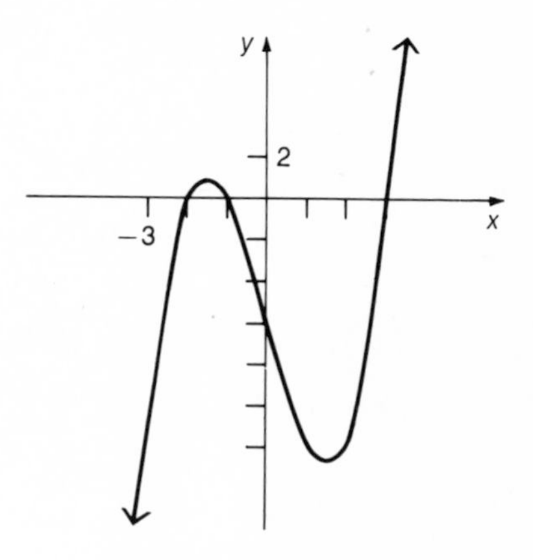

5.

$P(x) = x^4 - 5x^2 + 6$

	1	0	−5	0	6
−3	1	−3	4	−12	42
−2	1	−2	−1	2	2
−1	1	−1	−4	4	2
1	1	1	−4	−4	2
2	1	2	−1	−2	2
$3/2$	1	$3/2$	$-11/4$	$-33/8$	$-3/16$

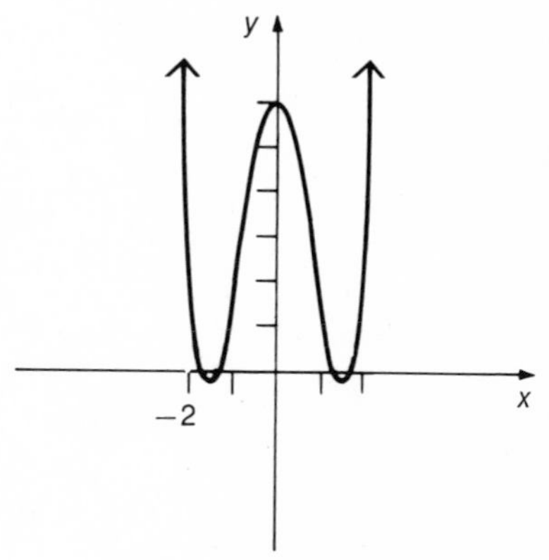

7.

$P(x) = 6x^3 + 11x^2 - x - 6$

	6	11	−1	−6
−3	6	−7	20	−66
−2	6	−1	1	−8
−1	6	5	−6	0
1	6	17	16	10

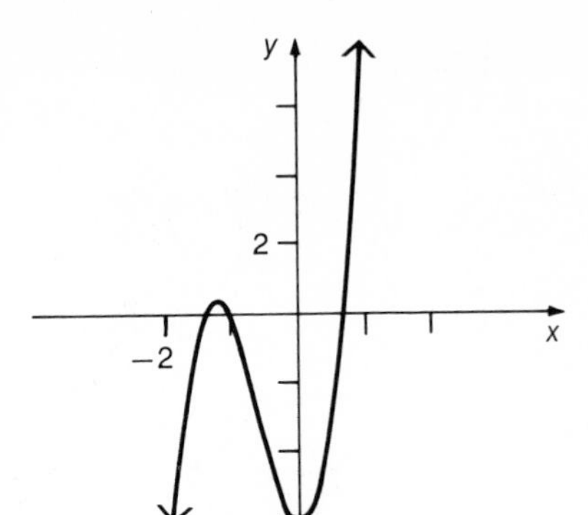

9.

$P(x) = x^4 + x^3 - 2$

	1	1	0	0	−2
−3	1	−2	6	−18	52
−2	1	−1	2	−4	6
−1	1	0	0	0	−2
1	1	2	2	2	0
2	1	3	6	12	22
$-\frac{1}{2}$	1	$\frac{1}{2}$	$-\frac{1}{4}$	$\frac{1}{8}$	$-\frac{33}{16}$
$\frac{1}{2}$	1	$\frac{3}{2}$	$\frac{3}{4}$	$\frac{3}{8}$	$-\frac{29}{16}$

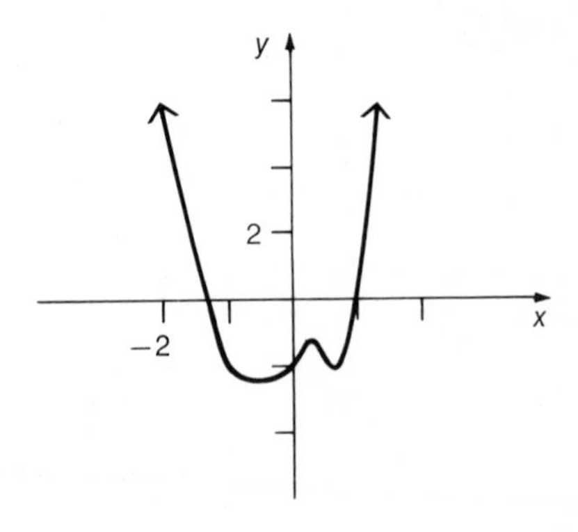

11.

$P(x) = 8x^4 - 2x^3 - 47x^2 - 52x - 15$

	8	−2	−47	−52	−15
−3	8	−26	31	−145	420
−2	8	−18	−11	−30	45
−1	8	−10	−37	−15	0
1	8	6	−41	−93	−108
2	8	14	−19	−90	−195
3	8	22	19	5	0
$-\frac{1}{2}$	8	−6	−44	−30	0
$-\frac{5}{4}$	8	−12	−32	−12	0

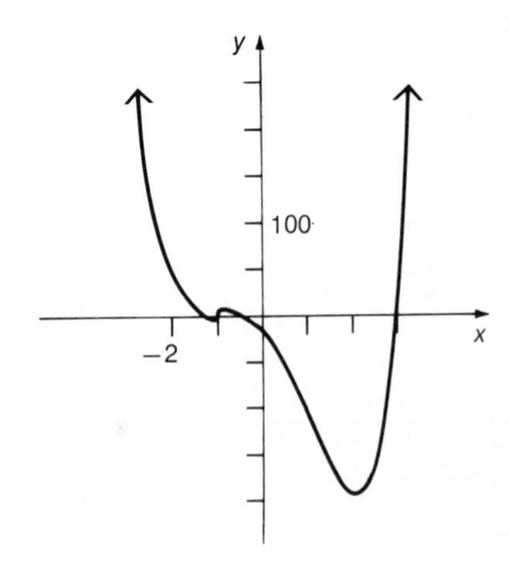

8.8 Exercises (page 208)

1.

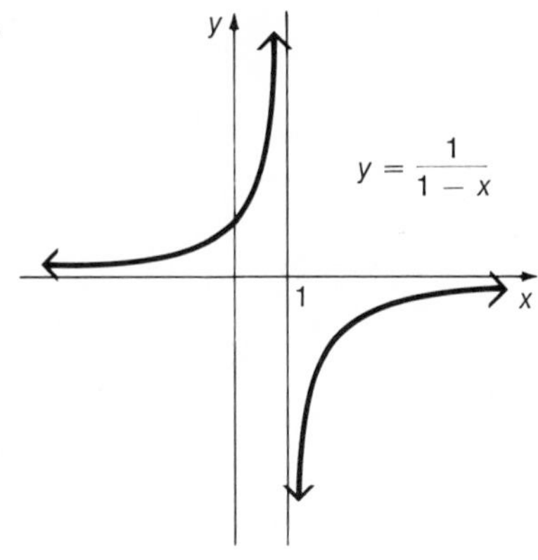

3.

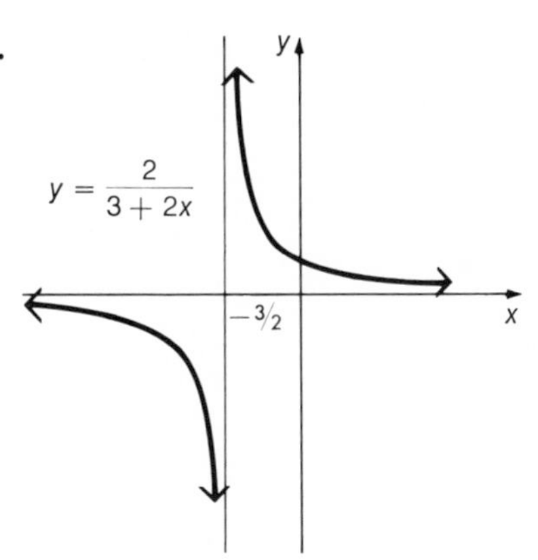

5.

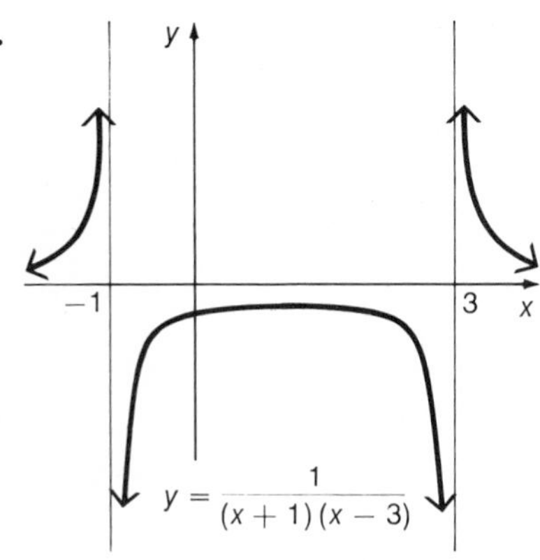

7.

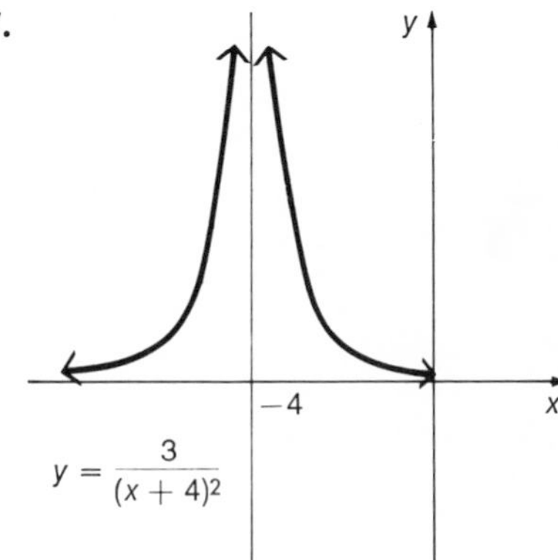

9.

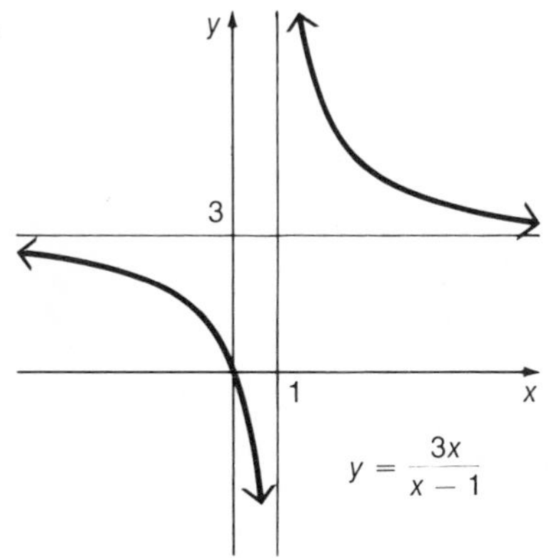

11.

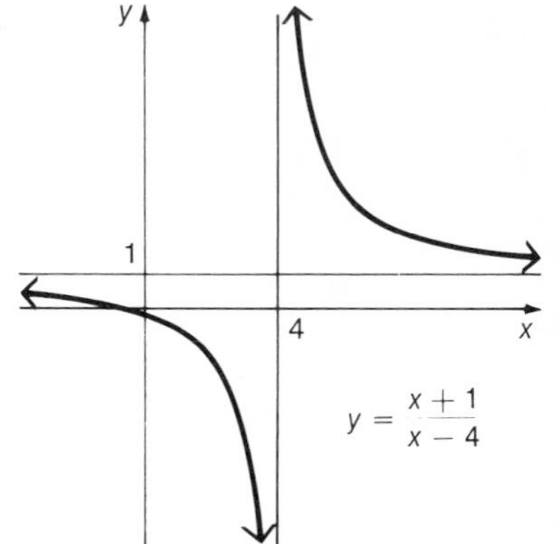

13.

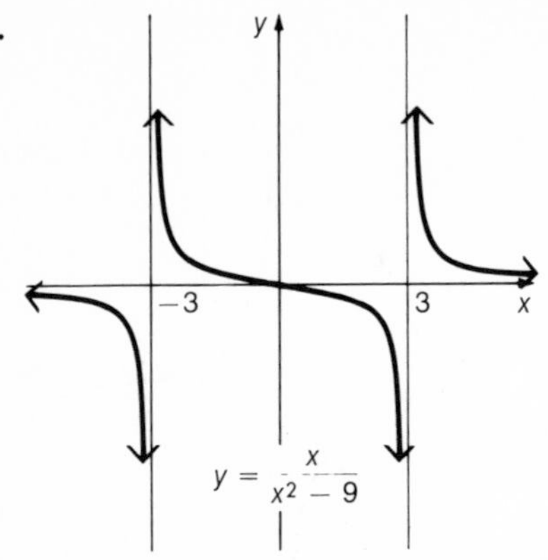

15.

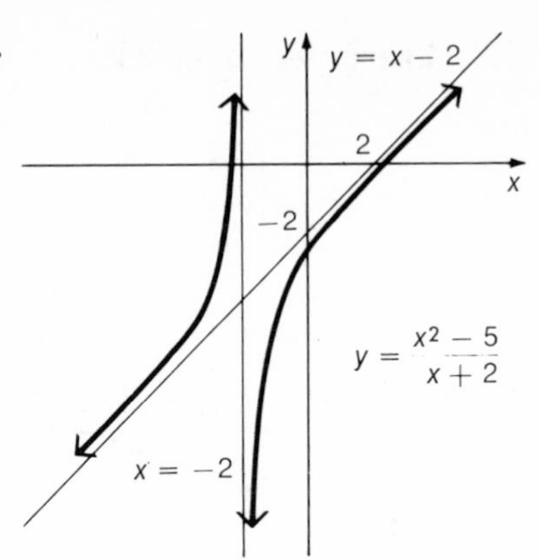

9.1 Exercises (page 213)

1. (a) 5040 (b) 60 (c) 720 (d) 12 **3.** 720 **5.** 120 30,240 **7.** 30
9. 120 **11.** 4 **13.** 552 **15.** $26 \cdot 25 \cdot 24 \cdot 10 \cdot 9 \cdot 8$, $2(26 \cdot 25 \cdot 24 \cdot 10 \cdot 9 \cdot 8)$

9.2 Exercises (page 217)

1. (a) 6 (b) 6 (c) 56 (d) 45 **3.** 27,405 **5.** 220, 220 **7.** 1326 **9.** 10
11. 28 **13.** (a) 84 (b) 10 (c) 40 (d) 28

9.3 Exercises (page 222)

1. $\{H\}$ **3.** $\{HHH, HHT, HTH, HTT, THH, THT, TTH, TTT\}$
5. $\{FFF, FFT, FTF, FTT, TFF, TFT, TTF, TTT\}$
7. (a) $E_1 = \{HH, TT\}$, $P(E_1) = \frac{1}{2}$ (b) $E_2 = \{HH, HT, TH\}$, $P(E_2) = \frac{3}{4}$
9. (a) $E_1 = \{2 \text{ and } 4\}$; $P(E_1) = \frac{1}{10}$

(b) $E_2 = \{1 \text{ and } 3, 1 \text{ and } 5, 3 \text{ and } 5\}$; $P(E_2) = \frac{3}{10}$

(c) $E_3 = \varnothing$; $P(E_3) = 0$

(d) $E_4 = \{1 \text{ and } 2, 1 \text{ and } 4, 2 \text{ and } 3, 2 \text{ and } 5, 3 \text{ and } 4, 4 \text{ and } 5\}$; $P(E_4) = \frac{3}{5}$

11. (a) $\frac{1}{5}$ (b) $\frac{8}{15}$ (c) 0 (d) 1 to 4 (e) 7 to 8 **13.** 1 to 4 **15.** $\frac{2}{5}$
17. (a) $\frac{1}{9}$ (b) $\frac{1}{36}$ (c) $\frac{35}{36}$ (d) $\frac{1}{2}$ (e) $\frac{1}{6}$

9.4 Exercises (page 227)

1. (a) $\frac{1}{2}$ (b) $\frac{7}{10}$ (c) $\frac{2}{5}$ **3.** (a) $\frac{1}{6}$ (b) $\frac{1}{3}$ (c) $\frac{1}{6}$ **5.** (a) $\frac{1}{8}$ (b) $\frac{3}{8}$ (c) $\frac{1}{4}$
7. (a) $\frac{1}{17}$ (b) $\frac{25}{102}$ (c) $\frac{1}{221}$ (d) $\frac{13}{102}$ (e) $\frac{25}{204}$ **9.** (a) $\frac{2}{3}$ (b) $(\frac{2}{3})^3$
11. (a) $\frac{9}{29}$ (b) $\frac{197}{203}$ **13.** $\frac{55}{221}$

9.5 Exercises (page 232)

1. $x^6 + 6x^5y + 15x^4y^2 + 20x^3y^3 + 15x^2y^4 + 6xy^5 + y^6$
3. $p^5 - 5p^4q + 10p^3q^2 - 10p^2q^3 + 5pq^4 - q^5$
5. $16x^4 + 32x^3t^3 + 24x^2t^6 + 8xt^9 + t^{12}$
7. $\frac{m^5}{32} - \frac{15m^4n}{16} + \frac{45m^3n^2}{4} - \frac{135m^2n^3}{2} + \frac{405mn^4}{2} - 243n^5$

9. $7920m^8p^4$ **11.** $\frac{4845}{65536}p^8q^{16}$ **13.** 2.5937 **15.** 245.9374

9.6 Exercises (page 235)

1. $C(7, 5)\left(\frac{1}{2}\right)^7 = \frac{21}{2^7}$ **3.** (a) $\left(\frac{1}{6}\right)^{12}$ (b) $C(12, 6)\left(\frac{1}{6}\right)^6\left(\frac{5}{6}\right)^6$ (c) $C(12, 1)\left(\frac{1}{6}\right)\left(\frac{5}{6}\right)^{11}$
5. (a) $(.95)^{20}$ (b) $(.95)^{20} + C(20, 1)(.05)(.95)^{19} + C(20, 2)(.05)^2(.95)^{18}$

INDEX

Absolute value, 19
properties, 21ff
Addition identity axiom, 5
Addition inverse axiom, 6
Addition property of equality, 9
Additive identity, 5
for complex numbers, 185
Additive inverse, 6
for complex numbers, 185
of a matrix, 136
Algebraic expression, 24
Area of a triangle, 157
Arithmetic progression, 71
sum of n terms, 73
Arithmetic sequence, 72
Associative axioms, 5
Asymptotes, 95, 206
Augmented matrix, 169
Axiom(s)
of completeness, 17
of equality, 5
field, 5–6
of logarithms, 111
of mathematical induction, 237
of order, 16

Base, 24
Base of a logarithm, 110
Binary operation, 5
Binomial, 25
Binomial theorem, 230
Bound
lower, 16
upper, 16

Cancellation property of addition, 10
Cancellation property of multiplication, 10
Carbon 14 dating, 124
Cartesian coordinate system, 51
Center of a circle, 89
Characteristic, 114
Circle, 89
center of, 89
radius of, 89
Closure axioms, 6
Closure axiom of order, 16
Coefficient, 24
Coefficient matrix, 169
Column matrix, 135

Combinations, 214
Common difference, 72
Common logarithms, 113
Common ratio, 126
Commutative axioms, 5
Complement, 219
Completeness, axiom of, 17
Completing the square, 76
Complex fraction, 37
Complex numbers, 184
  additive identity of, 185
  additive inverses of, 185
  conjugates of, 186
  equality of, 184
  field of, 186
  multiplicative identity of, 185
  multiplicative inverses of, 185
  negatives of, 185
  quotient of, 186
  product of, 184
  subtraction of, 185
  sum of, 184
Composite function, 51
Compound sentence, 20
Conditional probability, 226
Conics
  circle, 89
  ellipse, 92
  hyperbola, 94
  parabola, 76
Conic sections, 98
  summary of, 98
Constant of variation, 100
Conjugate, 186
Coordinate system, Cartesian, 51
Cramer's rule, 165

Deductive reasoning, 236
Degree, 25
Dependent events, 227
Dependent system, 160
Descartes, René, 52
Descartes' rule of signs, 201
Determinants, 148
  expansion of, 150
  properties of, 152
Difference of two cubes, 33
Difference of two squares, 33
Dimensions of a matrix, 134
Direct variation, 100
Discriminant, 81
Distance formula, 89
Distributive axiom, 6
  extended, 6
Division, 6
  algorithm, 29, 191
  of polynomials, 28
  synthetic, 30
Domain of a relation, 50

Elements of a matrix, 134
Elimination method, 161
Ellipse, 92
  foci of, 92
Empty set, 2
Equations
  absolute value, 64
  equivalent, 61
  exponential, 118
  first degree, 159
  linear, 159
  logarithmic, 118
  quadratic, 82
  quadratic in form, 82
  second degree, 82
Equal matrices, 135
Equality of complex numbers, 184
Equality test for rational numbers, 11
Event, 218
Events
  independent, 227
  mutually exclusive, 224
Expansion of a determinant, 150

Exponential, 24
  equations, 118
  function, 108
Exponent, 24
Exponents
  integer, 39
  properties of, 25
  rational, 42
  real, 108
Expressions, rational, 35
Extended distributive axiom, 6
Extraneous roots, 84

Factor theorem, 190, 192
Factoring, 33
Field, 5
  axioms, 5
  complete ordered, 17
  of complex numbers, 186
First degree equations, 159
First degree functions, 54
Focus of an ellipse, 92
Forgetting curve, 122
Functions, 51
  composite, 51
  exponential, 108
  first degree, 54
  graphical test for, 52
  greatest integer, 69
  linear, 54
  logarithmic, 110
  nearest integer, 69
  periodic, 70
  polynomial, 183
  quadratic, 75
  rational, 206
  rational zeros of, 197
  second degree, 75ff
  sequence, 71
  step, 69
  zeros of, 79
Fundamental theorem
  of algebra, 193
  of rational expressions, 35
  of rational numbers, 11

General binomial expansion, 230
Geometric means, 129
Geometric progression, 126
  infinite, 131
  sum of n terms, 127
Geometric sequence, 126
Greatest integer function, 69
Greatest lower bound, 16

Half-plane, 65
Horizontal asymptote, 206
Hyperbola, 94
  asymptotes, 95

Identity
  additive, 5
  matrix, 144
  multiplicative, 5
Identity axioms, 5
Imaginary number, 184
Inconsistent equations, 160
Independent events, 227
Index of a radical, 43
Inductive reasoning, 236
Inequalities, 63
  absolute value, 64
  linear, 65
  quadratic, 86
Infinite geometric sequences, 129
Infinite geometric progression, 131
Integers, 3
Integer exponents, 39
Interpolation, 115
Intersection, 2

Inverse
additive, 6
multiplicative, 6
Inverse axioms, 6
Inverse relations, 103
Inverse variation, 100
Irrational numbers, 3

Joint variation, 101

Leading coefficient, 183
Least upper bound, 16
Less than, definition of, 15
Limits, 130
Linear
equations, 159
functions, 54
interpolation, 115
relations, 54
Logarithm, 110
characteristic of, 114
common, 113
mantissa of, 114
natural, 120
Logarithmic
equations, 118
functions, 110
Lower bound, 16

Mantissa, 114
Mathematical induction, 237
Mathematical system, 4
Matrices, 134
additive inverses of, 136
augmented, 169
coefficients of, 169
column, 135
determinants of, 148
dimensions of, 134
elements of, 134
Matrices (cont.)
equal, 135
identity, 144
multiplication of, 137
multiplicative inverses of, 144
negatives of, 136
order of, 134
product of, 137
properties of, 146
row, 135
scalar, 140
square, 134
sum of, 135
zero, 143
Monomial, 25
Multiplication
property of equality, 9
property of zero, 10
Multiplicative identity, 5
axiom, 5
for complex numbers, 185
Multiplicative inverse, 6
axiom, 6
for complex numbers, 185
of a matrix, 144
Mutually exclusive events, 224

Natural logarithms, 120
Natural numbers, 2
Nearest integer function, 69
Negative, 6
of a matrix, 136
Nonlinear first degree relations, 68
Nonlinear system of equations, 175
*n*th root, 42
Null set, 2
Numbers
complex, 3, 184
imaginary, 3, 184
integers, 3
irrational, 3
natural, 2

Numbers (cont.)
pure imaginary, 184
rational, 3
real, 3
whole, 3

Oblique asymptote, 208
Order, axioms of, 16
Order of a matrix, 134
Ordered pair, 50
components of, 50
Ordered triple, 160

Parabola, 76
vertex of, 76
Perfect square trinomial, 33
Periodic function, 70
Permutations, 212
Point-slope form, 59
Polynomial function, 183
Polynomials
products of, 25
differences of, 25
sums of, 25
quotients of, 28
monomial, 25
trinomial, 25
binomial, 25
degrees of, 25
over the rationals, 24
over the integers, 24
over the reals, 24
in x, 24
Power, 24
Probability, 218
conditional, 226
Product
of complex numbers, 184
of two matrices, 137
Progression, geometric, 71, 126

Quadrant, 52
Quadratic formula, 81, 188
Quadratic function, 75
Quadratic inequalities, 86
Quotient of complex numbers, 186

Radical, 43
Radicals
index of, 46
rationalizing, 46
simplifying, 46
Radicand, 43
Radius of a circle, 89
Range, 50
Rational exponents, 42
Rational expression
domain of, 35
properties of, 35
Rational expressions, 35
Rational function, 206
Rational numbers, 3
equality test for, 11
fundamental theorem of, 11
Rationalizing, 46
Real exponents, 108
Real numbers, 3
Reciprocal, 6
Reflexive axiom, 5
Relation, 4, 50
domain of, 50
first degree, 50
inverse, 103
linear, 54
nonlinear, 68
range of, 50
second degree, 75ff
Remainder theorem, 196
Repeated trials, 233
Roots, properties of, 43
Row matrix, 135

Sample space, 218
Scalar, 140
Second degree equations, 82
Second degree functions, 75ff
Sequence, 71
 arithmetic, 72
 geometric, 126
 terms of, 71
Set, 1
 braces, 1
 complement of, 3
 elements of, 1
 empty, 2
 members of, 1
 null, 2
 replacement, 3
 subset of, 2
 universal, 2
 well-defined, 1
Sets
 disjoint, 2
 equal, 2
 equivalent, 2
 intersection of, 2
 union of, 2
Slope, 57
Slope-intercept form, 58
Square matrix, 134
Step function, 69
Subset, 2
Substitution axiom, 5
Substitution method, 176
Subtraction, 6
 for complex numbers, 185
Successive events axiom, 211
Sum
 of infinite geometric progression, 131
 of two complex numbers, 184
 of two cubes, 33
 of two matrices, 135
Symmetric axiom, 5
Symmetry, 94
Synthetic division, 30
System of equations, 159
 dependent, 160
 inconsistent, 160
 nonlinear, 175
System of inequalities, 159
System, mathematical, 4
 real number, 5

Term, 24
Theorem, 9
Transitive axiom, 5
Trichotomy axiom, 16
Trinomial, 25
 perfect square, 33

Union, 2
Universe of discourse, 2
Upper bound, 16

Variable, 3
 domain of, 3
Variation
 direct, 100
 inverse, 100
 joint, 101
Vertex, 76, 78
Vertical asymptote, 206

Whole numbers, 3

x-axis, 52
x-intercept, 54

y-axis, 52
y-intercept, 54

Zero-factor theorem, 79
Zero matrix, 143
Zero of a function, 79

80 79 78 77 76 75 74 73
1 2 3 4 5 6 7 8 9 10 11 12 13 14 15 16 17 18 19 20 21 22 23 24 25